Differential Neural Networks for Robust Nonlinear Control

Identification, State Estimation and Trajectory Tracking

Differential Neural Networks for Robust Nonlinear Control

Identification, State Estimation and Trajectory Tracking

Alexander S. Poznyak
Edgar N. Sanchez
Wen Yu

CINVESTAV-IPN, Mexico

World Scientific
New Jersey • London • Singapore • Hong Kong

Published by

World Scientific Publishing Co. Pte. Ltd.

P O Box 128, Farrer Road, Singapore 912805

USA office: Suite 1B, 1060 Main Street, River Edge, NJ 07661

UK office: 57 Shelton Street, Covent Garden, London WC2H 9HE

British Library Cataloguing-in-Publication Data
A catalogue record for this book is available from the British Library.

DIFFERENTIAL NEURAL NETWORKS FOR ROBUST NONLINEAR CONTROL

ISBN 981-02-4624-2

Printed in Singapore by Uto-Print

To our children
Poline and Ivan,
Zulia Mayari, Ana Maria and Edgar Camilo,
Huijia and Lisa.

0.1 Abstract

This book deals with *Continuous Time Dynamic Neural Networks Theory* applied to solution of basic problems arising in Robust Control Theory including identification, state space estimation (based on neuro observers) and trajectory tracking. The plants to be identified and controlled are assumed to be a priory unknown but belonging to a given class containing internal unmodelled dynamics and external perturbations as well. The error stability analysis and the corresponding error bounds for different problems are presented. The high effectiveness of the suggested approach is illustrated by its application to various controlled physical systems (robotic, chaotic, chemical and etc.).

0.2 Preface

Due to the big enthusiasm generated by successful applications, the use of *static (feedforward) neural networks* in automatic control is well established. Although they have been used successfully, the major disadvantage is a slow learning rate. Furthermore, they do not have memory and their outputs are uniquely determined by the current value of their inputs and weights. This is a high contrast to biological neural systems which always have feedback in their operation such as the cerebellum and its associated circuitry, and the reverberating circuit, which is the basis for many of the nervous system activities.

Most of the existing results on nonlinear control are based on static (feedforward) neural networks. On the contrary, there are just a few publications related to *Dynamic Neural Networks* for Automatic Control applications, even if they offer a better structure for representing dynamic nonlinear systems.

As a natural extension of the static neural networks capability to approximate nonlinear functions, the dynamic neural networks can be used to approximate the behavior of nonlinear systems. There are some results in this directions, but their requirements are quite restrictive.

In the summer of 1994, the first two authors of this book were interested by exploring the applicability of *Dynamic Neural Networks* functioning in continuous time for Identification and Robust Control of Nonlinear Systems. The third author became involved in the summer of 1996.

Four years late, we have developed results on weights learning, identification, estimation and control based on the dynamic neural networks. Here this class of networks is named by *Differential Neural Networks* to emphasize the fact that the considered dynamic neural networks as well as the dynamic systems with incomplete information to be controlled are functioning in continuous time. These results have been published in a variety of journals and conferences. The authors wish to put together all these results within a common frame as a book.

The main aim of this book is to develop a systematic analysis for the applications of dynamic neural networks for identification, estimation and control of a wide class of nonlinear systems. The principal tool used to establish this analysis is a

Lyapunov like technique. The applicability of the results, for both identification and robust control, is illustrated by different technical examples such as: chaotic systems, robotics and chemical processes.

The book could be used for self learning as well as a textbook. The level of competence expected for the reader is that covered in the courses of differential equations, the nonlinear systems analysis, in particular, the Lyapunov methodology, and some elements of the optimization theory.

0.3 Acknowledgments

The authors thank the financial supports of CONACyT, Mexico, projects 1386A9206, 0652A9506 and 28070A, as well as former students Efrain Alcorta, Jose P. Perez, Orlando Palma, Antonio Heredia and Juan Reyes-Reyes. They thank H. Sira-Ramirez, Universidad de los Andes, Venezuela, for helping to develop the application of sliding modes technique to learning with Differential Neural Networks.

The helpful review of Dr.Vladimir Kharitonov is greatly appreciated. Thanks are also due to anonymous reviewers of our publications, on the topics matter of this book, for their constructive criticism and helpful comments.

We want to thank the editors for their effective cooperation and great care making possible the publication of this book.

Last, but not least, we thank the time and dedication of our wives Tatyana, Maria de Lourdes, and Xiaoou. Without them this book would not be possible.

Alexander S. Poznyak
Edgar N. Sanchez
Wen Yu

Mexico, January of 2000

Contents

List of Figures

0.4 Introduction

Undoubtedly, since their strong rebirth in the last decade, *Artificial Neural Networks (ANN)* are playing an increasing role in Engineering. For some years, they have been seen as providing considerable promise for application in the nonlinear control. This promise is based on their theoretical capability to approximate arbitrary well continuous nonlinear mappings.

By large, the application of neural networks to automatic control is usually for building a model of the plant and, on the basis of this model, to design a control law. The main neural network structure in use is the static one or, in other word, is the feedforward type: the input-output information process, performed by the neural network, can be represented as a nonlinear algebraic mapping.

On the basis of *Static Neural Networks (SNN)* capability to approximate any nonlinear continuous function, a natural extension is to approximate the input-output behavior of nonlinear systems by *Dynamic Neural Networks (DNN)*: their information process is described by differential equations for continuous time or by difference equations for discrete time. The existing results about this extension require quite restrictive conditions such as: an open loop stability or a time belonging to a close set.

This book is intended to familiarize the reader with the new field of the dynamic neural networks applications for robust nonlinear control, that is, it develops a systematic analysis for identification, state estimation and trajectory tracking of nonlinear systems by means of Differential (Dynamic Continuous Time) Neural Networks . The main tool for this analysis is the *Lyapunov like approach.*

The book is aimed to graduate students, but the practitioner engineer can profit from it for self learning. A background in differential equations, nonlinear systems analysis, in particular Lyapunov approach, and optimization techniques is strongly recommended. The reader could read the appendices or use some of the given references to cover these topics. Therefore the book should be very useful for a wide spectrum of researchers and engineers interested in the growing field of the neurocontrol, mainly based on differential neural networks.

0.4.1 Guide for the Readers

The structure of the book scheme we developed consists of two main parts:

The Identifier (or the state estimator). A differential neural network is used to build a model of the plant. We consider two cases:

a) the dimension of the neural network state consides with one of the nonlinear system; so the neural networks becomes an identifier.

b) the nonlinear system output depends linearly on the states. The neural network allows to estimate the system state by means of a neural observer implementation.

The controller. Based of the model, implemented by the neural identifier or observer, the local optimal control law is developed , which at each time minimizes the tracking error with respect to a nonlinear reference model under the fixed prehistory of this process; it also minimize the required input energy.

Additionally, in order to perform a better identification, we developed two new algorithms to adapt on-line the neural network weights. These algorithms are based on the sliding modes technique and the gradient like contribution.

The book consists of four principal parts:

- An introductory chapter (Chapter 1) reviewing the basic concepts about neural networks.
- A part related to the *neural identification* and *estimation* (Chapters 2, 3, 4).
- A part dealing with the *passivation* and the *neurocontrol* (Chapters 5 and 6).
- The last part related to its applications (Chapters 7, 8, 9 and 10).

The content of each chapter is as follows.

Chapter One: *Neural Networks Structures.* The development and the structures of Neural Networks are briefly reviewed. We first take a look to biological ones. Then the different structures classifying them as static or dynamic neural networks are discussed. In the introduction, the importance of autonomous or intelligent systems is established, and the role which neural networks could play to implement such a system for the control aims is discussed. Regarding to biological neural networks, the main phenomena, taking place in them,.are briefly described. A brief review of the different neural networks structures such as the single layer, the multi-layer perceptron, the radial basis functions, the recurrent and the differential ones, is also presented. Finally, the applications of neural networks to robust control are discussed.

Chapter Two: *Nonlinear System Identification.* The on-line nonlinear system identification, by means of a differential neural network with the same space state dimension as the system, is analyzed. It is assumed that the system space state dimension completely measurable. Based of the Lyapunov-like analysis, the stability conditions for the identification error are determined. For the identification analysis an algebraic Riccati equation is ued. The new learning law ensures the identification error convergence to zero (model matching) or to a bounded zone (with unmodelled dynamics). As our main contributions, a new on-line learning law for differential neural network weights is developed and the theorem, giving a bound for the identification error which turns out to be proportional to the a priory uncertainty bound, is established. To identify on-line a nonlinear system from a given class a new stable learning law for a differential *multilayer* neural network is also proposed. By means of a Lyapunov-like analysis the stable learning algorithms for the hidden layer as well as for the output layer is determined. An algebraic Riccati equation is used to give a bound for the identification error. The new learning is similar with the back-propagation for multilayer perceptrons. With this updating law we can assure that the identification error is globally asymptotically stable (GAS). The applicability of these results is illustrate by several numerical examples.

Chapter Three: *Sliding Mode Learning.* The identification of continuous, uncertain nonlinear systems in presence of bounded disturbances is implemented using dynamic neural networks. The proposed neural identifier guarantees a bound for the state estimation error, which turns out to be a linear combinations of the internal and external uncertainties levels. The neural network weights are updated on-line by a learning algorithm based on the sliding mode technique. To the best of authors awareness, this is the first time when such a learning scheme is proposed for differential neural networks. The numerical simulations illustrate its effectiveness even for highly nonlinear systems in the presence of important disturbances.

Chapter Four: *Neural State Estimation.* A dynamic neural network solution of the state estimation is discussed. The proposed adaptive robust neuro-observer has an extended Luneburger structure. Its weights are learned on-line by a new gradient-like algorithm. The gain matrix is calculated by solving a matrix optimization problem and an inverted solution of a differential matrix Riccati equation. In the case when the normal nonlinear system is a priory unknown, the state observation using dynamic recurrent neural network, for continuous time, uncertain nonlinear systems, subjected to external and internal disturbances of bounded power, is discussed. The design of a suboptimal neuro-observer is proposed to achieve a perspectives accuracy of the estimation error, which is defined as the weighted squares of its semi-norm. This error turns out to be a linear combination of the power levels of the external disturbances and internal uncertainties. The numerical simulations of the proposed robust observer illustrate its effectiveness in the presence of the unmodelled uncertainties of a high level.

Chapter Five: *Passivation via Neuro Control.* An adaptive technique is suggested to provide the passivity property for a class of partially known SISO nonlinear systems. A simple differencial neural network (DifNN), containing only two neurons, is used to identify the unknown nonlinear system. By means of a Lyapunov-like analysis a new learning law is derived for this DifNN guarantying both a successful identification and passivation effects. Based on this adaptive DifNN model an adaptive feedback controller, serving for wide class of nonlinear systems with a priory incomplete model description, is designed. Two typical examples illustrate the effectiveness of the suggested approach.

Chapter Six: *Nonlinear System Tracking.* If the state measurements of a nonlinear system are available and its structure is estimated by a dynamic neural identifier or neuro-observer, to track a reference nonlinear model an optimal control law can be developed. To do that, first a neuro identifier is considered and, using the on-line adapted parameter of the corresponding differential neural network, an optimal control law is implemented. It minimizes the input energy and the tracking error between the designed DifNN and a given reference model. Then, assuming that not all the system states are measurable, the above discussed neuro-observer is implemented . The optimal control law has the same structure as before, but with the space states replaced by their estimates. In both cases a bound for the trajectory error is guaranteed. So, the control scheme is based on the proposed neuro-observer and, as a result, the final structure is composed by two parts: the neuro-observer and the tracking controller. Some simulation results conclude this chapter.

Chapter Seven: *Neural Control for Chaos.* Control for a wide class of continuous time nonlinear systems with unknown dynamic description (model) can be implemented using a dynamic neural approach. This class includes a wide group of chaotic systems which are assumed to have unpredictable behavior but whose state can be measured. The proposed control structure has to main parts: a neural identifier and a neural controller. The weights of the neural identifier are updated on-line by a learning algorithm based on the sliding mode technique. The controller assures tracking of a reference model. Bounds for both the identification and the tracking errors are established. So, in this chapter identification and control of unknown chaotic dynamical systems are consider. Our aim is to regulate the unknown chaos to a fixed points or a stable periodic orbits. This is realized by following two contributions: first, a dynamic neural network is used as identifier. The weights of the neural networks are updated by the sliding mode technique. This neuro-identifier guarantees the boundedness of identification error. Secondly, we derive a local optimal controller via the neuro-identifier to remove the chaos in a system. The controller proposed in this chapter is effective for many chaotic systems including Lorenz system, Duffing equation and Chua's circuit.

Chapter Eight: *Neuro Control for Robot Manipulator.* The neuro tracking problem for a robot manipulator with two degrees of mobility and with unknown load, friction and the parameters of the mechanical system , subject to variations within a given interval, is tackled. The design of the neuro robust nonlinear controller is proposed such a way that a certain accuracy of the tracking is achieved. The suggested neuro controller has a direct linearization part and a locally optimal compensator. Compared with sliding mode type and linear state feedback controllers, numerical simulations of this robust controller illustrate its effectiveness.

Chapter Nine: *Identification of Chemical Processes.* The identification problem for multicomponent nonstationary ozonization processes with incomplete observable states is addressed. The corresponding mathematical model containing unknown parameters is used to simplify the initial nonlinear model and to derive its observability conditions. To estimate the current concentration of each component, a dynamic neuro observer is suggested. Based on the obtained neuro observer outputs, the continuous time version of LS-algorithm, supplied by special projection procedure, is applied to construct the estimates of unknown chemical reaction constants. Simulation results related to the identification of ozonization process illustrate the applicability of the suggested approach.

Chapter Ten: *Neuro Control for a Multicomponent Distillation Column.* Control of a multicomponent non-ideal distillation column is proposed by using a dynamic neural network approach. The holdup, liquid and vapor flow rates are assumed to be time-varying, that is, the non-ideal conditions are considered. The control scheme is composed of two parts: a dynamic neural observer and a neuro controller for output trajectory tracking. Bounds for both the state estimation and the tracking errors are guaranteed. The trajectory to be tracked is generated by a reference model, which could be nonlinear. The controller structure which we propose is composed of two parts: the neuro-identifier and the local optimal controller. Numerical simulations, concerning a 5 components distillation column with 15 trays, illustrate the high effectiveness of the approach suggested in this chapter.

Three appendices end the book containing some auxiliary mathematical results:

Appendix A deals with some useful mathematical facts;

Appendix B contains the basis required to understand the Lyapunov-like approach used to derive the results obtain within this book;

Appendix C discusses some definitions and properties concerning to the locally optimization technique required to obtain the mentioned optimal control law.

0.5 Notations

":= " this symbol means "equal by definition";

$x_t \in \Re^n$ is the state vector of the system at time $t \in R^+ := \{t : t \geq 0\}$;

$\widehat{x}_t \in \Re^n$ is the state of the neural network;

x^* is the state of nonlinear reference model;

$u_t \in \Re^q$ is a given control action;

$y_t \in \Re^m$ is the output vector;

$f(x_t, u_t, t) : \Re^{n+q+1} \to \Re^n$ is a vector valued nonlinear function describing the system dynamics;

$\varphi(x^*, t) : \Re^{n+1} \to \Re^n$ is nonlinear reference model;

$C \in \Re^{n \times m}$ is the unknown output matrix;

$\xi_{1,t}$, $\xi_{2,t}$ are vector-functions representing external perturbations;

Υ_1, Υ_2 are the "bounded power" of $\xi_{1,t}$, $\xi_{2,t}$;

$A \in \Re^{n \times n}$ is a Hurtwitz (stable) matrix;

$W_{1,t} \in \Re^{n \times k}$ is the weights matrix for nonlinear state feedback;

$W_{2,t} \in \Re^{n \times r}$ is the input weights matrix;

W_1^* and W_2^* are the initial values for $W_{1,t}$ and $W_{2,t}$;

$\overline{W}_1$ and $\overline{W}_2$ are wieghted upper bounds for $W_{1,t}$ and $W_{2,t}$;

$\widetilde{W}_1$ and $\widetilde{W}_2$ are the weight estimation error of $W_{1,t}$ and $W_{2,t}$;

$K_t \in \Re^{n \times m}$ is the observer gain matrix;

$\phi(\cdot)$ is a diagonal matrix function;

$\sigma(\cdot)$ and $\gamma(.)$ are $n-$dimensional vector functions;

$\widetilde{\sigma}_t := \sigma(x_t) - \sigma(\widehat{x}_t)$, $\widetilde{\phi}_t := \phi(x_t) - \phi(\widehat{x}_t)$;

Δ_t is the identification error;

$\triangle f$ is the modeling error reflecting the effect of unmodelled dynamics;

L_i is the Liptshitz constant for the function $f(x) : R^n \xrightarrow{f} R^m$

$$\|f(x) - f(y)\| \leq L_i \|x - y\|, \quad \forall x, y \in \Re^n, \quad L_i \in [0, \infty);$$

$\overline{\lim\limits_{t\to\infty}}$ is the upper limit:

$$\overline{\lim_{t\to\infty}} x_t := \limsup_{t\to\infty} x_t = \lim_{t\to\infty} \sup_{n\geq t} x_n;$$

$\|\cdot\|$ is the Euclidian norm for vectors, and for any matrix A it is defined as

$$\|A\| := \sqrt{\lambda_{\max}(A^T A)};$$

$\lambda_{\max}(.)$ is the maximum eigenvalue of the respective matrix;

$\|\cdot\|_W$ is the weighted Euclidian norm of the vector $x \in R^n$:

$$\|\cdot\|_W := \sqrt{\sum_{i=1}^{n} w_i x_i^2};$$

$\|\cdot\|_Q$ is the semi-norms of a function, defined as

$$\|x\|_Q := \sqrt{\overline{\lim_{T\to\infty}} \frac{1}{T} \int_0^T x_t^T Q x_t dt};$$

$[\bullet]^+$ is the pseudoinverse matrix in Moor-Penrose sense, satisfing:

$$A^+ A A^+ = A^+, \quad A A^+ A = A$$

and

$$\forall x \in \Re^n (x \neq 0), \quad x^+ = \frac{x^T}{\|x\|}.$$

■the end of the proof.

Part I

Theoretical Study

1

Neural Networks Structures

In this chapter, we briefly review the structures of Neural Networks. We first take a look to biological ones. Then the different structures classifying them as static or dynamic neural networks are discussed. In the introduction, the importance of autonomous or intelligent systems is established, and the role of neural networks could play to implement such a system is also discussed. Regarding biological neural networks, we describe briefly the main phenomena which take place in them. A brief survey of the different neural networks structures: single layer, multi-layer perceptron, radial basis functions, and recurrent ones are also present. Finally, the applications of neural networks to control are discuss.

1.1 Introduction

The ultimate goal of control engineering is to implement an automatic system which could operate with increasing independence from human actions in an unstructured and uncertain environment [22]. Such a system may be named as an *autonomous* or *intelligent* one. It would need only to be presented with a goal and would achieve its objective by continuous interaction with its environment through feedback about its behavior. It would continue to adapt and perform tasks with increasing efficiency under changing and unpredictable conditions. It would be also very useful when direct human interaction could be hazardous, prone to failures, or impossible.

Biological systems are a possible framework for the design of such an autonomous system. They provide several clues for the development of *robust* (highly stable) learning and adaptation algorithms required for this kind of systems. Biological systems process information differently that conventional control schemes; they are model free and are quite successful for dealing with uncertainty and complexity. They do not require the development of a mathematical model to execute complex tasks. Indeed, they can learn to perform new tasks and easily adapt to changing environments. If the fundamental principles of computation imbedded in the nervous systems are understood, an entirely new generation of control methods could be developed far beyond the capabilities of the present techniques based on explicit mathematical model. These new methods could serve to implement the ultimate intelligent systems.

A control system has the ability to learn if it acquires information, during operation, about the unknown features of the plant and its environment such that the overall performance is improved. By enhancing the controller with learning, it is possible to expand the operation region and ultimately implement autonomous systems.

One class of models which has the potential to implement this learning is *the artificial neural network*. Indeed, the neural morphology of the nervous system is quite complex to analyze. Nevertheless, simplified analogies have developed, which could be applied to engineering applications. Based on these simplified understandings, artificial neural networks structures have been developed.

This chapter reviews briefly the biological neuron, and the artificial neural network structures. It also presents a brief review of neural networks applications to control.

1.2 Biological Neural Networks

In this section, we describe briefly the main phenomena which take place in *biological neural networks*. Being the nervous system, in particular the brain, an explosive research field, there exists an enormous bibliography about it. As a guide, we mention just a few: a good introduction to neurosciences is [1]; excellent textbooks are [68], [69], and a more recent reference is [2]. Very good short reviews about neurons, neural networks and the brain can be found in [22] and [16].

Neurons, or nerve cells, are the building blocks of the nervous system. Although they have the same general organization and biochemical apparatus of other cells, they posses unique features. They have a distinctive shape, an outer membrane capable of generating electric impulses, and a unique structure: the synapse to transfer information from one neuron to other neurons. Even, if there not exit two identical

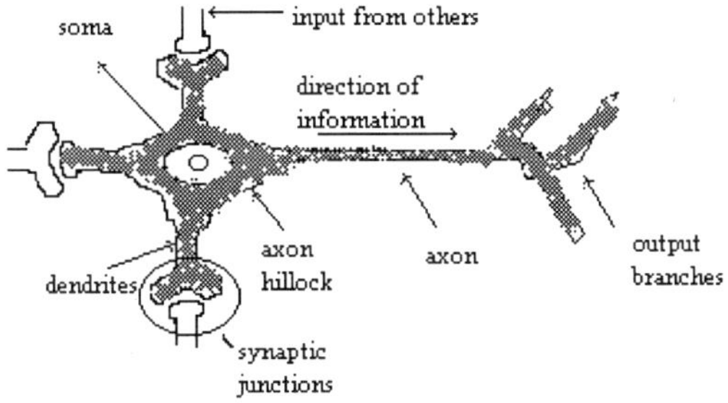

FIGURE 1.1. Biological Neuron Scheme.

neurons, it is possible to distinguish three regions on this specialized cell:

- the cell body,
- the dendrites,
- and the axon.

The *cell body*, or *soma*, provides the support functions and structure of the cell; it collects and process information received from other neurons. The *axon* extends away from the cell body and provides the path over which information travel to other neurons. The *dendrites* are tube like extensions that branch repeatedly and form a bushy tree around the cell body; they provide the main path on which the neuron receives coming information. A nerve impulse is triggered, at the origin of the axon, by the cell body in response to received information; the impulse sweeps along the axon until it reaches the end. The junction point of an axon with a dendrite of another neuron is called a *synapse*, which consists of two parts: the knob like axon terminal and the receptor region. There, information is conveyed from neuron to neuron by means of chemical transmitter, which are released by arriving nerve impulses. Figure 1.1 shows a scheme of the main neuron components.

An isolated neuron, which can be thought, as a multi-input, single-output (MISO) systems is shown in Figure 1.2

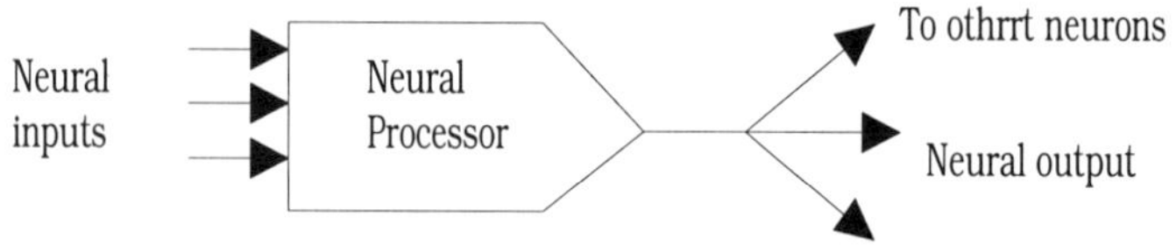

FIGURE 1.2. Biological Neuron Model.

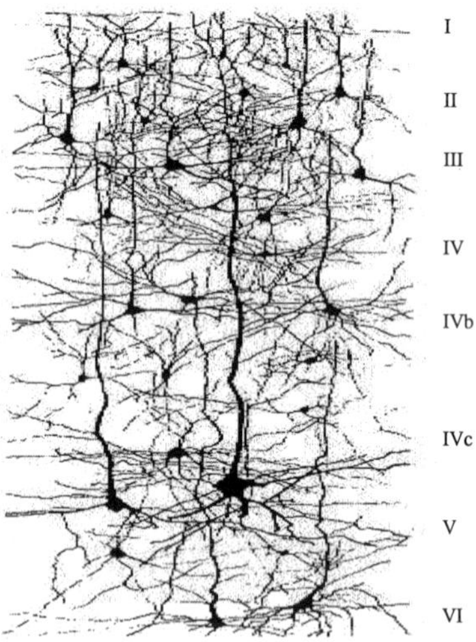

FIGURE 1.3. A biological neural network.

The massive interconnection of neurons constitutes a biological neural network as presented in Figure 1.3

Looking at the neuron in more detail, we can think of it as a tiny battery. In fact, neurons are filled and surrounded by fluids which contain dissolved chemicals; the fluid inside is in a big contrast from the one at the outside. Inside and around the cell body or soma are calcium (Ca^{++}), chloride (CL^{-}), potassium (K^{+}) and sodium (Na^{+}). K^{+} ions are concentrated inside the neuron and Na^{+}ones outside it; these ions are responsible for generating the nerve impulse. In an unexcited state there is only a minimal ion current through the membrane, and the voltage inside with respect to the outside (membrane potential) rests constant at about -70 mV. If the cell body is stimulated by a voltage greater than a certain threshold, a current of ions is established: Na^+into the cell body and K^+out of it, changing the cell body internal state by the increasing of the membrane potential. This ionic current is triggered by changes of the neurone membrane conductance in response to synapses. The presence of conductance changes opens channels which allows the ionic exchange. The changing of the cell body internal state starts a nerve impulse at the origin of the axon by locally lowering the voltage difference across the axon membrane; immediately ahead (in the direction of the nerve impulse propagation) of the altered region, the membrane conductance changes and Na^+ ions current enters changing the axon membrane potential to about + 30 mV. Soon after, about 0.5 ms, the sodium channels close, and another channels are open for Ka^+ions flow out to restore the axon potential to its resting value (-70mV) at that particular place. The sharp positive and then again negative potential constitutes the nerve impulse, also known as the activation potential. This potential wave propagates along the axon until it reaches the end. In summary the nerve impulse are traveling positive potential changes generated by the ionic current. Neurons have a refractory period; the axon can not immediately be started to transmit another nerve impulse. Generally this period lasts about 3- 5 ms. After this period a new nerve impulse is generated; so a nerve impulse train is produced. A nerve impulse is schematic presented in Figure 1.4.

When the nerve impulse reach the synaptic junction, it causes a release of neurotransmitters, which cross the synapse between the two neurons and alter the activity of the next neurone. Transmitter substances may have an excitatory or inhibitory effect; excitatory inputs will increase the neuron firing rate , and inhibitory ones will decrease this rate. A scheme of synapse is shown in Figure 1.5. In term of information processing the synapse performs a nerve impulse train frequency to voltage conversion. Each cell body receive numerous excitatory and inhibitory inputs, which are added in a spatio-temporal way by means of a weighted average. If this average exceeds a threshold it is converted in a nerve impulse train at the axon origin. Not all neurons generate nerve impulses; some of them, i.e. those in the retina, spread

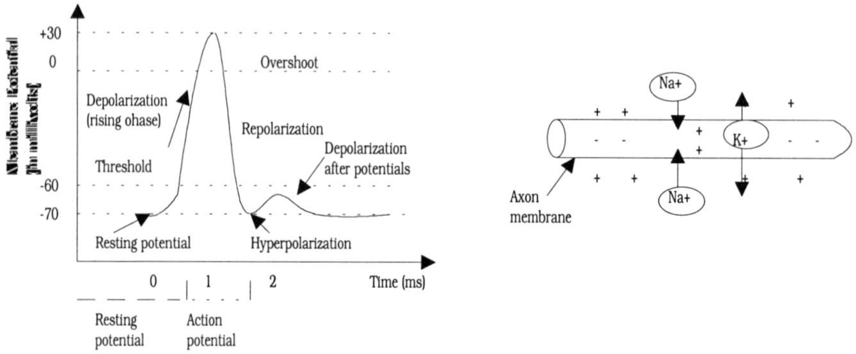

FIGURE 1.4. Nerve Impulse.

grades change in the membrane potentials along the axon. These changes cause the release of neurotransmitter at synapses.

The nature of biological neural networks is constituted by the density of neurons and their massive interconnection; these networks conforms the *brain* and the nervous system. The brain is a dense highly interconnected neural network; its neuron are continuously processing and sending information. The main components of the human brain are presented in 1.6.

The brain is covered by the cerebral cortex, the convoluted outer layer; in higher mammals, it is extensively folded in order to fit inside a skull of reasonably size. The size and complexity of the cerebral cortex conform a critical difference between higher mammals, in particular humans beings, and lower animals. Different cortex regions are specialized for complex tasks such as language processing, visual information analysis, and other aspect of behavior which constitutes intelligence. Figure 1.7 shows the main components of the cerebral cortex and some of the specialized areas.

The major neural structures placed within or below the cerebral cortex of a human brain are the cerebellum, the spinal cord, the brain stem, the thalamus, and the limbic system. The cerebellum is involved with sensory-motor coordination; the spinal cord transmits information up and down between the other components of

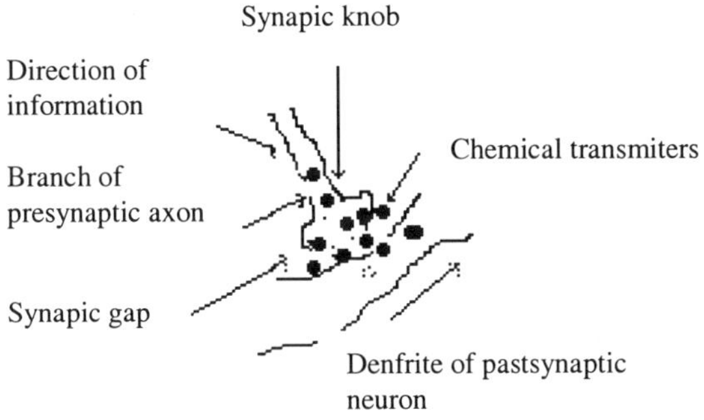

FIGURE 1.5. Synapse.

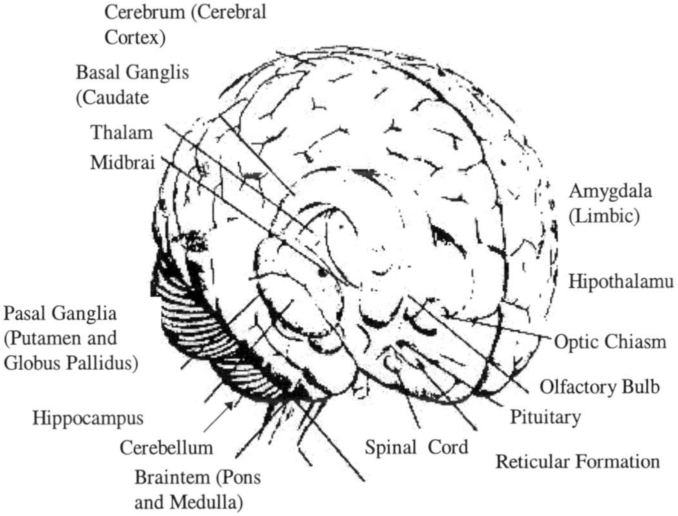

FIGURE 1.6. Human Brain Major Structures.

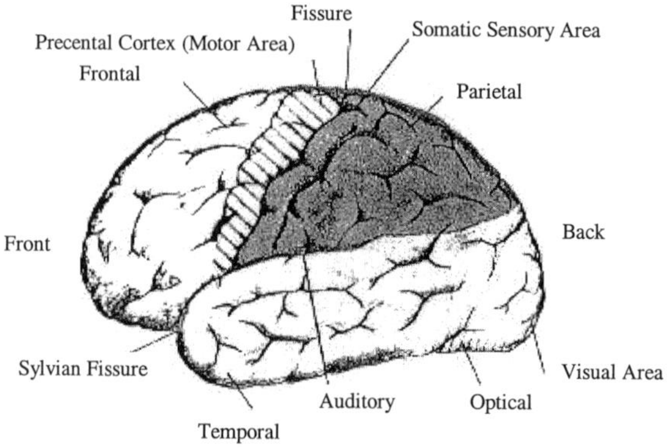

FIGURE 1.7. Cerebral Cortex.

the nervous system and the brain; the brain stem is concerned with respiration, heart rhythm, and gastrointestinal functions; the thalamus functions as a relay station for the projection of the major sensory systems into the cerebral cortex; the symbic system (amygdala, hippocampus and adjacent regions) are mainly involved with smell. In higher mammals, the hippocampus has taken on new roles as memory codification.

Very recent research results have allowed to determine which neurons or group of neurons are involved in particular intelligent tasks such as: mathematical calculations, positioning, and memory [70]. A hippocampus group of neurons responsible for memory codification is shown in Figure 1.8.

1.3 Neuron Model

An artificial neural network (ANN) is a massively parallel distributed processor, inspired from biological neural networks, which can store experimental knowledge and makes it available for use [71]. It has some similarities with the brain, such as:

a) knowledge is acquired through a learning process;

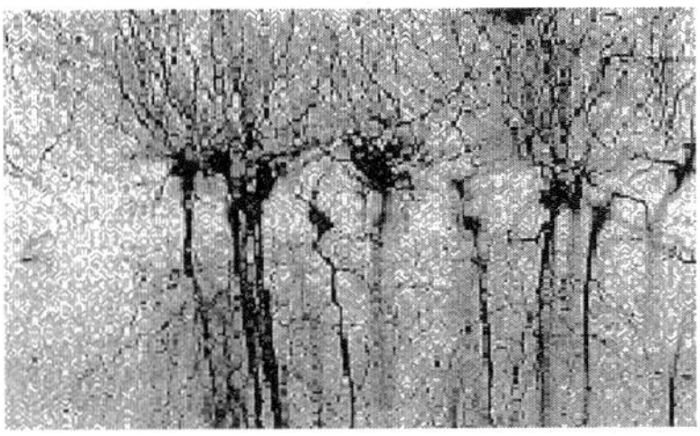

FIGURE 1.8. A hippocampus group of neurons responsible for memory codification.

b) interneuron connectivity named as synaptic weights are used to store this knowledge.

The procedure for *the learning process* is known as a *learning algorithm.* Its function is to modify the synaptic weights of the networks in order to attain a prespecified goal. The weights modification provides the traditional method for neural networks design and implementation.

The neuron is the fundamental unit for the operation of a neural network. 1.9 presents a neuron scheme. There are three basic elements:

1. A set of *synapsis links*, with each element characterized by its own weight.

2. An *adder* for summing the inputs signal components, multiplied by the respective synapsis weight.

3. A *nonlinear activation function* transforming the adder output into the output of the neuron.

An *external threshold* is also applied to lower the input to the activation function. In mathematical terms, the $i-th$ neuron can be described as:

$$\boxed{\begin{array}{l} v_i = \sum_{j=1}^{n} w_{ij} u_j \\ y_i = \varphi\left(v_i - \rho_i\right) \end{array}}$$

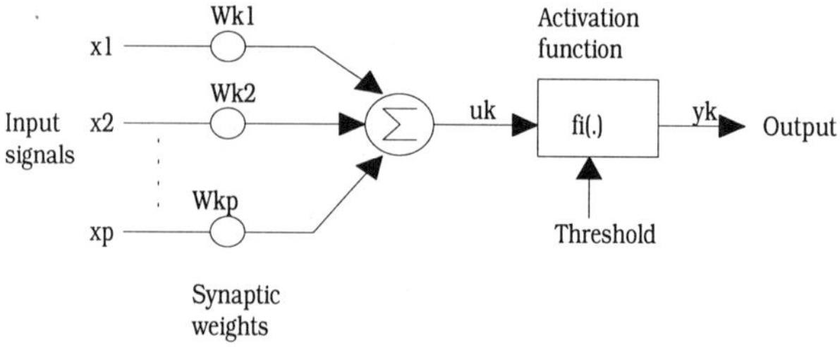

FIGURE 1.9. Nonlinear model of a neuron.

where:

u_j is the j-th component of the input,

w_{ij} is the weight connecting the j-th input component to neuron i,

v_i is the output of the adder,

ρ_i is the threshold,

$\varphi(.)$ is the nonlinear activation function,

y_i is the output of neuron i.

This scheme can be simplified as shown in Figure 1.10, which is characterized by:

- input nodes, which supply the input signal to the neuron,
- the neuron is represented a single node, named a computation one,
- communication links interconnecting the input nodes and the computation ones.

It allows to simplify the scheme drawing for the different neural networks structures.

1.4 Neural Networks Structures

The way in which the neurons of a neural network are interconnected determines its structure. For the purposes of identification and control, the structures the most used are:

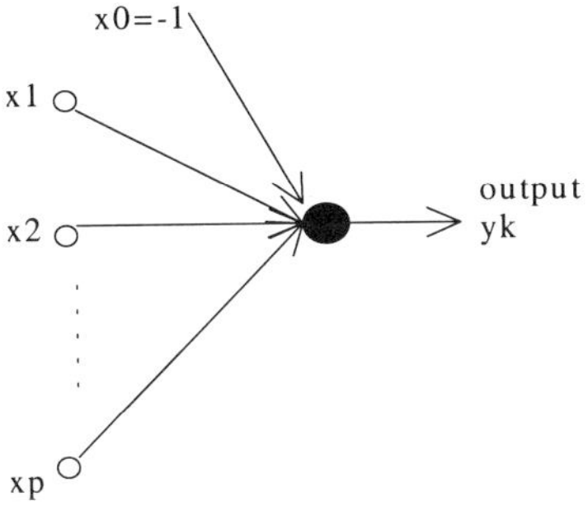

1.

FIGURE 1.10. Simplified scheme.

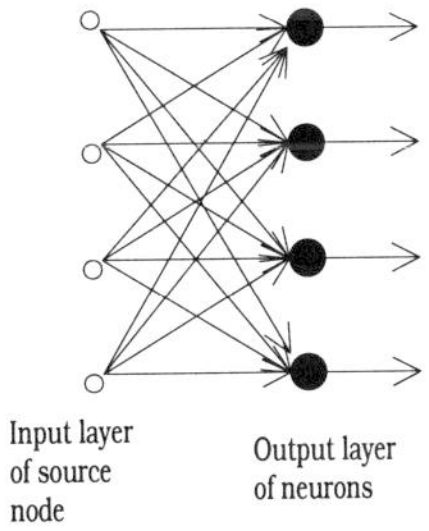

FIGURE 1.11. Single-Layer Fedforward Network.

1. Single-layer Feedforward Networks;

2. Multilayer Feedforward Networks;

3. Radial Basis Networks;

4. Dynamic (differential) or Recurrent Neural Networks;

1.4.1 Single-Layer Feedforward Networks

It is the simplest form of feedforward networks. It has just one layer of neurons as shown in Figure 1.11. The best known is the so called Perceptron. Basically it consists of a single neuron with adjustable synaptic weights and threshold.

The learning algorithm to adjust the weights of this neural networks first appeared in [62],[65]. There, it is proved that if the information vectors, used to train the perceptron are taken from two linearly separable classes, then the perceptron learning algorithm converges and defines the decision surface as an hyperplane separating the two classes. The respective convergence proof is known as the perceptron convergence theorem.

Basic to the perceptron is the McCulloch-Pitts model[38], which activation function $\varphi(.)$ is a hard limiter.

The purpose of the perceptron is to classify the input signal, with components:

$$u_1, u_2, ...u_n$$

into one of two classes: C_1 or C_2. The classification decision rule is to assign the point corresponding to input $u_1, u_2, ...u_n$ to class C_1 if the perceptron output (y) is equal to $+1$ and to class C_2 if it is -1.

Usually the threshold (ρ) is treated as a synaptic weight connected to a fixed input equal to -1, so the input vector is defined as:

$$(u(k))^T = (-1, u_1(k), u_2(k) ...u_n(k))$$

where k stands for the k-th example. The output adder is calculated by:

$$v(k) = w^T(k)\, u(k) = u^T(k)\, w(k) = \sum_{i=1}^{n} u_i(k) w_i(k) - w_0(k)$$

where w is the vector of synaptic weights.

For any k, on the n-dimensional space, the equation $w^T u$, with coordinates

$$u_1, u_2, ...u_n$$

defines an hyperplane separating the inputs in the two classes: C_1 and C_2. If these classes are linearly separable, then there exists a vector w such that:

$$w^T u \geq 0, \ \forall u \in C_1$$

and

$$w^T u < 0, \ \forall u \in C_2$$

The learning algorithm adapts the weight vector as follows:

1.

$$w(k+1) = w(k) \ \text{ if } w^T u \geq 0, \forall u \in C_1 \ \text{ or if } w^T u < 0, \ for \ \ u \in C_2,$$

2.

$$w(k+1) = w(k) - \eta(k) u(k) \ \text{ if } w^T u \geq 0, \ for \ \ u \in C_2,$$

or

$$w(k+1) = w(k) + \eta(k) u(k) \ \text{ if } w^T u < 0, \ for \ \ u \in C_1.$$

The convergence of this algorithm can be demonstrated by a contradiction argument [71].

If we look only at the adder output $v(k)$ and define the error as:

$$e(k) = d(k) - v(k)$$

then a least mean square algorithm can be derived to minimize this error. The derivation of this algorithm is based on the gradient descent method. The resulting learning law for the weights is as follows:

$$\boxed{w(k+1) = w(k) + \eta e(k) u(k)}$$

Because the error depends linearly on the weights, this algorithm assures a global minimum.

Based on the LMS algorithm, B. Widrow and collaborators (see [83] and [84]) proposed the Adaline (adaptive linear element), whose scheme is presented in Figure 1.12. It consists of an adder, a hard limiter and the LMS algorithm to adjust the weights.

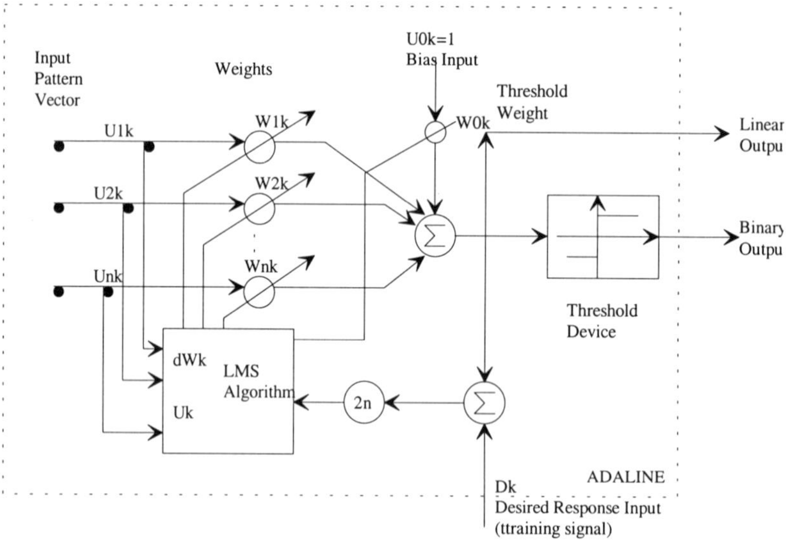

FIGURE 1.12. Adaline scheme.

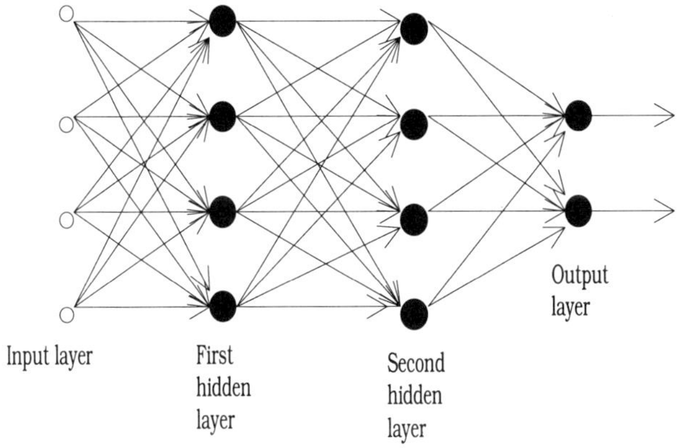

FIGURE 1.13. Mulilayer Perceptron.

1.4.2 Multilayer Feedforward Neural Networks

They distinguish themselves by the presence of one or more hidden layer (Figure 1.13) whose computation nodes are called *hidden neurons.* Typically the neurons in each layer have as their inputs the output signals of the preceding layer. If each neuron in each layer is connected to every neurone in the adjacent forward layer, then the neural network is named as fully connected, on the opposite case, it is called partly connected.

A multilayer perceptron has three distinctive characteristics:

1. The activation function of each neuron is smooth as opposed to the hard limit used in the single layer perceptron. Usually, this nonlinear function is a sigmoidal one defined as:

$$\boxed{\varphi_i(v_i) = 1/(1 + e^{-v_i})}$$

1. The network contains one or more layers of hidden neurons

2. The network exhibits a high degree of connectivity.

Multilayer perceptron derives its computing power through the combination of these characteristics and its ability to learn from experience. However the presence of distributed nonlinearities and the high connectivity of the network make the theoretical analysis difficult. Research interest in this kind of network dates back to [65] and to Madalines, constructed with many Adalines elements on the first layer and a variety of logic devices in the second layer[85].

Backpropagation Algorithm

The learning algorithm used to adjust the synaptic weights of a multilayer perceptron is known as back-propagation. The basic idea was first described in [86]. Subsequently it was rediscovered in [64]; similar generalization of the algorithm were derived in [52] and [39].This algorithm provides a compositionally efficient method for the training of multi-layer perceptron. Even if it does not give a solution for all problems, it put to rest the criticism about learning in multilayer neural networks, which could be inferred from [41].

The error at the output of neuron j, element of the output layer, is given as:

$$e_j(k) = d_j(k) - y_j(k)$$

where:

d_j is the desired output.

y_j is the neuron output

k indicates the k -th example

The instantaneous sum of the squared output errors is given by:

$$\mathcal{E}(k) = \frac{1}{2}\sum_{j=1}^{l} e_j(k)$$

where l is the number of neurons of the output layer.

The average squared error is obtained by summing $\mathcal{E}(k)$ over all the examples (an epoch) and then normalizing with respect to the epoch size

$$\mathcal{E}_{av} = \frac{1}{N} \sum_{k=1}^{N} \mathcal{E}(k)$$

with N the number of examples, which forms an epoch.

Using the gradient decent, as for the LMS algorithm, the weight connecting neuron i to neuron j is updated as:

$$\boxed{\Delta w_{ji}(k) = w_{ji}(k+1) - w_{ji}(k) = -\eta \frac{\partial \mathcal{E}(k)}{\partial w_{ji}(k)}}$$

The correction term $\Delta w_{ji}(k)$ is known as the delta rule. The term $\frac{\partial \mathcal{E}(k)}{\partial w_{ji}(k)}$ can be calculated as

$$\frac{\partial \mathcal{E}(k)}{\partial w_{ji}(k)} = \frac{\partial \mathcal{E}(k)}{\partial e_j(k)} \frac{\partial e_j(k)}{\partial y_j(k)} \frac{\partial y_j(k)}{\partial v_j(k)} \frac{\partial v_j(k)}{\partial w_{ji}(k)}$$

The partial derivatives are given by

$$\begin{aligned} \frac{\partial \mathcal{E}(k)}{\partial e_j(k)} &= e_j(k), \ \frac{\partial e_j(k)}{\partial y_j(k)} = -1 \\ \frac{\partial y_j(k)}{\partial v_j(k)} &= \varphi_j'(v_j(k)) \ with \ \varphi_j'(\beta) = \frac{\partial \varphi_j(\beta)}{\partial \beta} \\ \frac{\partial v_j(k)}{\partial w_{ji}(k)} &= y_i(k) \end{aligned}$$

So, the delta rule can be rewritten as

$$\Delta w_{ji}(k) = \eta \delta_j(k) y_i(k)$$

with

$$\delta_j(k) = -\frac{\partial \mathcal{E}(k)}{\partial y_j(k)} \frac{\partial y_j(k)}{\partial v_j(k)}.$$

Two cases can be distinguished: the neuron j is in the output layer or it is in a hidden layer.

Case 1.

If the neuron j is located in the output layer is straightforward to calculate $\delta_j(k)$

Case 2.

When neuron j is located in a hidden layer, there is not specified desired response for that neuron. So, its error signal has to be derived recursively in terms of the error for all the neurons to which it is connected. In this case it is possible to establish the following equation:

$$\delta_j(k) = \frac{\partial y_j(k)}{\partial v_j(k)} \sum_{n=1}^{m} \delta_n(k) w_{nj}(k)$$

where n indicates the n-th neuron to which neuron j is connected and m the total number of these neurons.

This is a brief exposition of *backpropagation* taken from[71], where a complete derivation is presented. The backpropagation algorithm has become the most popular one for training of the multilayer perceptron. It is compositionally very efficient and it is able to classify information non-linearly separable. The algorithm is a gradient technique, implementing only one step search in the direction of the minimum, which could be a local one, and not an optimization one. So, it is not possible to demonstrate its convergence to a global optimum.

Function Approximation

A multilayer perceptron trained with the backpropagation algorithm is able to perform a general nonlinear input-output mapping from $\Re^n$ (dimension of the input space) to $\Re^l$ (dimension of the output space). Research interest in the capabilities of the multilayer perceptron to approximate an arbitrary continuous function was first mentioned in [30]. It was Cybenko who first demonstrated that a single hidden layer is sufficient to uniformly approximate any continuous function with support in a unit hypercube [10] and [11]. In 1989, two additional papers [21],[31] were published on proofs of the multilayer perceptron as universal approximate of continuous functions.

The capability of multilayer perceptron to approximate arbitrary continuous functions is established in the following theorem

Theorem 1.1 *(Cybenko [10], [11]) Let $\sigma(.)$ be a stationary, bounded, and monotone increasing function. Let I_n denote the n-dimensional unit hypercube. Let $C(I_n)$ the space of continuous functions on I_n. Then for any $f \in C(I_n)$ and $\varepsilon > 0$, there exist an integer m and real constants α_i, ρ_i and w_{ij}, with $i = 1, ...m$ and $j = 1...n$, such that defining $F(u_1, u_2...u_n)$ as*

$$F(u_1, u_2...u_n) = \sum_{i=1}^{m} \alpha_i \varphi \left(\sum_{j=1}^{n} w_{ij} u_j - \rho_i \right);$$

it is an approximate realization of $f(.)$, that is,

$$|F(u_1, u_2...u_n) - f(u_1, u_2...u_n)| < \varepsilon, \forall (u_1, u_2...u_n) \in I_n$$

This theorem is directly applied to multilayer perceptron, with the following characteristics:

1. The input nodes: $u_1, u_2...u_n$.

2. A hidden layer of m neurons completely connected to the input.

3. The activation function $(\varphi(.))$, for the hidden neurons, is a constant, bounded, and is monotonically increasing.

4. The network output is a linear combinations of the hidden neurons output.

This property is very useful for applications of multilayer perceptron to control.

1.4.3 Radial Basis Function Neural Networks

Radial Basis Function (RBF) neural networks have three entirely different layers.

1. The input layer made up of input nodes.

2. The hidden layer, with a high enough number of nodes (neurons). Each of these nodes performs a nonlinear transformation of the input, by means of Radial Basis Functions.

3. The output layer, which is a linear combinations of the hidden inputs neurons.

Radial Basis Functions were first introduced for the solution of multivariate interpolation problems; early works on this approach is surveyed in [53]. The first application of radial basis functions to neural networks design is reported in [6]. Major contributions to the theory, design, and application of these functions to neural networks are [43], [61], [49].

A Radial Basis Function is a multi-dimensional one, which depends on the distance between its input vector $u \in \Re^n$ and a previously defined center $c \in \Re^n$. Usually, this distance is calculated as:

$$d = \sqrt{(u - c)^T (u - c)}$$

There exist different Radial Basis Functions, i.e,:

- $G(d) = d$ (piece-wise linear function),
- $G(d) = d^3$ (cubic function),
- $G(d) = e^{-\frac{d^2}{\sigma^2}}$ (Gaussian function),
- $G(d) = \left(\frac{d^2}{\sigma^2}\right)^{\frac{1}{2}}$ (multi-quadratic function).

When a RBF neural network is used for classification, it solves this problem by transforming it into a high dimensional space. The justification for doing so is given by the Cover's theorem[15], which establishes that a classification problem is more likely to be linearly separable in a high dimensional space than in a low dimensional one.

The other theoretical justification for RBF neural network is regularization theory for the solution of ill posed problems. This theory was introduced by Tikhonov in

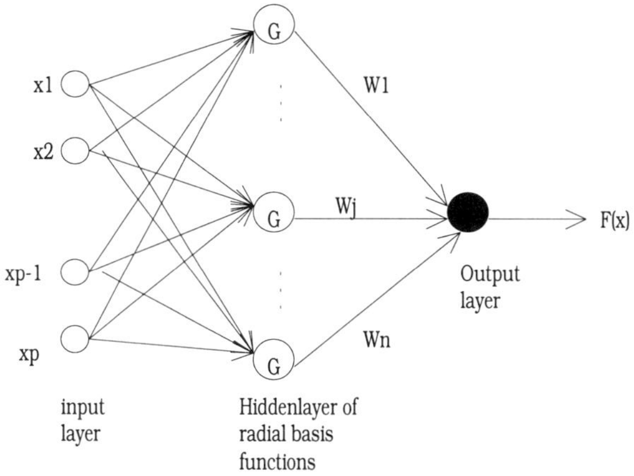

FIGURE 1.14. A general scheme for RBF neural networks.

1963 (see [82]). For approximation problems, the basic idea is to stabilize the solution by means of an auxiliary non-negative functional that embeds prior information, and thereby transforms an ill posed problem into a well posed one. In [49], the regularization theory is applied to analysis properties of RBF neural networks.

A general scheme for RBF neural networks is presented in Figure 1.14

Its mathematical model is given by:

$$\boxed{y = F(u) = \sum_{i=1}^{m} w_i G\left(\|u - c_i\|_{r_i}^{2} \right)}$$

where:

w_i are the weights,

$G(.)$ is the radial basis function,

$u^T = (u_1, u_2 .... u_n)$ is the input,

c_i are the centers., $c_i \in \Re^n$,

r_i are the radii.$r_i \in \Re$,

m is the number of neurons

and

$$\|u - c_i\|_{r_i}^2 = \frac{(u - c_i)^T (u - c_i)}{r_i}$$

Defining

$$R_i = \begin{pmatrix} \frac{1}{\sqrt{r_i}} & 0 & \dots & 0 & \dots & 0 \\ 0 & \frac{1}{\sqrt{r_i}} & \dots & 0 & \dots & 0 \\ . & . & \dots & . & \dots & . \\ 0 & . & \dots & \frac{1}{\sqrt{r_i}} & \dots & 0 \\ . & . & \dots & . & \dots & . \\ 0 & . & \dots & . & \dots & \frac{1}{\sqrt{r_i}} \end{pmatrix}$$

and it follows

$$\Omega_i^{-1} = R_i^T R_i,$$

then

$$\|u - c_i\|_{r_i}^2 = (u - c_i)^T R_i^T R_i (u - c_i) = (u - c_i)^T \Omega_i^{-1} (u - c_i)$$

The matrix Ω_i^{-1} could be interpreted as a covariance one. If

$$R_i = \begin{pmatrix} \frac{1}{\sqrt{r_{1i}}} & 0 & \dots & 0 & \dots & 0 \\ 0 & \frac{1}{\sqrt{r_{2i}}} & \dots & 0 & \dots & 0 \\ . & . & \dots & . & \dots & . \\ 0 & . & \dots & \frac{1}{\sqrt{r_{ii}}} & \dots & 0 \\ . & . & \dots & . & \dots & . \\ 0 & . & \dots & . & \dots & \frac{1}{\sqrt{r_{pi}}} \end{pmatrix}$$

then an elliptic base is obtained.

One of the most popular RBF are the Gaussian functions:

$$\begin{aligned} G\left(\|u - c_i\|_{r_i}^2\right) &= \exp\left(-\|u - c_i\|_{r_i}^2\right) = \exp\left(-(u - c_i)^T R_i^T R_i (u - c_i)\right) \\ &= \exp\left(-(u - c_i)^T \Omega_i^{-1} (u - c_i)\right) \end{aligned}$$

They have the property of being factorizables.

Gaussian RBF can also approximate any continuous nonlinear function (See [51] and [67]). This property does them very attractive for applications in control.

Learning Algorithm

Assuming that there are N training examples, the error is defined as:

$$\mathcal{E} = \frac{1}{2}\sum_{k=1}^{N} e^{2}(k)$$

with

$$e(k) = d(k) - \sum_{i=1}^{m} w_i G\left(\|u(k) - c_i\|_{r_i}^{2}\right)$$

The quadratic error $\mathcal{E}$ can be minimized with respect to w_i, c_i, and/or r_i.This minimization can be performed for each example or per epoch. In the following, the first case is analyzed in detail.

1. **Weights Adaptation.**

   If only the weights are changed then we have:

$$\frac{\partial \mathcal{E}(k)}{\partial w_i(k)} - \frac{\partial}{\partial w_i(k)}\left(\frac{1}{2}e^{2}(k)\right)$$
$$= e(k)\frac{\partial}{\partial w_i(k)}\left(d(k) - \sum_{i=1}^{m} w_i(k) G\left(\|u(k) - c_i\|_{r_i}^{2}\right)\right)$$
$$\frac{\partial \mathcal{E}(k)}{\partial w_i(k)} = -e(k) G\left(\|u(k) - c_i\|_{r_i}^{2}\right)$$

   By the use of the gradient descent, the weights are adapted as

$$w_i(k+1) = w_i(k) - \eta_w \frac{\partial \mathcal{E}(k)}{\partial w_i(k)} \quad for\ \imath = 1, .., m.$$

2. **Center Adaptation**

We consider this time that the centers are also updated, then:

$$\begin{aligned}\frac{\partial \mathcal{E}(k)}{\partial c_i(k)} &= \frac{\partial}{\partial c_i(k)}\left(\frac{1}{2}e^2(k)\right)\\ &= -w_i(k)\,e(k)\frac{\partial}{c_i(k)}G\left(\|u(k)-c_i(k)\|_{r_i}^2\right)\\ &\quad \frac{\partial}{\partial c_i(k)}G\left(\|u(k)-c_i(k)\|_{r_i}^2\right)\\ &= G'\left(\|u(k)-c_i(k)\|_{r_i}^2\right)\frac{\partial}{\partial c_i(k)}\left(\|u(k)-c_i(k)\|_{r_i}^2\right)\end{aligned}$$

where

$$G'(\beta) = \frac{\partial G(\beta)}{\partial \beta}$$

Considering that

$$\frac{\partial}{\partial c_i(k)}\left(\|u(k)-c_i(k)\|_{r_i}^2\right) = -2R_i^T(k)\,R_i(k)\,(u(k)-c_i(k))$$

we obtain

$$\frac{\partial \mathcal{E}(k)}{\partial c_i(k)} = 2w_i(k)\,e(k)\,G'\left(\|u(k)-c_i(k)\|_{r_i}^2\right)R_i^T(k)\,R_i(k)\,(u(k)-c_i(k))$$

So, the centers are updated as

$$c_i(k+1) = c_i(k) - \eta_r\frac{\partial \mathcal{E}(k)}{\partial c_i(k)} \quad for\ \ i = 1,..,m$$

3. **Radii Adaptation**

Finally, we consider that all three parameters are updated.

$$\begin{aligned}\frac{\partial \mathcal{E}(k)}{\partial \Omega_i^{-1}(k)} &= \frac{\partial}{\partial \Omega_i^{-1}(k)}\left(\tfrac{1}{2}e^2(k)\right)\\ &= -w_i(k)\,e(k)\frac{\partial}{\partial \Omega_i^{-1}(k)}G\left(\|u(k)-c_i(k)\|_{r_i(k)}^2\right)\\ &\quad \frac{\partial}{\partial \Omega_i^{-1}(k)}G\left(\|u(k)-c_i(k)\|_{r_i(k)}^2\right)\\ &= G'\left(\|u(k)-c_i(k)\|_{r_i(k)}^2\right)\frac{\partial}{\partial \Omega_i^{-1}(k)}\left(\|u(k)-c_i(k)\|_{r_i(k)}^2\right)\end{aligned}$$

Now, taking in account that

$$\frac{\partial}{\partial \Omega_i^{-1}(k)} \left(\left\| u(k) - c_i(k) \right\|_{r_i(k)}^2 \right) = (u(k) - c_i(k))(u(k) - c_i(k))^T = Q_i(k)$$

we derive:

$$\frac{\partial \mathcal{E}(k)}{\partial \Omega_i^{-1}(k)} = -w_i(k) e(k) G' \left(\left\| u(k) - c_i(k) \right\|_{r_i(k)}^2 \right) Q_i(k)$$

and the radii are updated as

$$\Omega_i^{-1}(k+1) = \Omega_i^{-1}(k) - \eta_\Omega \frac{\partial \mathcal{E}(k)}{\partial \Omega_i^{-1}(k)} \; for \; i = 1, .., m.$$

It is worth noting that $\mathcal{E}$ is convex only with respect to w_i. Usually the following relation is selected

$$\eta_\Omega < \eta_r < \eta_w$$

So the descent step of the weights is bigger than the centers one, and this last step is bigger than the radii one. The weights adaptation helps to minimize the error between the output of the network and the desired output. To adapt the centers assures the clustering of the input information. The radii are adapted to reach a certain degree of overlapping between the radial basis functions.

In the case of a per epoch minimization the respective partial derivative are:

$$\begin{aligned}
\frac{\partial \mathcal{E}}{\partial w_i} &= -\sum_{k=1}^{N} e(k) G \left(\left\| u(k) - c_i \right\|_{r_i}^2 \right) \\
\frac{\partial \mathcal{E}}{\partial c_i} &= 2w_i \sum_{k=1}^{N} e(k) G' \left(\left\| u(k) - c_i \right\|_{r_i}^2 \right) R_i^T(k) R_i(k) (u(k) - c_i) \\
\frac{\partial \mathcal{E}}{\partial \Omega_i^{-1}} &= -w_i(k) \sum_{k=1}^{N} e(k) G' \left(\left\| u(k) - c_i \right\|_{r_i}^2 \right) Q_i(k) \\
Q_i(k) &= (u(k) - c_i)(u(k) - c_i)^T
\end{aligned}$$

The respective parameter adaptation is given by a similar gradient decent formula.

All the structures of neural networks, tackled here, can be classified as static; they perform a nonlinear static transformation from their inputs to their outputs.

1.4.4 Recurrent Neural Networks

A common approach for encoding temporal information using static neural networks is to include delay inputs and outputs. However, this representation is limited, since it can only encode a finite number of previous measured outputs and imposed inputs; moreover, it tends to require prohibitively large amounts of memory, thereby hindering its use for all but relatively low order dynamical systems. As an very efficient and promising alternative, the international research community has been exploring the use of recurrent or dynamic neural networks.

Recurrent or dynamic neural networks distinguish themselves from static neural networks in that they have at least one feedback loop. One of the first surveys of structures, learning algorithms and applications of this kind of neural networks is given in [25] There, it is signaled that, neural networks, whose structures include feedback, are present from the very earliest development of artificial neural networks; in fact, in [38],McCulloch and Pitts developed models for feedforward networks, which have time dependence and time delays; however, these networks were implemented with threshold logic neurons. Then, they extended their network to those with dynamic memory; these networks had feedback. Later, these networks were modeled as finite automata with a regular language in [36] , which is usually referenced as the first work on this kind of automata.

The feedback loops involve the use, for discrete time, of branches composed of unit delay elements denoted by q^{-1}, such that: $u(k-1) = q^{-1}u(k)$, with k indicating the k-th sample in time. Figure 1.15 shows a discrete time recurrent neural networks.

The feedback loops result in a nonlinear dynamical behavior due to the nonlinear activation function of the neurons. So the term dynamic neural network, which is even least well known, describes better this structure of neural networks. Due to these facts, we name them as *dynamic neural networks.*

This kind of neural networks allows better understanding of biological structures. They can offer also great computational advantages. In fact it is well known that a static infinite order linear adder is equivalent to a single pole feedback linear system. (see Figure 1.15). From Figure 1.15 , we find that the output is:

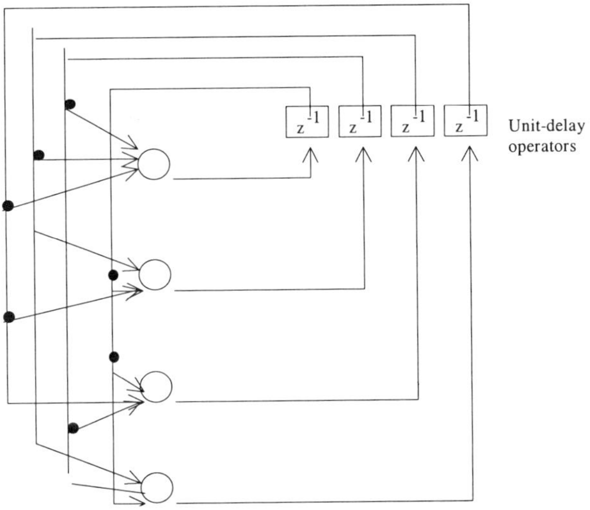

FIGURE 1.15. Discrete Time recurrent Neural Networks.

$$v(k) = u(k) + u(k-1) + ...u(k-n) = \sum_{i=0}^{n} u(k-i), \; n \to \infty$$

The linear system is described by:

$$\begin{aligned} v(k) &= v(k-1) + u(k) \\ \frac{v(k)}{u(k)} &= \frac{1}{1-q^{-1}} = 1 + q^{-1} + q^{-2}...q^{-n} \\ v(k) &= u(k) + u(k-1) + ...u(k-n) \end{aligned}$$

It is clear that the two structures are equivalent, but from a computational point of view, a system with feedback is equivalent to a large, possible infinity, static structure. This property is very interesting for identification and control, and open the road for applications of dynamic neural networks to these fields.

Neural structures with feedback are particularly appropied for system identification and control. They are important because most of the system to be modelled and controlled are indeed nonlinear dynamic ones.

A very well known dynamic neural network is the Hopfield one. There exist two versions: the discrete-time one and the continuous time one. Both were proposed by Hopfield; the former was introduced in [26], and the latest in[27] .

The discrete-time Hopfield neural network can be represented by the following mathematical model:

$$\boxed{x_i\,(k+1) = sgn\left[\sum_{j=1}^{n} w_{ij}x_j\,(k) + u_i - \rho_i\right]}$$

where:

x_i is the state of the i-th neuron,

n is the number of neurons,

u_i is the input to the i-th neuron,

ρ_i is the threshold of the i-th neuron,

w_{ik} is the synaptic weight connecting neuron j to neuron i.

This model can be rewritten as a matrix equation:

$$\begin{aligned}
x\,(k+1) &= \Gamma\left(Wx\,(k) + u - \rho\right) \\
&= \Gamma(v\,(k)), v\,(kA) = Wx\,(k) + u - \rho \\
x^T &= \left(x_{1,}x_{2,\ldots}x_{i,\ldots}x_n\right), \\
u^T &= \left(u_{1,}u_{2,\ldots}u_{i,\ldots}u_n\right), \rho^T = \left(\rho_{1,}\rho_{2,\ldots}\rho_{i,\ldots}\rho_n\right) \\
v^T &= \left(v_{1,}v_{2,\ldots}v_{i,\ldots}v_n\right), \\
(\Gamma(v))^T &= \left(sign\,(v_1)\,, sign\,(v_2)\ldots sign\,(v_i)\ldots sign\,(v_n)\right) \\
W &= \begin{pmatrix}
w_{11} & w_{12} & \cdots & w_{1i} & \cdots & w_{1n} \\
w_{21} & w_{22} & \cdots & w_{i2} & \cdots & w_{2n} \\
. & . & \cdots & . & \cdots & . \\
w_{i1} & w_{i2} & \cdots & w_{ii} & \cdots & w_{in} \\
. & . & \cdots & . & \cdots & . \\
w_{n1} & w_{n2} & \cdots & w_{ni} & \cdots & w_{nn}
\end{pmatrix}
\end{aligned}$$

where W is defined as the weight matrix.

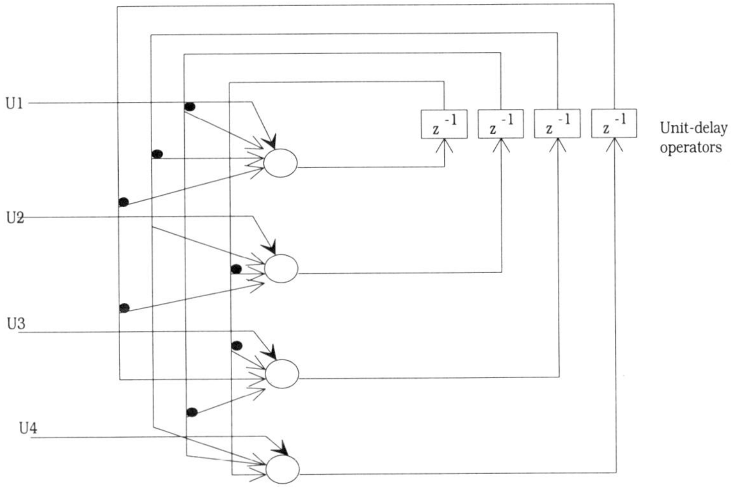

FIGURE 1.16. A diagram of this kind of recurrent neural network.

Due to the sign function, the state of each neuron can only take the value ± 1. Using an energy function [26], it is possible to find conditions guaranteeing that the neural networks converges to a steady state given by:

$$x_i = sgn\left[\sum_{j=1}^{n} w_{ij} x_j(k) + u_i - \rho_i\right]$$

For convergence, it is required to have a symmetric weight matrix, with no-negative diagonal elements. Figure 1.16 shows a diagram of this kind of recurrent neural network. An alike discrete-time dynamic neural network is the so called Brain-in- the-Box one [5].

1.4.5 Differential Neural Networks

The continuous time Hopfield Neural Network or, on our terminology, the *Differential Neural Networks*, can be described by an electric circuit, which is based on a RC network connecting nonlinear amplifiers [35]. The i-th amplifiers are characterized by their input-output functions: $x_i = \varphi_i(v_i)$, with $x_i$and v_i the respective input and

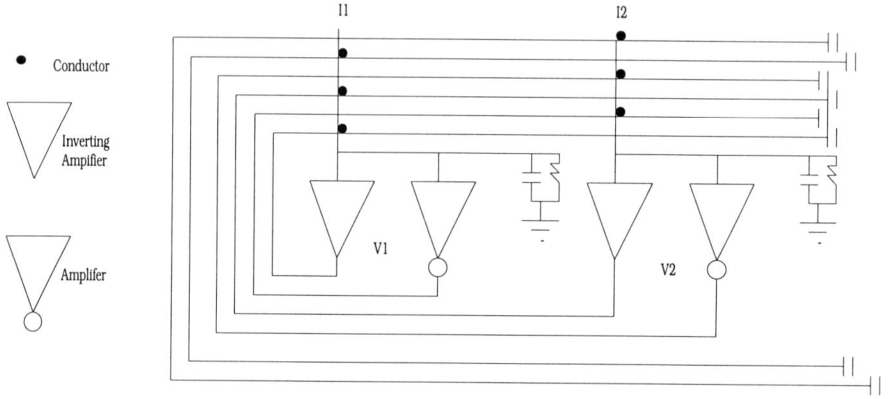

FIGURE 1.17. Hopfield Neural network

output voltages; additionally the following biological interpretation can be done: x_i is the neuron soma voltage, x_i is the neuron output voltage, and $\varphi_i(.)$ is the activation function. This neural network is presented in Figure 1.17. Writing at the input of the i-th amplifier, then we obtain:

$$
\begin{aligned}
C_i \frac{dv_i}{dt} &= \sum_{j=1}^{n} T_{ij} x_j - \frac{v_i}{R_i} + u_i \\
&= \sum_{j=1}^{n} T_{ij} \varphi_j(v_j) - \frac{v_i}{R_i} + u_i, x_i = \varphi_i(v_i), i = 1...n \\
\frac{1}{R_i} &= \frac{1}{\rho_i} + \sum_{j=1}^{n} \frac{1}{R_{ij}}, T_{ij} = \frac{1}{R_{ij}}
\end{aligned}
$$

where, in biological terms:

C_i is the input capacitance of the neuron membrane,

R_i is the neuron transmembrane resistance,

T_{ij} is the conductance between neuron j and neuron i,

n is the number of neurons.

So, this neural network can be seen as a nonlinear system , whose state is v_i. This system can be written in the matrix equation:

$$\boxed{\frac{dv}{dt} = Av + W_1\varphi(v) + W_2u}$$

where

$$\begin{aligned}
x &= \varphi(v) \\
v^T &= v^T = (v_1, v_{2,\ldots} v_{i,\ldots} v_n) \\
(\varphi(v))^T &= (\varphi(v_1), \varphi(v_2) \ldots \varphi(v_i) \ldots \varphi(v_n)) \\
x^T &= (x_1, x_{2,\ldots} x_{i,\ldots} x_n) \\
u^T &= (u_{1,} u_{2,\ldots} u_{i,\ldots} u_n)
\end{aligned}$$

and

$$A = \begin{pmatrix}
-\frac{1}{R_1C_1} & 0 & \ldots & 0 & \ldots & 0 \\
0 & -\frac{1}{R_2C_2} & \ldots & 0 & \ldots & 0 \\
. & . & \ldots & . & \ldots & . \\
0 & 0 & \ldots & -\frac{1}{R_iC_i} & \ldots & 0 \\
. & . & \ldots & . & \ldots & . \\
0 & 0 & \ldots & 0 & \ldots & -\frac{1}{R_nC_n}
\end{pmatrix}$$

$$W_1 = \begin{pmatrix}
\frac{T_{11}}{C_1} & \frac{T_{12}}{C_1} & \ldots & \frac{T_{1i}}{C_1} & \ldots & \frac{T_{1n}}{C_1} \\
\frac{T_{21}}{C_2} & \frac{T_{21}}{C_2} & \ldots & \frac{T_{2i}}{C_2} & \ldots & \frac{T_{2n}}{C_2} \\
. & . & \ldots & . & \ldots & . \\
\frac{T_{i1}}{C_i} & \frac{T_{i2}}{C_i} & \ldots & \frac{T_{ii}}{C_i} & \ldots & \frac{T_{in}}{C_i} \\
. & . & \ldots & . & \ldots & . \\
\frac{T_{n1}}{C_n} & \frac{T_{n2}}{C_n} & \ldots & \frac{T_{ni}}{C_n} & \ldots & \frac{T_{nn}}{C_n}
\end{pmatrix}$$

$$W_2 = \begin{pmatrix}
\frac{1}{C_1} & 0 & \ldots & 0 & \ldots & 0 \\
0 & \frac{1}{C_2} & \ldots & 0 & \ldots & 0 \\
. & . & \ldots & . & \ldots & . \\
0 & 0 & \ldots & \frac{1}{C_i} & \ldots & 0 \\
. & . & \ldots & . & \ldots & . \\
0 & 0 & \ldots & 0 & \ldots & \frac{1}{C_n}
\end{pmatrix}$$

Using as the state, the neuron output voltage, then

$$\frac{dx_i}{dt} = \frac{\psi_i(x_i)}{C_i}\left(\sum_{j=1}^{n} T_{ij}x_j - \frac{\varphi_i^{-1}(x_i)}{R_i C_i} + u_i\right)$$

$$\psi_i(x_i) = \frac{d\varphi_i(v_i)}{dv_i}|_{u_i=\varphi_i^{-1}(x_i)},\ i = 1,..,n$$

In [35], starting from this state space representation and using the gradient method, the energy function, originally proposed in [26], is derived. Then by means of the invariance principle, the conditions for convergence of this neural network to the equilibrium points are established; these conditions require the weight matrix (W_1) to be symmetric.

Considering again the Differential Neural Networks, taking the state $x = v$, and taking in account an expanded input (u^*), which includes the nonlinear feedback term, it possible to derive the following mathematical model :

$$\frac{dx_i}{dt} = -a_i x_i + b_i \sum_{j=1}^{2n} w_{ij} u_j^*,\ i = 1...n$$

$$(u^*)^T = (\varphi(x_1), \varphi(x_2)...\varphi(x_i)...\varphi(x_n), u_{1,} u_{2,...} u_{i,...} u_n)$$

where a_i, b_i are real constants, which allow this mathematical transformation.

Recurrent high-order neural networks(RHONN) are expansion of this first order one. Their properties have been analyzed in [54], [9] [17], [32] and [33]. For example, for the second order case, the product $y_j y_k$ is included in the input. Pursuing along this line, high order interactions such as triplets $(y_j y_k y_l)$, cuadruplets $(y_j y_k y_l y_m)$, etc can be considered.

Let take in account a RHONN with n neurons and m inputs. The state of each neuron is given by the differential equation:

$$\frac{dx_i}{dt} = -a_i x_i + b_i \sum_{k=1}^{l} w_{ik} \prod_{j\in I_k} \left(u_j^*\right)^{d_j(k)},\ l = m + n$$

where $\{I_1, I_2, ...I_l\}$ is a collection of l not-ordered subsets of $\{1, 2, ...l\}$, and

$$d_j(k) \in Z^+.$$

Nonlinear System Approximation

Is it possible to approximate nonlinear system behaviors by dynamic neural networks? There exist some results concerning this question. They may be classified in two groups: The first one, as a natural extension, is based on the function approximation properties of static neural networks [72], [20], and is limited for time belonging to a closed set. The second one uses the operator representation of the system to derive conditions for the validity of its approximation by a dynamic neural network; it has been extensively analyzed by I. W. Sandberg, both for continuous and discrete time [73], [74], [75].

The results in [73], [74] are limited to unidimensional maps; they are based on the concept of approximately-finite memory, which means that for maps G taking a subset S of B into B, where B is a collection of bounded $\Re$ valued maps on either $[0, \infty)$ or $\{0, 1, ...\}$,that given $\gamma > 0$ there is a $\Delta > 0$ such that

$$\left|(Gs)(t) - (GW_{t,\Delta}s)(t)\right| < \gamma, \ \ s \in S$$

where $W_{t,\Delta}$ is the window map defined by

$$\begin{aligned} (W_{t,\Delta}s)(\tau) &= s(\tau), \qquad t - \Delta \leq \tau \leq t \\ &= 0, \qquad \text{otherwise} \end{aligned}$$

If the system fulfills this condition, then it can be approximated arbitrary well by a dynamic neural network with the following structure: the input signal is first passed trough a parallel connection of l unidimensional linear systems and then applied to a nonlinear static map from $\Re^m$ to $\Re$, which could include sigmoid functions. (see Fig.1.18).

This result is extended to multidimensional nonlinear systems in [75]. The basic concept is *the myopic map* . In this case, first the set of continuous maps from $\Re^m$ to $\Re^n$ $(C(\Re^m, \Re^n))$ is considered. Then the map of bounded function $C_b(\Re^m, \Re^n)$ contained in $C(\Re^m, \Re^n)$ is taken into account. Let S be a nonempty subset of $C_b(\Re^m, \Re^n)$, and w be a continuous $\Re$ valued function defined on $\Re^m$ such that

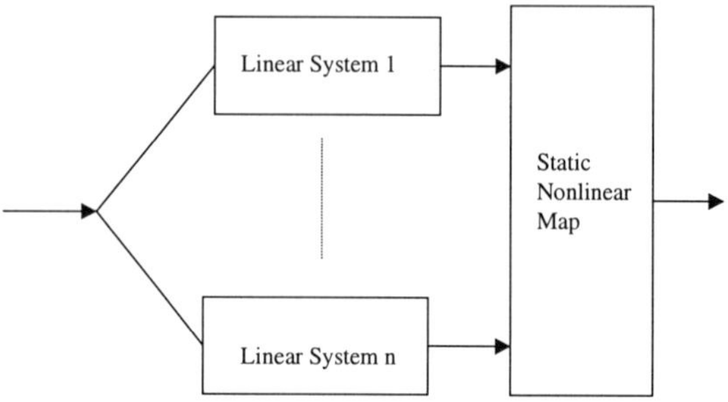

FIGURE 1.18. Nonlinear static map

$w(\alpha) \neq 0$ for all α and

$$\lim_{|\alpha| \to \infty} w(\alpha) = 0$$

Additionally, let G map S to the set of $\Re$ valued functions on $\Re^m$. It is said that G is myopic on S with respect to w is given an $\epsilon > 0$ there is $\delta > 0$ such that

$$\sup_{\alpha \in \Re^m} \|w(\alpha)[x(\alpha) - y(\alpha)]\| < \delta$$

implies

$$|(Gx)(0) - (Gy)(0)| < \epsilon, \quad \forall x, y \in S$$

Roughly speaking, if G is myopic, the value of $(Gx)(\alpha)$ is always relatively independent of the values of x at points remote from α. If the nonlinear system fulfills the myopic condition, then it can be approximated arbitrary well by a similar structure used for the unidimensional map. The only difference is that this time the linear systems have a multidimensional input.

Both concepts: the approximately finite memory and the myopic map implicitly mean that the nonlinear system, to be approximated, is open loop stable.

Digressing from these approaches, in [55], a dynamic neural network with the same state dimension as the one of the nonlinear system is considered, and the stability conditions of the approximation error are determined by means of a Lyapunov-like analysis. The obtained result does not require the system to be stable or time to belong to a closed interval.

This brief review is by no means complete. There exist important neural networks structures such as: competitive one and those root in statistics [71]. However, they have been less applied in identification and control of nonlinear systems.

1.5 Neural Networks in Control

In reference to neural networks in control, the following characteristics and properties are important:

1. *Nonlinear systems.* Neural networks offer a great promise in the realm of nonlinear control. This stems from their theoretical capability to approximate arbitrary nonlinear functions.

2. *Parallel distributed processing.* Neural networks have a highly parallel structure, which allows immediately parallel implementations.

3. *Learning and adaptation.* Neural networks are training using data from the system under study. An adequately trained neural network has the ability to generalize for inputs not appearing in the training data. Moreover, they can also be adapted on-line.

4. *Multivariable systems.* Neural networks have the ability to process many inputs and outputs; they are readily applicable to multivariable systems.

A modelling structure with all these features has a great promise for application in nonlinear identification and control. The compilation book [42] provides a broad overview of neural networks in control as well as the survey [28]. Due to the property of neural networks to implement input-output mappings, they are very suitable for applications on identification and control of nonlinear systems.

1.5.1 *Identification*

Neural networks have the potential to be applied to model nonlinear systems. An important question is that of systems identifiability [37], i.e. can the dynamic systems under consideration be adequately represented within a given particular model structure?. Identifiability on neural networks is related to uniqueness of the weights and to whether two networks with different para meters can produce identical input/output behavior. Results on this subject are in [76] for static neural networks, and in [4] for dynamic ones.

In order to represent nonlinear systems by neural networks, a straightforward approach is to augment the networks inputs with signals corresponding to its inputs and outputs. Assuming that the nonlinear system is described by [28]:

$$\begin{aligned} y(k+1) &= g(y(k), y(k-1) \ldots y(k-n), u(k), u(k-1) \ldots u(k-m)) \\ y, u &\in \Re,\ m \leq n \end{aligned}$$

This model does not consider explicitly disturbances. For a method including disturbances, see [12]. Special cases of this model have been considered in [45], where y depends linearly on either its past values or on those of u.

An obvious approach to model the system is to select the input-output structure of the neural network to be the same as that of the system. Denoting the output of the neural network as y_{nn}, then there exist two possible implementations.

a) series parallel model

In this case, the system outputs are used as the inputs of the neural network.

$$y_{nn}(k+1) = g(y(k), y(k-1) \ldots y(k-n), u(k), u(k-1) \ldots u(k-m))$$

Because there is not recurrence in the equation, this correspond to a static neural network.

b) *parallel model*

Here past outputs of the neural network are used as component of its input.

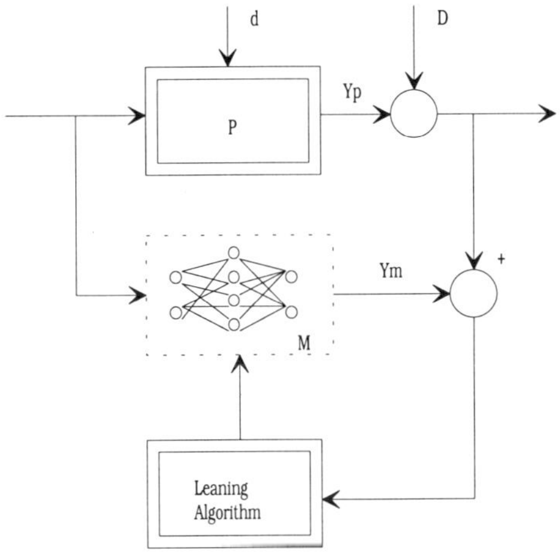

FIGURE 1.19. Identification scheme based on static neural netwotk.

$$y_{nn}(k+1) = g\left(y_{nn}(k), y_{nn}(k-1) ... y_{nn}(k-n), u(k), u(k-1) ... u(k-m)\right)$$

Because of the recurrence of y_{nn} in this equation, this correspond to a dynamic neural network.

Once the neural network structure defined, the next step is to determine the learning algorithm. For the case of static neural networks, the algorithms above mentioned can be used; however, in case of dynamic neural networks, new learning algorithm are required. Based on the backpropagation algorithm, in [45] a dynamic version is introduced. A generic identification scheme based on static neural network is presented in Figure 1.19.

Inverse models of dynamic systems are important for some control structures. Static neural networks can be used to implement inverse models. The simplest approach is to introduce a training signal to the system and to use the system output as the input to the neural network, whose output is compare with the training signal; the respective error is used to train the neural network [57]. The input-output relation of the neural network modelling the plant inverse is:

$$u(k) = g^{-1}(r(k+1), y(k), y(k-1)...y(k-n), u(k-1)...u(k-m))$$

However this simple method presents two drawbacks:

a) The training signal must be chosen to sample over a wide range of systems input, and the actual operational inputs may be hard to define a priori.

b) If the nonlinear system is not one-one, then an incorrect inverse can be obtained.

To overcome these problem, an improved structure known as specialized inverse learning is proposed in [57] In this approach the network inverse model precedes the system and receives as input a training signal which spans the desired operational output of the controlled system.

This input-output approach is not easily extended to multi-input, multi-output approach. For these cases is better to use a state space representation. Given the nonlinear system:

$$x(k+1) = f(x(k), u(k))$$

$$y(k) = h(x(k))$$

$$x \in \Re^n, u \in \Re^m, y \in \Re^l, k = 0, 1...$$

the following neural network identifier is proposed in [77]

$$\begin{aligned} x_{nn}(k+1) &= W_1\varphi(V_a x_{nn}(k) + V_b u(k) + \rho_x) + Ke(k) \\ y_{nn}(k) &= W_2\varphi\left(V_c x_{nn}(k) + V_d u(k) + \rho_y\right) \\ e(k) &= y(k) - y_{nn}(k),\ x(0) = x_{nn}(0) \\ W_1 &\in \Re^{nxnx}, V_a \in \Re^{nxxn}, V_b \in \Re^{nxxm}, \rho_x \in \Re^{nx} \\ W_2 &\in \Re^{nxny}, V_c \in \Re^{nyxn}, V_d \in \Re^{nyxm}, \rho_y \in \Re^{ny} \end{aligned}$$

The nonlinear mapping φ is implemented as a static multilayer feedforward neural network. Nevertheless the global neural network constitutes a dynamic one. The authors introduce a modified dynamic backpropagation algorithm and by means of linear matrices inequalities (LMI) determine stability conditions for the neural identifier. However is not possible to establish global stability but only local one. Recently, the authors published the extension of this approach to the continuous time case [78].

It is worth mentioning, that the learning procedure can be implemented also on-line.

In [33] a RHONN is consider, which is rewritten as:

$$\begin{aligned}\frac{dx_{nn,i}}{dt} &= -a_i x_{nn,i} + b_i \sum_{k=1}^{l} w_{ik} z_k, \; l = m+n, \; i = 1, ..., n \\ z^T &= (z_1, z_2, ...z_i...z_l) \\ z_i &= \left(\prod_{j \in I_i} \left(u_j^*\right)^{d_j(i)} \right)\end{aligned}$$

Taking

$$\theta_i = b_i \left(w_{i1}, w_{i2}...w_{ii}, ...w_{in}\right)^T$$

then

$$\frac{dx_{nn,i}}{dt} = -a_i x_{nn,i} + \theta_i^T z$$

Given the nonlinear system

$$\begin{aligned}\frac{dx_i}{dt} &= f_i(x, u), \; i = 1...n, \\ x(t) &\in \Re^n, \; u(t) \in \Re^m \; \forall t \in [0, \infty)\end{aligned}$$

First it is assumed that there exist a set of parameters θ^*, such that the neural networks models exactly the nonlinear system. Under this condition, it is possible to write the system as:

$$\frac{dx_i}{dt} = -a_i x_i + \theta_i^{*T} z$$

Defining the identification error as

$$e_i = x_{nn,i} - x_i$$

the authors are able to demonstrate that the learning law

$$\frac{d\theta_i}{dt} = -\eta_i z e_i$$

assures

$$\lim_{I \to \infty} e_i = 0$$

In order to handle the case of not exact modelling the learning law has to be modified, using techniques from robust nonlinear adaptive control.

In [63] a particular RHONN is used. This neural network is described by:

$$\frac{dx_{nn}}{dt} = Ax_{nn} + BW_1 \varphi(x_{nn}) + BW_2 \varphi'(x_{nn}) u$$

$$A, B, W_1, W_2, \varphi'(x_{nn}) \in \Re^{nxn}$$

$$x_{nn}, u, \varphi(x_{nn}) \in \Re^n$$

where:

A, B are diagonal matrices with elements $a_i < 0, b_i$ respectively,

W_1 is a synaptic weights matrix,

W_2 is a diagonal matrix of synaptic weights,

$\varphi(x_{nn})$ has elements

$$\varphi_i(x_{nn,i}) = \frac{k}{1 + e^{-\beta x_{nn,i}}}$$

$\varphi'(x_{nn})$ is a diagonal matrix with elements

$$\varphi_i'(x_{nn,i}) = \frac{k}{1 + e^{-\beta x_{nn,i}}} + \lambda$$

Using the Lyapunov method (see Appendix B), they derive the following learning law:

$$\begin{aligned} \frac{dw_{1,ij}}{dt} &= -b_i p_i \varphi_j (x_j) e_i \\ \frac{dw_{2,ii}}{dt} &= -b_i p_i \varphi_i'(x_{nn,i}) u_i e_i \end{aligned}$$

where p_i are the element of the diagonal matrix $P = P^T > 0$ which is the solution of the Lyapunov Matrix equation

$$PA + AP = -I$$

Then they modified this learning law in order to do it robust in presence of singular perturbed systems.

These two schemes using RHONN requires that both the system and the neural networks to start from identical initial conditions and the time has to belong to a closed interval. In a recent publication [34], these conditions are relaxed by means of a quite elaborated learning law.

1.5.2 Control

A large number of control structures has been proposed. For a recent and complete review, see [3]. It is beyond the scope of this chapter to provide a full survey of all architectures used. We give particular emphasis to those structures which are well established.

Supervised Control

There exist plants where a human closes the control loop, due to the enormous difficulties to implement an automatic controller using standard techniques. For some of these cases, it is desirable to design an automatic controller which mimics the human actions. A neural network is able to clone the human actions. Training the

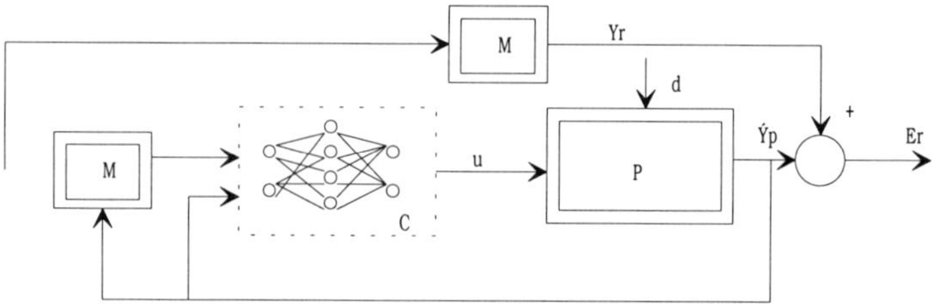

FIGURE 1.20. Model reference neurocontrol.

neural network is similar to learn a model as described above. The neural network inputs correspond to sensory information perceived by the human, and the outputs correspond to the human control actions. An example is presented in [24]

Direct Inverse Control

In this structure, an inverse model of the plant is directly utilized. This inverse model is cascaded with the plant, so the composed system results in an identity map between the desired response and the plant one. It heavily relies on the quality of the inverse model. The absence of feedback dismisses the robustness of this structure. This problem can be overcomed by the use of on-line learning of the neural network parameters implementing the inverse model.

Model Reference Control

The desired behavior of the closed loop system is specified through a stable reference model (M), usually a linear one, which is defined by its input-output relation $\{r(k), y^m(k)\}$The control goal is to force the plant output $\{y(k)\}$ to match the output of the reference model asymptotically:

$$\lim_{k\to\infty} \left\| y^m(k) - y(k) \right\| \leq \epsilon, \ \epsilon \geq 0$$

where ϵ is a specified constant.

Figure 1.20 shows the structure for this controller. The error between the two outputs is used to train the neuro controller [45]. This approach is related to the training of a inverse plant model. If the reference model is the identity mapping the two approaches coincide.

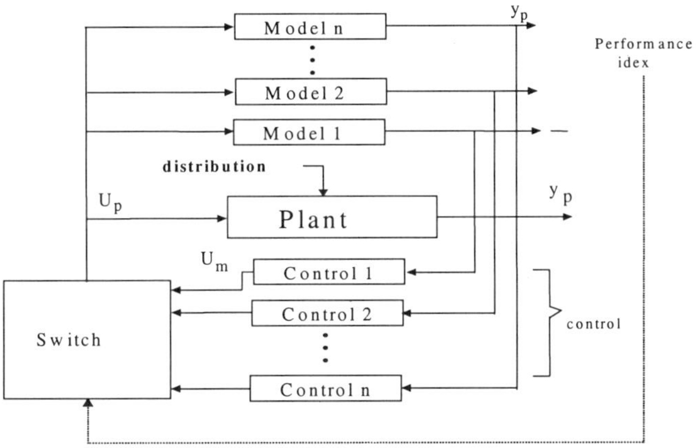

FIGURE 1.21. Scheme of multiple models control

Based on the results discussed in [45], the methodology of multiple models, switching and tuning, as shown in Fig.1.21.

It is extended to intelligent control in [46], [47]. There two applications, one for aircraft control and the other to robotics, are discussed. Even is the results are encouraging, stability analysis of the scheme presented in Fig. 2a, when neural networks are used for modeling and/or control, is quite complicated; only recently it has been possible to establish preliminary results [13], which extended existing ones for the case of linear systems [48]. It is also worth noting that all these results concerning multiple models, switching and tuning are only valid for single input, single output (SISO) systems.

One very interesting scheme of neuro control, which has the structure of reference model but not use the inverse of the plant is presented in [59]. There, on the basis of recurrent neural networks, an identifier and a controller are developed to ensure that the nonlinear plant tracks the model of reference. Both, the identifier and the controller are trained of-line by means of an extended Kalman filter, which had been proposed by several authors for training of feedforward neural networks. [79], [18], [58], [58]. This scheme had been successfully tested, in simulations, to control

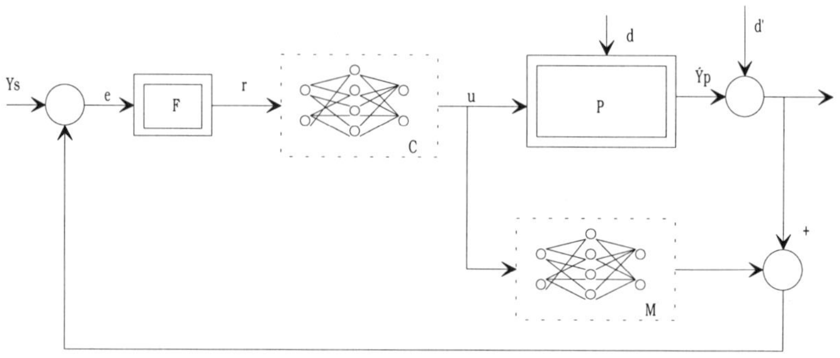

FIGURE 1.22. Internal model neurocontrol.

complex nonlinear systems such as engine idle speed control [59].

Internal Model Control (IMC)

In this structure, a system forward and inverse model are used directly within the feedback loop [23]. Robustness and stability analysis for the IMC has been developed [44]; moreover, IMC extends to nonlinear systems [19].

In this structure, a system model (M) is connected in parallel with the plant. The difference between the system and the model outputs is used as feedback signal, which is processed by a controller(C); this controller is implemented as the inverse model of the plant.

IMC realization by neural networks is straightforward [29]. The system model and its inverse are implemented using neural networks model as shown in Figure 1.22.

It is worth noting that IMC is limited to open-loop stable plants.

Predictive Control

In this structure, a neural network model gives prediction of the future plant response over an horizon. These predictions are sent to an optimization module in order to minimize a performance criterion The result of this optimization is a suitable control action (u')

This control signal u' is selected to minimize the index.

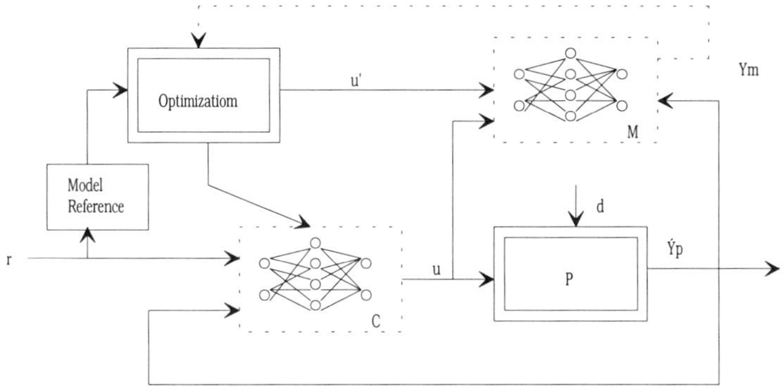

FIGURE 1.23. Predictive neurocontrol.

$$J = \sum_{j=N_1}^{N_2} \left(y^m (k+j) - y_{nn} (k+j)\right)^2 + \sum_{j=1}^{N_2} \lambda_j \left(u'(k+j-1) - u'(k+j-1)\right)^2$$

subject to the constraint of the dynamical model for the system.

Constants N_1 and N_2 define the optimization horizon. The values of λ weight the control actions. As shown in Figure 1.23, it is possible to train a second neural network to reproduce the actions given by the optimization module. Once this second neural network is trained, the plant model and optimization routine are not needed anylonger.

Optimal Control

The N-stage optimal control is stated as follows. Given a nonlinear system

$$\begin{aligned} x(k+1) &= f(x(k), u(k), k), x(0) = c_0 \\ x(k) &\in \Re^n, \; u(k) \in \Re^m, \; k = 0, 1, ...N-1 \end{aligned}$$

consider a performance index of the form

$$J = l_N(x_N) + \sum_{k=0}^{N-1} l_k(x_k, u_k)$$

with l_k a positive real function for $k = 0, 1, ...N$. The problem is to find a sequence $u(k)$ that minimizes J.

It is possible to implement this control laws by means of a feedforward neural network. In this case, the control is parametrized by this neural network through its weights (w_k):

$$u(k) = h(x_k, w_k)$$

An algorithm to implement this control law is presented in [77]. A similar approach has been developed for optimal tracking [60].

Adaptive Control

The difference between indirect and direct adaptive control is based on their structures. In indirect methods, first system identification from input-output measurement of the plant is done and then a controller is adapted based on the identified model. In direct methods the controller is learned directly without having a model from the plant. Neural networks can be used for both methods.

By far the major part of neural adaptive control is based on the indirect method. First a neural network model for the system is derived on-line from plant measurements. and then one of the control structures above mentioned is implemented upon this adaptive neural model. One of the first result on nonlinear adaptive neural network is [45], where, on the basis of specific neural models, an indirect adaptive model reference controller is implemented. In [67], an IMC adaptive control is developed using RBF neural networks. Applications to robot control, where static neural networks are used to estimate adoptively part of the robot dynamics are presented in [40]. Based on the system model identified on-line by a dynamic neural network, explained above, in [63], a control law is built in order to assure the tracking of a linear reference model.

Regarding direct adaptive neural control, in [80], a direct adaptive controller is developed using RBF neural networks. In [77], a combination of both methods is used; in fact a dynamic neural network model of the plant is adapted on-line, as well as a dynamic neuro controller.

Reinforcement Learning

This structure can be classified as a direct adaptive control one. Without having a model for the plant, a so called critic evaluates the performance of the plant; the reinforcement learning controller is rewarded or punished depending on the outcome of trials with the system. Reinforcement learning neural controller was first illustrated in [7].

The Q-Learning method is close to dynamic programming; it applies when no model is available for the plant, and in fact it is a direct adaptive optimal control strategy [81], [50] and [56]. Considering a finite state finite action Markov decision problem, the controller observes at each k the state $x(k)$, select a control action $u(k)$, and receives a reward $r(k)$. The objective is to find a control law that maximizes, at each k the expected discount sum of the rewards.

$$E\left\{\sum_{j=1}^{\infty}\gamma^{j}r\left(k+j\right)\right\}, 0<\gamma<1$$

with γ the discount factor.

Figure 1.24 shows an scheme of this control structure.

1.6 Conclusions

In this chapter, we have briefly reviewed the basic concepts of biological neural networks. Then we have presented the relevant artificial neural networks structures from the point of view of automatic control. We have given our reasons to name recurrent neural networks as dynamic ones. The importance of this kind of neural networks for identification and control was illustrated by means of an example.

After presenting the main artificial neural networks structures, we have discussed nonlinear system identification by both static and dynamic neural networks. We

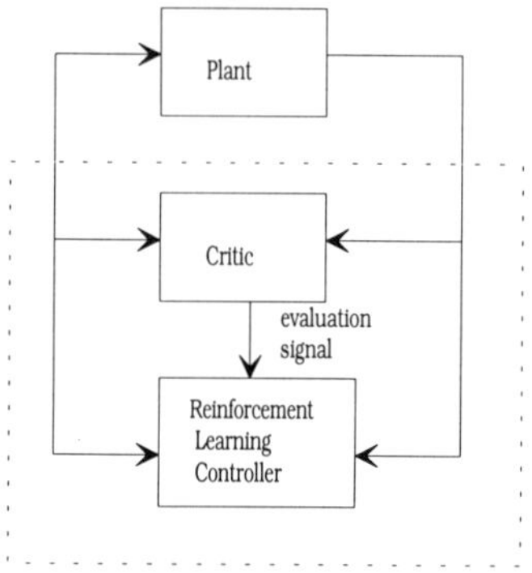

FIGURE 1.24. Reinforcement Learning control.

signal the main advantages and disadvantages of each one of this schemes. Then, the main existing schemes of neuro control are presented.

This chapter gives the main fundamentals concepts allowing to understand our results on identification, state estimation and trajectory tracking of nonlinear systems using Differential Neural Networks.

1.7 References

[1] Editors of Scientific American, *The Brain*, Scientific American, New York, 1979.

[2] M. A. Arbib, *The Handbook of Brain Theory and Neural Networks*, The MIT Press, Cambridge, MA, USA, 1995.

[3] M. Agarwal, "A systematic classification of neural network based control", .*IEEE Control Systems Magazine*, vol. 17, pp 75-93, 1997.

[4] F. Albertini and E. D. Sontag, "For neural networks function determines form", *Neural Networks*, vol. 6, pp 975-990, 1993.

[5] J. A. Anderson, J. W. Silverstein, S. A. Ritz and R. S. Jones, "Distinctive features, categorical perceptions, and probability learning: Some applications of a neural model", *Psychological Review*, vol. 84, pp413-451, 1977.

[6] D. S. Broomhead, and D. Lowe, "Multivariable functional approximation and adaptive networks", *Complex Systems*, vol.2, pp 321-355, 1988.

[7] A. G. Barto, R. S. Sutton., and C. W. Anderson, "Neuronlike adaptive elements that can solve difficult leasrning control problems", *IEEE Trans. on System, Man, and Cybernetics,* vol. 13, pp 834-846, 1983.

[8] A. E. Bryson and Y. C. Ho, *Applied Optimal Control*, Blaisdel, MA, USA, 1969.

[9] P. Baldi, "Neural networks, orientations of the hypercube and algebraic threshold units", *IEEE Trans. on Information Theory*, vol. 34, pp 523-530, 11988.

[10] G. Cybenko, "Approximation by superposition of a sigmoidal function", *Technical Report*, University of Illinois, Urbana, 1988,

[11] G. Cybenko, "Approximation by Superposition of a Sigmoidal Function", *Mathematics of Control, Signals, and Systems*, vol. 2, pp 303-314, 1989.

[12] S. Chen, S. A. Billings, C. F. Cowan, and P. M. Grant, "Practical identification of NARMAX models using radial basis functions", *Intl. Journal of Control*, vol. 52, pp 1327-1350, 1990.

[13] L. Chen and K. S. Narendra, "Nonlinear adaptive control using neural networks and multiple models, *2000 American Control Conference*, Chicago, Ill., USA, June 2000.

[14] A. Cichoki and R. Unbehaven, *Neural Networks for Optimization and Signal Processing*, J. Wiley and Sons, 1993.

[15] T. M. Covers, "Geometrical and statistical properties of systems of linear inequalities with applications in pattern recognition", *IEEE Tran. on Electronics Computers*, Vol. 14, pp 326-334, 1965.

[16] J. Dayhoff, *Neural Networks Architectures: An Introduction*, Van Nostrand Reinhold, New York, 1990.

[17] A. Dempo, O. Farotini, and T. Kailath, "High-order absolutely stable neural networks", *IEEE Trans. on Circuits and Systems*, vol.38, No 1, 1991.

[18] S. C. Douglas and T. H. Y. Meng, "Linearized least squares training of multilayer feedforward neural networks" *Proc. of Interntl. Joint Conf. on Neural Neyworks*, pp 133-140, Seattle, WA, USA, 1991.

[19] G. C. Economu, M. Morari, and B. O. Palson,"Internal model control. 5.Extension to nonlinear systems", *Ind. Eng. Chem. Process De.Dev.*, vol. 25, pp 403-411, 1986.

[20] K. Funahashi and Y. Nakamura, "Approximation of dynamical systems by continuous time recurrent neural networks", *Neural Networks*, vol. 6, pp 801-806, 1993.

[21] K. Funahashi, "On the approximate realization of continuous mapping by neural networks", *Neural Networks*, vol. 2, pp 183-192, 1989.

[22] M. M. Gupta and D. N. Rao, Editors, *Neuro-Control Systems, Theory and Applications*, IEEE Press, USA, New York, 1994.

[23] C. E. Garcia and M. Morari, "Internal model control -1. A unifying review and some new results", *Ind. Eng. Chem. Process De.Dev.*, vol. 21, pp 308-323, 1982.

[24] E. Grant and B. Zhang, "A neural net approach to supervised learning of pole placement", in *Proc. of 1989 IEEE Symposium on Itelligent Control*, 1989.

[25] C. L. Giles, G. M. Khun, and R. J. Williams, Eds., special issue on Dynamic Recurrent Neural Networks, *IEEE Trans. on Neural Networks*, vol. 5, March 1994.

[26] J. J. Hopfield, "Neural networks and physical systems with emergent collective computational abilities", *Proc. of the National Academy of Science, USA*, vol. 79, pp 2445-2558, 1982.

[27] J. J. Hopfield, "Neurons with grade response have collective computational properties like those of a two-state neurons", *Proc. of the National Academy of Science, USA*, vol. 81, pp 3088-3092, 1984.

[28] K. J. Hunt, D. Sbarbaro, R. Zbikowski and P.J. Gawthrop, "Neural Networks for Control Systems- A Survey", *Automatica*, vol. 28, pp. 1083-1112, 1992.

[29] K. J. Hunt and D. S. Sbarbaro, "Neural network for non-linear internal model control", *Proc. IEE-D*, vol.138, pp431-438, 1991, 1991.

[30] R. Hecht-Nielsen, "Kolgomorov's mapping neural networks existence theorem", *First IEEE International Conference on Neural Networks*, vol. 3, pp 11-14, San Diego, CA, 1987.

[31] K. Hornik, M. Stinchcombe, and H. White, "Multilayer Feedforward networks are universal approximators", *Neural Networks*, vol. 2, pp 359-366, 1989.

[32] Y. Kamp and M. Hasler, *Recursive Neural Networks for Associative Memory*, Wiley, New York, USA, 1990.

[33] E. B. Kosmatopoulos, M. M. Polycarpou, M. A.Christodoulou, and P. A. Ioannou,"High-order neural networks structures for identification of dynamical systems", *IEEE Trans. on Neural Networks*, vol. 6, pp 422-431, 1995.

[34] E. B. Kosmatopoulos, M. A. Christodoulou, and P. A. Ioannou, "Dynamic neural networks that ensure exponential identification error convergence", *Neural Networks*, vol. 10, pp299-314,1997

[35] H. K. Khalil, *Nonlinear Systems*, 2^{nd}edition, Prentice Hall, New York, USA, 1996.

[36] S. C. Kleene, "Representation of events in nerve nets and finite automata", in *Automata Studies*, C. E. Shanon and J. McCarthy, Eds., Princeton University Press, Princeton, N.J., USA, 1956

[37] L. Lung, *System Identification-Theory for the user*, Prentice Hall, New York, USA, 1987.

[38] W. S. McCullock and W. Pitss, "A logical calculus of the ideas immanent in nervous activity", *Bulletin of Mathematical Biophysics*, vol. 5, pp 115-133. 1943.

[39] Y. Lecun, "Une procedure d ´apprentissage pour reseau a seuil assymetrique", *Cognitiva*, vol. 85, pp 599-604, 1985.

[40] F. L. Lewis, K. Liu, and A. Yesilderek,"Neural net robot controller with guaranteed tracking performance", *IEEE Trans. on Neural Networks*, vol. 6, pp 703-715,1995.

[41] M. L. Minsky and S. A. Papert, *Perceptrons*, MIT Press, Cambridge, MA, 1969.

[42] W. T. Miller, R. S. Sutton, and P. J. Werbos, *Neural Networks for Control*, MIT Press, Cambridge, MA, USA, 1990.

[43] J. E. Moody and C. J. Darken, "Fast learning in networks of locally tuned processing units", *Neural Computations*, vol. 1, pp 281-294, 1989.

[44] M. Morari and E. Zafiriou, *Robust Process Control*, Prentice Hall New Jersey, USA, 1989.

[45] K. S. Narendra, and K. Parthasarathy, "Identification and control for dynamic systems using neural networks", *IEEE Trans. on Neural Networks*, vol. 1, pp 4-27, 1990.

[46] K. S. Narendra, and S. Mukhopadhyay, "Intelligent control using usinfg neural networks", *IEEE Control System Magazine*, vol. 12, pp 11-19, April 1992.

[47] K. S. Narendra, J. Balakrishnan, and M. Kemal, "Adaptation and learning using multiple models, switching, and tuning", *IEEE Control System Magazine*, vol. 15, pp 37-51, June, 1995.

[48] K. S. Narendra and J. Balakrishnan, "Adaptive control using multiple models", *IEEE Trans. on Automatic Control*, vol. 42, pp 171-187, 1997.

[49] T. Poggio and F. Girosi, "Networks for approximation and learning", *Proceedings of the IEEE*, vol. 78, pp 1481-1497, 1989.

[50] K. Najim and A. Poznyak, *Learning Automata: Theory and Applications.* Elsivier, Pergamon Press, Oxford, 1994.

[51] J. Park, and I. W. Sandberg, "Universal approximation using Radial Basis Function Networks", *Neural Computation*, vol.3, pp 246-257, 1991.

[52] D. B. Parker, "Learning-logic: casting the cortex of the human brain in Silicon", *Technical Report TR-47*, Center for Computational Research in Economics and Management Science, MIT, , 1985.

[53] M. J. D. Powell, "Radial basis functions for multivariable interpolation: a review", *IMA Conference on Algorithms for the Approximation of Functions and Data*, pp 143-167, Shrivenham, U.K., 1985.

[54] P. Paretto and J. J. Niez, "Long term memory storage capacity of multiconnected neural networks", *Biol. Cybern.*, vol. 54, pp 53-63, 1986.

[55] A. S. Poznyak and E. N. Sanchez, "Nonlinear system approximation by neural networks: error stability analywsis", *Intell. Automat. and Soft Compt.: An Intertl. Journ.*, vol. 1, pp 247-258, 1995.

[56] A. Poznyak and K. Najim, *Learning Automata and Stochastic Optimization*, Lecture notes in Control and Information sciences 225, Springer-Verlag, New York, 1997.

[57] D. Psaltis, A. Sideris, and A. A. Yamamura, "A multilayered neural network controller", *IEEE Control Systems Magazine*, vol. 8, pp17-21, 1988.

[58] G. V. Puskorius, and L. A. Feldkamp, "Decoupled extended Kalman filter for training multilayer perceptrons", *Proc. of Interntl. Joint Conf. on Neural Neyworks*, pp 771-777, Seattle, WA, USA, 1991.

[59] G. V. Puskorius and L. A. Feldkamp, "Neurocontrol of nonlinear dynamical systems with Kalman filter trained recurrent networks", *IEEE Trans. on Neural Networks*, vol. 5, pp 279-297, 1994.

[60] T. Parisini and R. Zoppoli, "Neural networks for feedback feedforward nonlinear control systems", *IEEE Trans. on Neural Networks*, vol. 5, pp 436-449,1994.

[61] S. Renals, "Radial basis function network for speech pattern classification", *Electronics Letters*, vol. 25, pp 437-439, 1989.

[62] F. Rosenblatt,"The perceptron: a probabilistic model for information storage and organization in the brain", *Psychological Review*, vol. 65, pp 386-408, 1958.

[63] G. A. Rovithakis and M. A. Christodoulou, "Adaptive control of unknown plants using dynamical neural networks", *IEEE Trans. on Systems, Man and Cybernetics*, vol.24, pp 400-412.

[64] D. E. Rumelhart, G. E. Hinton, and R. J. Williams, "Learning internal representation by error propagation ", in *Parallel Distributed Processing: Explorations in the Microstructure of Cognigtion*,D. E. Rumelhart and J. L. McClelland Eds., Cambridge, MA, USA, 1986.

[65] F. Rosenblatt, *Principles of Neurodynamics*, Spartan Books, Washington D.C., 1962.

[66] D.W. Rocks *et al.*, "Comparative analysis of backpropagation and the extended Kalman filter for training multilayer perceptrons", *IEEE Trans. on Pattern Analysis and Machine Intelligence*, vol. 14, pp 686-690, 1992.

[67] D. G. Sbarbaro, "Connectionist feedforward networks for control of nonlinear systems", *Ph. D. Thesis*, Faculty of Engineering, Glasgow University, 1992.

[68] G. Sheperd, *The Synaptic Organization of the Brain*, Oxford University Press, England, 1979.

[69] G. Sheperd, *Neurobiology*, Oxford University Press, England, 1983.

[70] Scientific and Technological Center, *Scientific News*, Embassy of France, Mexico, December 1997.

[71] S. Haykin, *Neural Networks. A Comprehensive Foundation*, IEEE Press, New York, 1994.

[72] E. Sontag, "Neural nets as systems models and controllers", in *Proc. 7th Yale workshop on Adaptive and Learning Systems*, pp 73-79, Yale University, USA, 1992.

[73] I. W. Sandberg, "Approximation theorems for discrete time systems", *IEEE Trans on Circuits and Systems*, vol. 38, pp 564-566, 1991.

[74] I. W. Sandberg, "Uniform approximation and the circle criterion", *IEEE Trans on Automatic Control*, vol. 38, pp 1450-1458, 1993.

[75] I. W. Sandberg, "Uniform approximation of multidimensional myopic maps", *IEEE Trans on Circuits and Systems-I*, vol. 44, pp 477-485, 1997.

[76] H. J. Sussmann, "Uniqueness of the weights for minimal feedforward nets with a given input-output map", *Neural Networks*, vol. 5, pp589-593, 1992.

[77] J. A. K. Suykens, J. Vanderwalle, and B. De Moor., *Artificial Neural Networks for Modelling and Control of Non-linear Systems* , Kluwer Academic Publishers, Boston, MA, USA, 1996.

[78] J. A. K. Suykens, J. Vanderwalle, and B. De Moor, "Nonlinear H_∞ control for continuous-time recurrent neural networks", *in Proc. 1997 European Control Conference*, Belgium, 1997.

[79] S. Singhai and L. Wu, "Training multilayer perceptrons with the extended Kalman algorithm", in *Advances in Neural Information Processing Systems I*, D. S. Touresky, Ed., Morgan Kaufman, San Mateo, CA, USA, 1989.

[80] J. J. Slotine and R. M. Sanner, "Neural networks for adaptive control and recursive identification: a theoretical framework", in *Essays on Control: Perspectives in the theory and its applications*, H.L.Tretelman and J.C.Willems, Editors, Birkhauser, Germany, 1993.

[81] R. S. Sutton, A. Barto, and R. Williams, "Reinforcement learning is direct adaptive optimal control", *IEEE Control Systems Magazine*, vol. 12, pp19-22, 1992.

[82] A. N. Tikhonov, "On solving incorrectly posed problems and methods of regularization", *Doklady Akademii Nauk*, vol. 151, pp501-504, Moscow, Russia (former USSR), 1963

[83] B. Widrow and M. E. Hoff, "Adaptive switching circuits", *IRE Wescon convention record*, pp 96-104, 1960.

[84] B. Widrow and M. A. Lher, "30 years of adaptive neural networks: Perceptron, madaline, and backpropagation", *Proceedings of the IEEE*, vol. 78, pp 1415-1442, 1990.

[85] B. Widrow, "Generalisation and information storage in networks of adaline neurons", in *Self-Organizing Systems*,(M.C. Yovitz, GF.T.Jacobi, and G.D.Goldstein, eds.), Sparta, Washington, D.C, USA, 1962.

[86] P. J. Werbos, "Beyond regression: new tools for prediction and analysis in the behavioral sciences", *Ph. D Thesis*, Harvard University, Cambridge, MA, USA, 1974.

2

Nonlinear System Identification: Differential Learning

The adaptive nonlinear identification using a dynamic neural network, with the same state space dimension as the system, is analyzed. The system space states are assumed to be completely measurable. The new learning law ensure the identification error convergence to zero (model matching) or to a bounded zone (with unmodeled dynamics). By means of a Lyapunov-like analysis we determine stability conditions for the identification error. For the identification analysis we use an algebraic Riccati equation. We also establish the theorems which give bounds for the identification error and state that they are proportional to the a priory uncertainty bound.

2.1 Introduction

Recently, there has been a big interest in applying neural networks to identification and control of nonlinear systems [19], [20]. Nonlinear system identification can be approached as the approximation of the system behavior by dynamic neural networks. In this direction there exists already some results. They may be classified in two groups:

- the first one, as a natural extension, is based on *the function approximation properties of static neural networks* [22, 6] and is limited for time belonging to a closed set;

- the second one uses *the operator representation of the system* to derive conditions for the validity of its approximation by a dynamic neural network.

The last one has been extensively analyzed by I. W. Sandberg, both for continuous and discrete time ([23, 24] and references therein). The structure proposed is constituted by the parallel connection of neurons, with no interaction between them; it is required that the nonlinear system fulfills the approximately-finite memory condition. In [1], a dynamic neural network, based on the Hopfield model, was proposed for nonlinear systems identification using operator representation; the approximation property was proposed as a conjecture. Using the fading memory condition this conjecture was partially proved in [25]. Both of them, the approximately-finite memory and the fading memory conditions, require the nonlinear system to be stable.

The above mentioned results give just conditions for the existence of a dynamic neural network, which minimizes the approximation error to the nonlinear systems behavior; they do not determine the number of neurons and/or the value of their weights to effectively obtain the minimum error. A recently result [18] solves the problem of the neuron number by means of recursively high-order neural networks. There this number is selected to be equal to the dimension of the nonlinear systems state, which has to be completely measurable. This measurability condition is relaxed in [24] to singular perturbed systems. For these results, the time is also required to belong to a closed set. In [11] a high order parallel neural networks can ensure that the identification error converges to zero, but they need the regressor vector is persistently exciting, that for closed-loop control is not reasonable.

There are not many stability analyses in neural control in spite of successful neural control applications reported and that, even for neural information storage application, energy function studies are used for proof of convergence to desired final values [19]. To the best of our knowledge, there are only a few results published regarding nonlinear systems control by dynamic neural networks. In our opinion, the most important are the ones by M.A.Christodoulou and coworkers [24, 29] and references therein. They utilize a particular version of recursively high-order neural networks. In [24] they identify the nonlinear systems by means of the dynamic neural network, then calculate the control law, based on the neural network model, to force the system to follow a linear model. The neural network weights are on-line adapted to minimize the identification error. Stability of the whole system is performed via a Lyapunov function; as above mentioned their approach can deal with singular perturbed systems. In [29], they develop a direct adaptive regulation scheme for affine in the control nonlinear systems; again they analyze stability using a Lyapunov function. In both papers, they illustrate the applicability of the respective approach by the speed control of a D.C. motor. Other results [16, 3] utilize a SISO affine control representation of the nonlinear system, which is approximated by a dynamic neural network. This neural network is linearized by an inner loop designed using differential geometry techniques [9], and the outer control law is implemented using a PID controller.

In this book we analyze both: nonlinear system identification and control. First, in this chapter the nonlinear system is identified by means of a dynamic neural network; then, in the chapter 6, we force the identified system to track a signal generated by a nonlinear model using a nonlinear controller. The identification error and tracking error stability analysis is performed by a Lyapunov like method. It is worth mentioning that the stability analysis methodology that we use is similar to the one introduced by A.N. Mitchel and coworkers for the robustness analysis of neural information storage [26, 27].

The neural networks could be qualified as static (feedforward) or as dynamic (recurrent or differential) ones. Most of publications in nonlinear system identification and control use *static neural nets*, which are implemented for the approximation of nonlinear function in the right side-hand of dynamic model equations [11]. The main drawback of these NN is that the weight updates do not utilize any information on the local data structure and the function approximation is sensitive to the training data [7]. *Dynamic neural nets* can successfully overcome this disadvantage as well as present adequate behavior in presence of unmodeled dynamics, because their structure incorporate feedback. They have powerful representation capabilities. One of best known dynamic neural nets was introduced by Hopfield [5]. The most dynamic neural network structured (studied, for example, in [18],[13] and [18]) have no any hidden layers and, as a result, the approximation capabilities of this networks turn out to be limited, due to the same reasons as for a single layer perceptrons. To overcome these shortcomings, at least there exist two ways:

1. to use high order neural networks (see [24] and [29]) which contain multiple nonlinear functions in order to approximate nonlinear dynamics; the learning

law for high order networks is similar to the ones used for the single layer cases;

2. to employ multilayer dynamic neural networks (see [13]) which contain additional hidden layers in order to improve the approximation capabilities; dynamic multilayer neural networks are like a multilayer perceptrons combined with a dynamic operator, so the original back propagation algorithm as well as its modifications turn out to be a reasonable learning law for them.

In this section we will follow to the second approach. In general, using traditional technique the approximation error can be easy done arbitrarily small for a big enough class of nonlinear functions, but the corresponding state identification error stability cannot be guaranteed [7]. As it is shown in [18], [13] and [29], the Lyapunov-like method turns out to be a good instrument to generate a learning law and to establish error stability conditions . All papers mentioned above deal with the simple structure of dynamic NN containing only single output layer. As well as in the static case, it is not easy to update the weights of dynamic multilayer neural networks. In this book we successfully solve this problem.

This section presents the material in the following way: the on-line learning of the neural network parameters is considered; then the identification error stability is analyzed. After all we extend these results to dynamic multilayer neural networks. We illustrate the applicability of these results by several examples, discuss the perspectives and made the conclusions.

2.2 Identification Error Stability Analysis for Simplest Differential Neural Networks without Hidden Layers

2.2.1 Nonlinear System and Differential Neural Network Model

The nonlinear system to be identified is given as:

$$\boxed{\begin{array}{c} \dot{x}_t = f(x_t, u_t, t) \\ x_t \in \Re^n, \; u_t \in \Re^q, \quad n \geq q \end{array}} \tag{2.1}$$

We assume the following *parallel* structure of neural networks (in [24, 29] the *series-parallel* structure are used) :

$$\boxed{\dot{\widehat{x}}_t = A\widehat{x}_t + W_{1,t}\sigma(\widehat{x}_t) + W_{2,t}\phi(\widehat{x}_t)\gamma(u_t)} \tag{2.2}$$

where

$\widehat{x}_t \in \Re^n$ is the state of the neural network,

$u_t \in \Re^q$ is an input measurable control action,

$W_{1,t} \in \Re^{n\times k}$ is the matrix for nonlinear state feedback,

$W_{2,t} \in \Re^{n\times r}$ is the input matrix,

$A \in \Re^{n\times n}$ is a Hurtwitz matrix.

The vector field $\sigma(x_t) : \Re^n \to \Re^k$ is assumed to have the elements increasing monotonically. The nonlinearity $\gamma(u_t)$ define the vector field from $\Re^q$ to $\Re^s$. The function $\phi(\cdot)$ is the transformation from $\Re^n$ to $\Re^{r\times s}$. The typical presentation of the elements $\sigma_i(\cdot)$ and $\phi_{ij}(.)$ are as sigmoid functions (see Figure 2.1), i.e.,

$$\begin{aligned} \sigma_i(x) &= a_i\left(1+e^{-b_i x_i}\right)^{-1} - c_i \\ \phi_{ij}(x) &= \bar{a}_{ij}\left(1+e^{-\sum_p \bar{b}_{ij}^p x_p}\right)^{-1} - \bar{c}_{ij} \end{aligned} \tag{2.3}$$

The neural network (2.2) can be classified as a Hopfield-type one. Because $\sigma(\cdot)$ and $\phi(\cdot)$ are chosen as sigmoid functions, the following assumption is fulfilled.

A2.1: The function $\sigma(\cdot)$ and $\phi(\cdot)$ satisfy sector conditions [3] (see Figure 2.1):

$$\begin{gathered} \widetilde{\sigma}_t^T \Lambda_\sigma \widetilde{\sigma}_t \le \Delta_t^T D_\sigma \Delta_t, \quad \sigma_t^T(x) Z_\sigma \sigma_t(x) \le x_t^T C_\sigma x_t \\ \gamma^T(u_t) \widetilde{\phi}_t^T \Lambda_\phi \widetilde{\phi}_t \gamma(u_t) \le \Delta_t^T D_\phi \Delta_t \left\|\gamma(u_t)\right\|^2 \\ \gamma^T(u_t) \phi_t^T(x) Z_\phi \phi_t(x) \gamma(u_t) \le x_t^T C_\phi x_t \left\|\gamma(u_t)\right\|^2 \end{gathered}$$

$$\Delta_t := \widehat{x}_t - x_t \tag{2.4}$$

where

$$\begin{aligned} \widetilde{\sigma}_t &:= \sigma(\widehat{x}_t) - \sigma(x_t) \\ \widetilde{\phi}_t &:= \phi(\widehat{x}_t) - \phi(x_t) \end{aligned} \tag{2.5}$$

Λ_σ, Λ_ϕ, D_σ, D_ϕ, Z_σ, Z_ϕ, C_σ, C_ϕ are known positive constants.

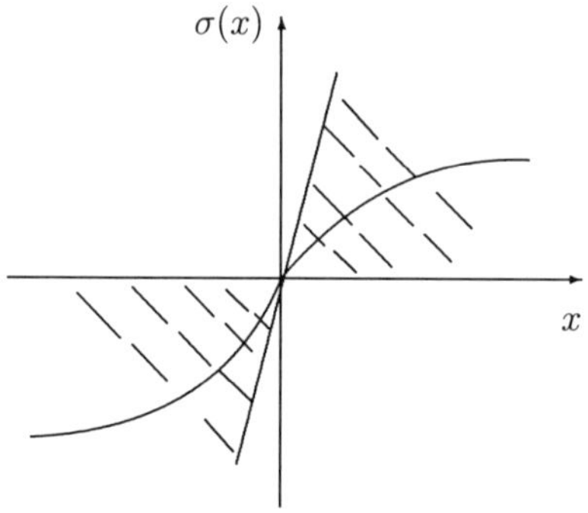

FIGURE 2.1. The shaded part satisfies the sector condition.

We assume the control input function $\gamma(\cdot)$ is bounded.

A2.2: The nonlinear function $\gamma(\cdot)$ is selected as

$$\|\gamma(u_t)\|^2 \leq \overline{u}$$

There exist two possibilities to fulfill this constraint:

1. consider bounded $\gamma(\cdot)$ function;

2. use bounded control actions $\{u_t\}$ ($\|u_t\| \leq u^+ < \infty$) and assume that the nonlinearity $\gamma(\cdot)$ is continuous.

Below we will assume that one of them is realized.

2.2.2 Exact Neural Network Matching with Known Linear Part

To illuminate the main features of the derived analysis method, let us first assume that we deal with the simplest situation when an exact neural network model of the plant is available. It means that for the known stable matrix A there exist weight matrices $W_1^*$and W_2^*, such that the given system can be completely presented by following neural network

$$\boxed{\dot{x}_t = Ax_t + W_1^*\sigma(x_t) + W_2^*\phi(x_t)\gamma(u_t)} \tag{2.6}$$

Here we not obligatory know the exact values of W_1^* and W_2^*, but we assume that their upper estimates are available, i.e.,

$$W_1^* \Lambda_1^{-1} W_1^{*T} \leq \overline{W}_1, \quad W_2^* \Lambda_2^{-1} W_2^{*T} \leq \overline{W}_2 \tag{2.7}$$

where Λ_1, Λ_2, $\overline{W}_1$ and $\overline{W}_2$ are priory known matrices.

Even the assumption that our plant is an exact matching of a simplest (without hidden layers) differential neural network is a very strong, from the methodological point of view we prefer to start with it. Sure, that these results will be used in the further chapters where the systems of more general structures will be investigated.

It is well known [3], if the matrix A is stable, the pair $\left(A, R^{\frac{1}{2}}\right)$ is controllable and the pair $\left(A, Q^{\frac{1}{2}}\right)$ is observable and the special frequency condition

$$S(\omega) \triangleq I - \left[R^{1/2}\right]^T \left[-i\omega I - A^{*T}\right]^{-1} Q \left[i\omega I - A^*\right]^{-1} \left[R^{1/2}\right] \geq 0 \tag{2.8}$$

are fulfilled, then the following matrix Riccati equation:

$$A^T P + PA + PRP + Q = 0 \tag{2.9}$$

has a positive solution $P = P^T > 0$. The following matrix inequality is sufficient to fulfill the frequency condition given above and simplifies it for the case when $R > 0$ (see Appendix A):

$$\frac{1}{4}\left(A^T R^{-1} - R^{-1} A\right) R \left(A^T R^{-1} - R^{-1} A\right)^T \leq A^T R^{-1} A - Q.$$

So, let us accept the following assumption:

A2.3: There exits a strictly positive defined matrix Q_0 such that for

$$R := \overline{W}_1 + \overline{W}_2$$

and

$$Q := Q_0 + D_\sigma + D_\phi \overline{u}$$

the matrix Riccati equation (6.43) has a positive solution.

We will develop a learning law to guarantee the stability of neural network as well as asymptotic stability for the identification error vector. The next theorem presents this algorithm and state its asymptotic properties.

Theorem 2.1 *Let us consider the unknown nonlinear system (2.1) and a model matching neural network (2.2) whose weights are adjusted as*

$$\boxed{\begin{array}{c} \dot{W}_{1,t} = -K_1 P \Delta_t \sigma(\widehat{x}_t)^T \\ \dot{W}_{2,t} = -K_2 P \Delta_t \gamma(u_t)^T \phi(\widehat{x}_t)^T \\ W_{1,0},\, W_{2,0} \text{ - the initial weight matrices} \end{array}} \tag{2.10}$$

where K_1 and K_2 are positive defined matrices, P is the solution of matrix Riccati equation given by (6.43). Assume also that the assumptions ***A2.1, A2.2*** *and* ***A2.3*** *hold. Then the weights dynamics are bounded:*

$$W_{1,t} \in L_\infty, \quad W_{2,t} \in L_\infty$$

and converge

$$\lim_{t\to\infty} \dot{W}_{1,t} = 0, \ \lim_{t\to\infty} \dot{W}_{2,t} = 0$$

We also can conclude that the identification process is asymptotically consistent, i.e.,

$$\lim_{t\to\infty} \Delta_t = 0 \tag{2.11}$$

Proof. From (2.6) and (2.2) the error equation can be expressed as

$$\dot{\Delta}_t = A\Delta_t + \widetilde{W}_{1,t}\sigma(\widehat{x}_t) + W_1^*\widetilde{\sigma}_t + \widetilde{W}_{2,t}\phi(\widehat{x}_t)\gamma(u_t) + W_2^*\widetilde{\phi}_t\gamma(u_t) \tag{2.12}$$

where

$$\widetilde{W}_{1,t} := W_{1,t} - W_1^* \tag{2.13}$$

and

$$\widetilde{W}_{2,t} := W_{2,t} - W_2^* \tag{2.14}$$

Define Lyapunov function candidate as

$$V_t := \Delta_t^T P \Delta_t + tr\left[\widetilde{W}_{1,t}^T K_1^{-1} \widetilde{W}_{1,t}\right] + tr\left[\widetilde{W}_{2,t}^T K_2^{-1} \widetilde{W}_{2,t}\right] \tag{2.15}$$

So, calculating its derivative, we obtain

$$\dot{V}_t = 2\Delta_t^T P \dot{\Delta}_t + 2tr\left[\dot{\widetilde{W}}_{1,t}^T K_1^{-1}\widetilde{W}_{1,t}\right] + 2tr\left[\dot{\widetilde{W}}_{2,t}^T K_2^{-1}\widetilde{W}_{2,t}\right] \tag{2.16}$$

Using (2.12), we get

$$\begin{aligned}\Delta_t^T P \dot{\Delta}_t = \Delta_t^T P A^* \Delta_t + \Delta_t^T P \left(W_1^* \widetilde{\sigma}_t + W_2^* \widetilde{\phi}_t \gamma(u_t)\right) \\ + \Delta_t^T P \left(\widetilde{W}_{1,t}\sigma(\widehat{x}_t) + \widetilde{W}_{2,t}\phi(\widehat{x}_t)\gamma(u_t)\right).\end{aligned} \tag{2.17}$$

As the term $\Delta_t^T P W_1^* \widetilde{\sigma}_t$ is a scalar, using **A2.1, A2.2** and matrix inequality (see Appendix A)

$$X^T Y + \left(X^T Y\right)^T \leq X^T \Lambda^{-1} X + Y^T \Lambda Y \tag{2.18}$$

which is valid for any $X, Y \in \Re^{n\times k}$ and for any positive defined matrix $0 < \Lambda = \Lambda^T \in \Re^{n\times n}$, we obtain

$$2\Delta_t^T P W_1^* \widetilde{\sigma}_t \leq \Delta_t^T P W_1^* \Lambda_1^{-1} W_1^{*T} P \Delta_t + \widetilde{\sigma}_t^T \Lambda \widetilde{\sigma}_t \leq \Delta_t^T \left(P\overline{W}_1 P + D_\sigma\right)\Delta_t \tag{2.19}$$

and

$$2\Delta_t^T P W_2^* \widetilde{\phi}_t u_t \leq \Delta_t^T \left(P\overline{W}_2 P + D_\phi \overline{u}\right)\Delta_t$$

If we select the adaptive law as (6.45) and taking into account that

$$\begin{aligned}\dot{\widetilde{W}}_{1,t} = \dot{W}_{1,t},\ \dot{\widetilde{W}}_{2,t} = \dot{W}_{2,t} \\ 2\sigma(\widehat{x}_t)^T \widetilde{W}_{1,t} P \Delta_t = 2tr\left[P\Delta_t \sigma(\widehat{x}_t)^T \widetilde{W}_{1,t}\right] \\ 2\gamma(u_t)^T \phi(\widehat{x}_t)^T \widetilde{W}_{2,t} P \Delta_t = 2tr\left[P\Delta_t \gamma(u_t)^T \phi(\widehat{x}_t)^T \widetilde{W}_{2,t}\right]\end{aligned}$$

(2.16) becomes

$$\begin{aligned}\tfrac{d}{dt}V_t \leq \Delta_t^T \left(PA + A^T P + P\left(\overline{W}_1 + \overline{W}_2\right)P + D_\sigma + D_\phi \overline{u} + Q_0\right)\Delta_t - \Delta_t^T Q_0 \Delta_t \\ + 2tr\left[\left(\dot{\widetilde{W}}_{1,t}^T K_1^{-1} + P\Delta_t \sigma(\widehat{x}_t)^T\right)\widetilde{W}_{1,t}\right] \\ + 2tr\left[\left(\dot{\widetilde{W}}_{2,t}^T K_2^{-1} + P\Delta_t \gamma(u_t)^T \phi(\widehat{x}_t)^T\right)\widetilde{W}_{2,t}\right] \\ = \Delta_t^T \left(PA + A^T P + P\left(\overline{W}_1 + \overline{W}_2\right)P + D_\sigma + D_\phi \overline{u} + Q_0\right)\Delta_t - \Delta_t^T Q_0 \Delta_t\end{aligned}$$

Using **A2.3,** we can conclude that

$$\frac{d}{dt}V_t \leq -\Delta_t^T Q_0 \Delta_t \leq 0$$

where $Q_0 > 0$. So, integrating both sides of this inequality from $t = 0$ up to $t = T$, finally, we obtain

$$\int_{t=0}^{T} \|\Delta_t\|_{Q_0}^2 \, dt \leq V_0 - V_T \leq V_0 < \infty$$

and, hence, we get

$$\int_{t=0}^{\infty} \|\Delta_t\|_{Q_0}^2 \, dt \leq V_0 < \infty$$

So, we can conclude that the process $\{\Delta_t\}$ is quadratically integrable ($\Delta_t \in L_2$) and bounded ($\Delta_t \in L_\infty$) (because $\{V_t\}$ is bounded process too), i.e.,

$$\Delta_t \in L_2 \cap L_\infty$$

In view of this fact, from the error equation (2.12) we also conclude that its derivative is also bounded, i.e.,

$$\dot{\Delta}_t \in L_\infty$$

Using Barlalat's Lemma (Appendix A) we derive (2.11). As the signals u_t, $\sigma(\widehat{x}_t)$, $\phi(\widehat{x}_t)$ and P are bounded, we can conclude that

$$\lim_{t\to\infty} \dot{W}_{1,t} = 0$$

and

$$\lim_{t\to\infty} \dot{W}_{2,t} = 0.$$

■

Remark 2.1 *The* ***series-parallel*** *structure corresponding to the following differential neural network*

$$\dot{\widehat{x}}_t = A\widehat{x}_t + W_{1,t}\sigma(x_t) + W_{2,t}\phi(x_t)\gamma(u_t)$$

(the nonlinearities $\sigma(x_t), \phi(x_t)$ are the functions of x_t but not $\hat{x}_t$ as it takes place in parallel structure, studied in this section) needs a Lyapunov equation

$$PA^* + A^{*T}P = -Q \tag{2.20}$$

which provides its solution P for the corresponding learning law [24]. But the ***parallel structure*** *needs the matrix Riccati equation (6.43). Sure, that the Lyapunov equation (2.20) can be considered as the partial case of the Riccati equation (6.43) when $R = 0$.*

2.2.3 Non-exact Neural Networks Modelling: Bounded Unmodeled Dynamics Case

In this paragraph, we consider the more realistic case when the dynamic neural networks does not match exactly the nonlinear system. We define the modelling error as

$$\Delta f(x_t, u_t, t) := f(x_t, u_t, t) - [Ax_t + W_1^*\sigma(x_t) + W_2^*\phi(x_t)\gamma(u_t)]$$

where A^*, W_1^* and W_2^* are fixed given (and known) matrices. W_1^* and W_2^* can be considered as the initial values for the weight matrices. It means that the nonlinear operator characterizing the given nonlinear system can be expressed as follows

$$\boxed{f(x_t, u_t, t) := Ax_t + W_1^*\sigma(x_t) + W_2^*\phi(x_t)\gamma(u_t) + \Delta f(x_t, u_t, t)}$$

As for the unmodeled dynamics $\Delta f(x_t, u_t, t)$ we assume that they are bounded as:

A2.4:

$$\|\Delta f(x_t, u_t, t)\| \leq \|\eta_\sigma\| + \left\|\eta_\phi \gamma(u_t)\right\|$$

where the matrices $\eta_\sigma \in \Re^n$, $\eta_\phi \in \Re^{n\times m}$ are assumed to be diagonal and satisfying

$$\|\eta_\sigma\|_{\Lambda_1} = \eta_\sigma^T \Lambda_1 \eta_\sigma \leq \overline{\eta}_\sigma, \quad \left\|\eta_\phi\right\|_{\Lambda_2} = \eta_\phi^T \Lambda_2 \eta_\phi \leq \overline{\eta}_\phi$$

Next assumption concerns to the uncertainties bounds involved in to the system description.

A2.5: There exits a strictly positive defined matrix Q_0 such that if the matrices R and Q are defined as

$$\begin{aligned} R &:= \overline{W}_1 + \overline{W}_2 + \Lambda_1^{-1} + \Lambda_2^{-1} \\ Q &:= Q_0 + D_\sigma + D_\phi \overline{u} \end{aligned} \tag{2.21}$$

the matrix Riccati equation (6.43) has a positive solution.

In the presence of any uncertainty or unmodeled dynamics, the weights learned according to the update law (6.45) derived in the previous subsection, can drift to infinity, i.e., may become unbounded! This fact has been stated by numerous computer simulations. This weights unboundedness effect is known as parameter drift. So, some modifications of this learning law are extremely recommended. *Dead zone* method may overcome the problem [12].

Next theorem states the main result concerning a new modified learning procedure and the corresponding identification error bound.

Theorem 2.2 *Let us consider the unknown nonlinear system (2.1) and parallel neural network (2.2) with modelling error as in (2.11) whose weights are adjusted as*

$$\boxed{\begin{aligned} \dot{W}_{1,t} &= -s_t K_1 P \Delta_t \sigma(\widehat{x}_t)^T \\ \dot{W}_{2,t} &= -s_t K_2 P \Delta_t \gamma(u_t)^T \phi(\widehat{x}_t)^T \\ W_{1,0} &= W_1^*,\ W_{2,0} = W_2^* \end{aligned}} \tag{2.22}$$

where ***dead-zone function*** s_t *is defined as*

$$s_t := \left[1 - \frac{\mu}{\|P^{1/2}\Delta\|}\right]_+, \quad [z]_+ = \begin{cases} z & z \geq 0 \\ 0 & z < 0 \end{cases} \tag{2.23}$$

$$\mu = \left(\overline{\eta}_\sigma + \overline{\eta}_\phi \overline{u}\right) / \lambda_{\min}\left(P^{-1/} Q_0 P^{-1/2}\right) \tag{2.24}$$

Here $\overline{u}$ *is defined by* ***A2.2****. Assuming also that* ***A2.4*** *and* ***A2.5*** *are verified, the following facts hold:*

a)

$$\Delta_t, W_{1,t}, W_{2,t} \in L_\infty \tag{2.25}$$

b) *the identification error* Δ_t *satisfies the following tracking performance*

$$\limsup_{T\to\infty} T^{-1} \int_0^T \Delta_t^T Q_0 \Delta_t s_t dt \le \overline{\eta}_\sigma + \overline{\eta}_\phi \overline{u} \tag{2.26}$$

Proof. From (2.1) and (2.2) we have

$$\begin{array}{c} \dot{\Delta}_t = A\Delta_t + \widetilde{W}_{1,t}\sigma(\widehat{x}_t) + \widetilde{W}_{2,t}\phi(\widehat{x}_t)\gamma(u_t) \\ +W_1^*\widetilde{\sigma}_t + W_2^*\widetilde{\phi}_t\gamma(u_t) + \Delta f(x_t, u_t, t) \end{array} \tag{2.27}$$

where $\widetilde{\sigma}_t$, $\widetilde{W}_{1,t}$ and $\widetilde{W}_{2,t}$ are defined as in (2.5), (2.13) and (2.14) correspondingly. If we select Lyapunov function as

$$V_t := \left[\left\|P^{1/2}\Delta\right\| - \mu\right]_+^2 + tr\left[\widetilde{W}_{1,t}^T K_1^{-1}\widetilde{W}_{1,t}\right] + tr\left[\widetilde{W}_{2,t}^T K_2^{-1}\widetilde{W}_{2,t}\right] \tag{2.28}$$

where $P = P^T > 0$, then, in view of Lemma 11.6 in Appendix A, we derive:

$$\begin{array}{c} \dot{V}_t \le 2\left[\left\|P^{1/2}\Delta\right\| - \mu\right]_+ \frac{\left(P^{1/2}\Delta\right)^T}{\left\|P^{1/2}\Delta\right\|} P^{1/2}\dot{\Delta} \\ +2tr\left[\dot{\widetilde{W}}_{1,t}^T K_1^{-1}\widetilde{W}_{1,t}\right] + 2tr\left[\dot{\widetilde{W}}_{2,t}^T K_2^{-1}\widetilde{W}_{2,t}\right] \\ = 2\left[1 - \mu\left\|P^{1/2}\Delta\right\|^{-1}\right]_+ \Delta P\dot{\Delta} \\ +2tr\left[\dot{\widetilde{W}}_{1,t}^T K_1^{-1}\widetilde{W}_{1,t}\right] + 2tr\left[\dot{\widetilde{W}}_{2,t}^T K_2^{-1}\widetilde{W}_{2,t}\right] \end{array} \tag{2.29}$$

If we define s_t as in (2.23), then (2.29) becomes

$$\dot{V}_t \le s_t 2\Delta^T P\dot{\Delta} + 2tr\left[\dot{\widetilde{W}}_{1,t}^T K_1^{-1}\widetilde{W}_{1,t}\right] + 2tr\left[\dot{\widetilde{W}}_{2,t}^T K_2^{-1}\widetilde{W}_{2,t}\right]$$

From (2.27) we have

$$\begin{array}{c} 2\Delta^T P\dot{\Delta} = 2\Delta^T P A^*\Delta + \\ 2\Delta^T P\left(\widetilde{W}_{1,t}\sigma(\widehat{x}_t) + \widetilde{W}_{2,t}\phi(\widehat{x}_t)\gamma(u_t) + W_1^*\widetilde{\sigma}_t + W_2^*\widetilde{\phi}_t\gamma(u_t) + \Delta f(x_t, u_t, t)\right) \end{array}$$

Using the matrix inequality (see Lemma 11.1 of Appendix A)

$$X^TY + Y^TX \leq X^T\Lambda X + Y^T\Lambda^{-1}Y \tag{2.30}$$

which is valid for any $X,\ Y \in R^{n\times m}$ and any $0 < \Lambda = \Lambda^T \in R^{n\times n}$, the modelling error effect may be estimated as

$$\begin{aligned} 2\Delta_t^T P\Delta f &\leq \Delta_t^T P\Lambda_1^{-1}P\Delta_t + \left\|\eta_\sigma\right\|_{\Lambda_1} + \left\|\eta_\phi \gamma(u_t)\right\|_{\Lambda_1} \\ &\leq \Delta_t^T P\Lambda_1^{-1}P\Delta_t + \overline{\eta}_\sigma + \overline{\eta}_\phi \overline{u} \end{aligned}$$

The terms $2\Delta^T PW_1^*\widetilde{\sigma}_t$ and $W_2^*\widetilde{\phi}_t\gamma(u_t)$ are estimated by the analogous way as

$$\begin{gathered} 2\Delta^T PW_1^*\widetilde{\sigma}_t \leq \Delta^T PW_1^*\Lambda_2^{-1}W_1^{*T}P\Delta + \widetilde{\sigma}_t^T\Lambda_2\widetilde{\sigma}_t \\ \leq \Delta^T P\overline{W}_1 P\Delta + \Delta^T D_\sigma \Delta = \Delta^T\left[P\overline{W}_1 P + D_\sigma\right]\Delta \\ 2\Delta^T PW_2^*\widetilde{\phi}_t\gamma(u_t) \leq \Delta^T P\overline{W}_2 P\Delta + \overline{u}\Delta^T D_\phi \Delta \\ = \Delta^T\left[P\overline{W}_2 P + \overline{u}D_\phi\right]\Delta \end{gathered}$$

Since $s_t \geq 0$, (2.29) can be rewritten as

$$\dot{V}_t \leq s_t\Delta_t^T L\Delta_t + L_{w1} + L_{w2} - \Delta_t^T Q_0 \Delta_t s_t + \left(\overline{\eta}_\sigma + \overline{\eta}_\phi \overline{u}\right) s_t \tag{2.31}$$

where

$$\begin{aligned} L &:= PA + A^TP + PRP + Q \\ L_{w1} &:= 2tr\left[\dot{\widetilde{W}}_{1,t}^T K_1^{-1}\widetilde{W}_{1,t}\right] + 2\Delta_t^T P\widetilde{W}_{1,t}\sigma s_t \\ L_{w2} &:= 2tr\left[\dot{\widetilde{W}}_{2,t}^T K_2^{-1}\widetilde{W}_{2,t}\right] + 2\Delta_t^T P\widetilde{W}_{2,t}\gamma(u)\, s_t \end{aligned}$$

R and Q are defined in (2.21). Using the updating law as in (2.22), and in view of

$$\dot{\widetilde{W}}_{1,t} = -\dot{W}_{1,t},\ \dot{\widetilde{W}}_{2,t} = -\dot{W}_{2,t}$$

we get

$$L_{w1} = 0,\ L_{w2} = 0$$

Finally, from (2.31) we derive that

$$\dot{V}_t \leq -s_t \left[\Delta_t^T Q_0 \Delta_t - \left(\overline{\eta}_\sigma + \overline{\eta}_\phi \overline{u} \right) \right] \tag{2.32}$$

The right-hand side of the last inequality can be estimated by the following way:

$$\dot{V}_t \leq -s_t \lambda_{\min} \left(P^{-1/} Q_0 P^{-1/2} \right) \left(\left\| P^{1/2} \Delta \right\|^2 - \mu \right) \leq 0 \tag{2.33}$$

where μ is defined as in (2.24). So, V_t is bounded. The statement a) (2.25) is proved.

Since $0 \leq s_t \leq 1$, from (2.33) we have

$$\dot{V}_t \leq -\Delta_t^T Q_0 \Delta_t s_t + \left(\overline{\eta}_\sigma + \overline{\eta}_\phi \overline{u} \right) s_t \leq -\Delta_t^T Q_0 \Delta_t s_t + \overline{\eta}_\sigma + \overline{\eta}_\phi \overline{u}$$

Integrating (2.32) from 0 up to T yields

$$V_T - V_0 \leq -\int_0^T \Delta^T Q_0 \Delta s_t dt + \overline{\eta}_\sigma + \overline{\eta}_\phi \overline{u}$$

So,

$$\int_0^T \Delta^T Q_0 \Delta s_t dt \leq V_0 - V_T + \left(\overline{\eta}_\sigma + \overline{\eta}_\phi \overline{u} \right) T \leq V_0 + \overline{\eta}_1 T$$

Because $W_{1,0} = W_1^*$ and $W_{2,0} = W_2^*$, V_0^0 and V_0 are bounded, (2.26) is obtained and b) is proved. ■

The estimate (2.26) for the performance index reflects the behavior of the identification error *"in average"*. In the next subsection we will present the results dealing with estimation of the maximum value of the identification error ("non average estimate").

2.2.4 Estimation of Maximum Value of Identification Error for Nonlinear Systems with Bounded Unmodeled Dynamics

The equation (2.4) can be written as

$$\dot{\Delta}_t = A_0 \Delta_t + h(\Delta_t, \widehat{x}_t, u_t) \tag{2.34}$$

where

$$h(\cdot) := \widetilde{W}_{1,t} \sigma(\widehat{x}_t) + \widetilde{W}_{2,t} \phi(\widehat{x}_t) \gamma(u_t) + W_1^* \widetilde{\sigma}_t$$
$$+ W_2^* \widetilde{\phi}_t \gamma(u_t) + \Delta f(x_t, u_t, t) + (A^* - A_0) \Delta_t$$

Here A_0 is a Hurtwitz matrix. In view of (2.13), (2.14), **A2.2** and **A2.4**, we have

$$\left(\widetilde{W}_{1,t}\sigma(\widehat{x}_t)+\widetilde{W}_{2,t}\phi(\widehat{x}_t)\gamma(u_t)\right)^T\Lambda_h\left(\widetilde{W}_{1,t}\sigma(\widehat{x}_t)+\widetilde{W}_{2,t}\phi(\widehat{x}_t)\gamma(u_t)\right)\leq\mu_1(\widehat{x}_t,u_t)$$

$$(\Delta f(x_t,u_t,t))^T\Lambda_h(\Delta f(x_t,u_t,t))\leq\mu_2(\widehat{x}_t,u_t)$$

where $\mu_1(\widehat{x}_t,u_t)$ and $\mu_2(\widehat{x}_t,u_t)$ are bounded functions (because we already proved that the weight matrices are bounded). From the assumption **A2.1**

$$\left(W_1^*\widetilde{\sigma}_t+W_2^*\widetilde{\phi}_t\gamma(u_t)\right)^T\Lambda_h\left(W_1^*\widetilde{\sigma}_t+W_2^*\widetilde{\phi}_t\gamma(u_t)\right)\leq\mu_3(\widehat{x}_t,u_t)\Delta_t^T\Lambda_\Delta\Delta_t$$

$$((A-A_0)\Delta_t)^T\Lambda_h((A-A_0)\Delta_t)\leq\mu_4(\widehat{x}_t,u_t)\Delta_t^T\Lambda_\Delta\Delta_t$$

where $\mu_3(\widehat{x}_t,u_t)$ and $\mu_4(\widehat{x}_t,u_t)$ are bounded functions too. So, $h(\cdot)$ satisfies the following sector conditions

$$h(\cdot)^T\Lambda_h h(\cdot)\leq\varepsilon_0(\widehat{x}_t,u_t)+\varepsilon_1(\widehat{x}_t,u_t)\Delta_t^T\Lambda_\Delta\Delta_t \tag{2.35}$$

where

$$\begin{aligned}\varepsilon_0(\widehat{x}_t,u_t)&=\mu_1(\widehat{x}_t,u_t)+\mu_2(\widehat{x}_t,u_t)\\ \varepsilon_1(\widehat{x}_t,u_t)&=\mu_3(\widehat{x}_t,u_t)+\mu_4(\widehat{x}_t,u_t)\end{aligned}$$

the functions $\varepsilon_0(\cdot)$ and $\varepsilon_1(\cdot)$ are uniformly bounded functions, i.e., there exist the constants ε^i such that

$$\varepsilon_i(\cdot)\leq\varepsilon^i<\infty,\ i=0,1,\ \forall x\in\Re^n,\ \forall u\in\Re^q$$

Similar with **A2.5** we can also select A_0 and Q_1 to satisfy following assumption.

A2.6: There exits a strictly positive defined matrix Q_1 such that if the matrices R, Q and A are defined as

$$R:=\Lambda_h^{-1},\quad Q:=\varepsilon^1\Lambda_\Delta+Q_1,\quad A:=A_0$$

the matrix Riccati equation (6.43) has a positive solution $P=P^T>0$.

Theorem 2.3 *Under the assumptions* ***A2.1-A2.6,*** *for the unknown nonlinear system (2.1) and parallel neural network (2.2), the following property holds:*

$$\limsup_{T\to\infty} \|\Delta(t)\| \leq \sqrt{\frac{\varepsilon^0}{\lambda_{\max}(P)\,\lambda_{\min}(R_p)}} \tag{2.36}$$

where

$$R_p = P^{-\frac{1}{2}} Q_1 P^{-\frac{1}{2}} \tag{2.37}$$

$\lambda_{min}(.)$ *is the minimum eigenvalue of the respective matrix.*

Proof. Let us consider the nonnegative defined scalar function

$$V_1(\Delta) = \Delta^T P \Delta \in \Re^+.$$

Computing its time derivative over the trajectories of equation (2.34), we obtain:

$$\dot{V}_1 = 2\Delta^T P \dot{\Delta} = 2\Delta_t^T P(A_0\Delta + h) = \Delta_t^T\left(P_i A_0 + A_0^T P_i\right)\Delta_t + 2\Delta^T P h \tag{2.38}$$

Using (2.35) we obtain:

$$2\Delta^T P h \leq h^T \Lambda_h h + \Delta^T P \Lambda_h^{-1} P \Delta \leq \Delta^T\left(P\Lambda_h^{-1}P + \varepsilon^1 \Lambda_\Delta\right)\Delta + \varepsilon^0 \tag{2.39}$$

Substituting inequalities (2.38) in (2.39), then we get:

$$\dot{V}(\Delta) \leq \Delta^T\left(PA_0 + A_0^T P + P\Lambda_h^{-1}P + \varepsilon^1\Lambda_\Delta + Q_1\right)\Delta - \Delta^T Q_1 \Delta + \varepsilon^0.$$

Taking into account **A2.6** we obtain:

$$\dot{V}(\Delta) \leq -\lambda_{\min}(R_p)\, V(\Delta) + \varepsilon^0 \tag{2.40}$$

where R_p is defined as (2.37). By solving (2.40), we derive:

$$V(\Delta) \leq V(\Delta(0))\, e^{-\lambda_{\min}(R_P)t} + \varepsilon^0 \int_0^t e^{-\lambda_{\min}(R_P)(t-\tau)} d\tau$$

which can be explicitly evaluated as:

$$V(\Delta) \leq V(\Delta(0))\, e^{-\lambda_{\min}(R_P)t} + \frac{\varepsilon^0}{\lambda_{\min}(R_P)}\left(1 - e^{-\lambda_{\min}(R_P)t}\right)$$

Because

$$V(\Delta) \geq \lambda_{\min}(P)\,\|\Delta\|^2$$

we obtain (2.36). ■

2.3 Multilayer Differential Neural Networks for Nonlinear System On-line Identification

In this section we start to study more complex differential neural networks containing some new possibilities to adopt its behavior. We deal with *multilayer dynamic neural network* and derive a new stable learning law for a big class of such neural networks. Their applications to nonlinear system on-line identification is talked. By means of a Lyapunov-like analysis we derive stability conditions for the weights of both hidden and output layers. An algebraic Riccati equation approach is used to establish a bound for the identification error.

2.3.1 Multilayer Structure of Differential Neural Networks

Let us consider the following dynamic neural network:

$$\boxed{\dot{\widehat{x}}_t = A\widehat{x}_t + W_{1,t}\sigma(V_{1,t}\widehat{x}_t) + W_{2,t}\phi(V_{2,t}\widehat{x}_t)\gamma(u_t)} \tag{2.41}$$

where

$\widehat{x}_t \in \Re^n$ is the state of the neural network,

$u_t \in \Re^q$ is an input measurable control action,

$W_{1,t} \in \Re^{n\times k}$ is the matrix of output weight,

$W_{2,t} \in \Re^{n\times r}$ is the matrix of output weight,

$V_1 \in \Re^{m\times n}$ is the weight describing hidden layers connections,

$V_2 \in \Re^{k\times n}$ is the weight describing hidden layers connections,

$A \in \Re^{n\times n}$ is a Hurtwitz matrix.

The vector field $\sigma(x_t) : \Re^m \to \Re^k$ is assumed to have the elements increasing monotonically as in (2.3). The nonlinearity $\gamma(u_t)$ define the vector field from $\Re^q$ to $\Re^s$. The function $\phi(\cdot)$ is the transformation from $\Re^k$ to $\Re^{r\times s}$.

The structure of this dynamic system is shown in Figure 2.2. The most simple structure without any hidden layers (containing only input and output layers) corresponds to the case when

$$p = q = n \text{ and } V_1 = V_2 = I \tag{2.42}$$

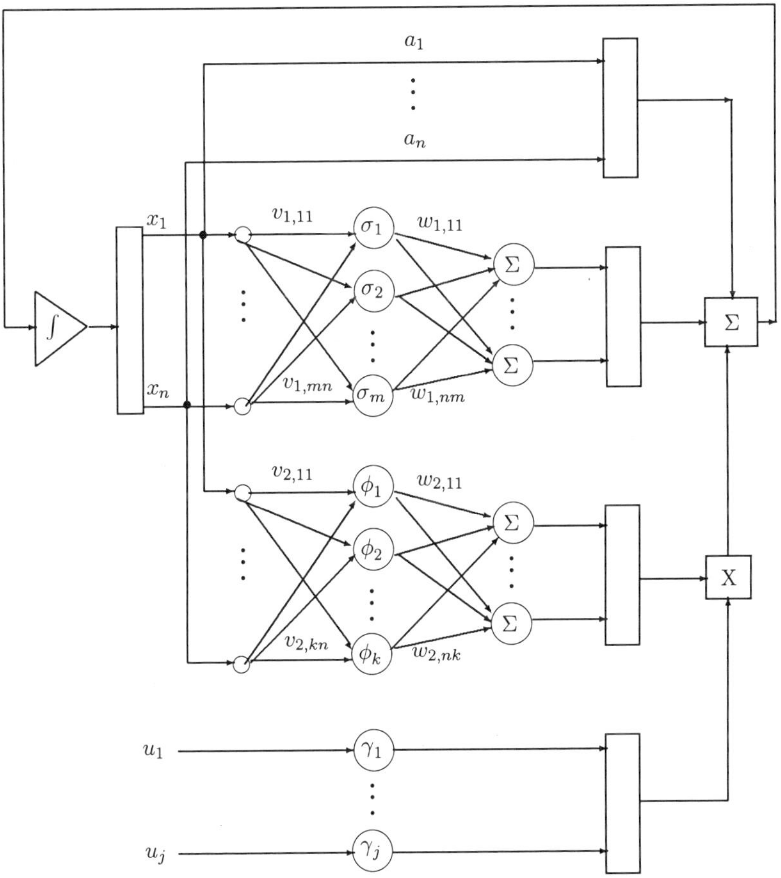

FIGURE 2.2. The general structure of the dynamic neural network.

and was studied in the previous section. This simple structure has been considered by many authors (see, for example, [24], [18], [13] and [18]). Below we will deal with dynamic neural networks of general type given by (2.41). The stable learning design for such dynamic neural networks is our main novelty presented in this chapter.

The nonlinear system to be identified is given as (2.1). Let us also assume that the control input is bounded (**A2.2**).

2.3.2 Complete Model Matching Case

Let us first consider the subclass of nonlinear systems (2.1)

$$\boxed{\begin{array}{c} \dot{x}_t = f(x_t, u_t, t) \\ x_t \in \Re^n,\ u_t \in \Re^q, \quad n \geq q \end{array}}$$

which can be exactly presented by a equation (2.41), i.e., an exact neural network model of the plant is available.

Mathematically this fact can be expressed as follows: there exist the matrix A^*, the weights W_1^*, W_2^*, V_1^* and V_2^* such that the nonlinear system (2.1) can be exactly described by the neural network structure (2.41) as

$$\boxed{\dot{x}_t = A^* x_t + W_1^* \sigma(V_1^* x_t) + W_2^* \phi(V_2^* x_t) \gamma(u_t)} \tag{2.43}$$

Here we consider the matrix $A = A^*$ and V_1^*, V_2^* as known, we do not know a priory the weights W_1^* and W_2^*. The upper boundness for the first weight matrices is assumed to be known, that is,

$$W_1^{*T} \Lambda_1^{-1} W_1^* \leq \overline{W}_1, \quad W_2^{*T} \Lambda_2^{-1} W_2^* \leq \overline{W}_2$$

where $\overline{W}_i$ and Λ_i are known positive define matrices.

Let us define the identification error as before:

$$\Delta_t := \widehat{x}_t - x_t \tag{2.44}$$

It is clear that all sigmoidal functions, commonly used in neural networks, satisfy the following conditions.

A2.7: *The differences of σ and ϕ fulfill the generalized Lipschitz condition given by*

$$\begin{gathered}\widetilde{\sigma}_t^T\Lambda_1\widetilde{\sigma}_t \leq \Delta_t^T\Lambda_\sigma\Delta_t, \quad \widetilde{\phi}_t^T\Lambda_2\widetilde{\phi}_t \leq \Delta_t^T\Lambda_\phi\Delta_t \\ \widetilde{\sigma}_t'^T\Lambda_1\widetilde{\sigma}_t' \leq \left(\widetilde{V}_{1,t}\widehat{x}_t\right)^T\Lambda_{v1}\left(\widetilde{V}_{1,t}\widehat{x}_t\right) \\ \widetilde{\phi}_t'^T\Lambda_2\widetilde{\phi}_t' \leq \left(\widetilde{V}_{2,t}\widehat{x}_t\right)^T\Lambda_{v2}\left(\widetilde{V}_{2,t}\widehat{x}_t\right)\end{gathered} \tag{2.45}$$

where

$$\begin{gathered}\widetilde{\sigma}_t := \sigma(V_1^*\widehat{x}_t) - \sigma(V_1^*x_t), \quad \widetilde{\phi}_t := \phi(V_2^*\widehat{x}_t) - \phi(V_2^*x_t) \\ \widetilde{\sigma}_t' = \sigma(V_{1,t}\widehat{x}_t) - \sigma(V_1^*\widehat{x}_t), \quad \widetilde{\phi}_t' = \phi(V_{2,t}\widehat{x}_t) - \phi(V_2^*\widehat{x}_t)\end{gathered}$$

and Λ_σ, Λ_ϕ, Λ_{v1}, Λ_{v2}, are known normalizing positive constant matrices.

Because the sigmoid functions σ and ϕ are differentiable and satisfy the Lipschitz condition, based on Lemma 11.5 (see Appendix A) we conclude that

$$\begin{gathered}\widetilde{\sigma}_t' := \sigma(V_{1,t}\widehat{x}_t) - \sigma(V_1^*\widehat{x}_t) = D_\sigma\widetilde{V}_{1,t}\widehat{x}_t + \nu_\sigma \\ \widetilde{\phi}_t'\gamma(u_t) := \phi(V_{2,t}\widehat{x}_t)\gamma(u_t) - \phi(V_2^*\widehat{x}_t)\gamma(u_t) \\ = \sum_{i=1}^{q}\left[\phi_i(V_2^*\widehat{x}_t) - \phi_i(V_{2,t}\widehat{x}_t)\right]\gamma_i(u_t) = \sum_{i=1}^{q}\left[D_{i\phi}\widetilde{V}_{2,t}\widehat{x}_t + \nu_{i\phi}\right]\gamma_i(u_t) \\ \gamma_i(u_t) \text{ is a scalar } (i\text{-th component of } \gamma(u_t)) \\ D_\sigma = \frac{\partial\sigma(Z)}{\partial Z}\Big|_{Z=V_{1,t}\widehat{x}_t} \in \Re^{m\times m}, \quad \|\nu_\sigma\|_{\Lambda_1}^2 \leq l_1\left\|\widetilde{V}_{1,t}\widehat{x}_t\right\|_{\Lambda_1}^2, \quad l_1 > 0 \\ D_{i\phi} = \frac{\partial\phi_i(Z)}{\partial Z}\Big|_{Z=V_{2,t}\widehat{x}_t} \in \Re^{m\times m}, \quad \|\nu_\phi\|_{\Lambda_2}^2 \leq l_2\left\|\widetilde{V}_{2,t}\widehat{x}_t\right\|_{\Lambda_2}^2, \quad l_2 > 0 \\ \widetilde{V}_{1,t} = V_{1,t} - V_1^*, \quad \widetilde{V}_{2,t} = V_{2,t} - V_2^*\end{gathered} \tag{2.46}$$

In the following, as before, we will use the fact that if the pair $(A, R^{1/2})$ is controllable, the pair $(Q^{1/2}, A)$ is observable, and the special *local frequency condition* or its "matrix equivalent" (2.8) is fulfilled (see also Appendix A), the matrix Riccati equation

$$A^TP + PA + \Gamma R\Gamma + Q = 0 \tag{2.47}$$

has a positive solution. In view of this fact we can accept the following additional assumption.

A2.8: *For a given matrix A there exists a strictly positive defined matrix Q_0 such that the matrix Riccati equation (2.47) with the matrices R and Q given by*

$$R = 2\overline{W}_1 + 2\overline{W}_2, \quad Q = Q_0 + \Lambda_\sigma + \Lambda_\phi \overline{u} \tag{2.48}$$

has a positive solution, $\overline{u}$ is defined as **A2.2**.

Next theorem presents the new learning procedure and states the fact of its stability that is the first main contributions of this study.

Theorem 2.4 *Let us consider the unknown nonlinear system (2.1) and a multilayer dynamic neural network (2.41) whose weights are adjusted as*

$$\boxed{\begin{array}{r}\dot{W}_{1,t} = -K_1 P \Delta_t \sigma^T + K_1 P \Delta_t \widehat{x}_t^T \left(V_{1,t} - V_1^*\right)^T D_\sigma \\ \dot{W}_{2,t} = -K_2 P \Delta_t \left(\phi \gamma(u_t)\right)^T + K_2 P \Delta_t \widehat{x}_t^T \left(V_{2,t} - V_2^*\right)^T \sum_{i=1}^{q} \left(\gamma_i(u_t) D_{i\phi}^T\right) \\ \dot{V}_{1,t} = -K_3 D_\sigma^T W_{1,t}^T P \Delta_t \widehat{x}_t^T - \frac{l_1}{2} K_3 \Lambda_1 \left(V_{1,t} - V_1^*\right) \widehat{x}_t \widehat{x}_t^T \\ \dot{V}_{2,t} = -K_4 \sum_{i=1}^{q} \left(\gamma_i(u_t) D_{i\phi}^T\right) W_{2,t}^T P \Delta_t \widehat{x}_t^T - \frac{q l_2 \overline{u}}{2} K_4 \Lambda_2 \left(V_{2,t} - V_2^*\right) \widehat{x}_t \widehat{x}_t^T\end{array}} \tag{2.49}$$

where $K_i \in R^{n \times n}$ $(i = 1 \cdots 4)$ are positive defined matrices, P is the solution of the matrix Riccati equation given by (2.47). The initial matrices are assumed to be bounded. Assuming also that the assumptions ***A2.2, A2.7-A2.8*** *hold, we conclude that the weights are bounded, i.e.,*

$$W_{1,t} \in L^\infty, \quad W_{2,t} \in L^\infty, \quad V_{1,t} \in L^\infty, \quad V_{2,t} \in L^\infty \tag{2.50}$$

and the identification process is globally asymptotically stable, that is,

$$\lim_{t \to \infty} \Delta_t = 0 \tag{2.51}$$

Proof. From (2.41) and (2.43) the error equation can be expressed as

$$\begin{array}{r}\dot{\Delta}_t = A\Delta_t + \widetilde{W}_{1,t}\sigma(V_{1,t}\widehat{x}_t) + \widetilde{W}_{2,t}\phi(V_{2,t}\widehat{x}_t)\gamma(u_t) \\ + W_1^* \widetilde{\sigma}_t + W_2^* \widetilde{\phi}_t \gamma(u_t) + W_1^* \widetilde{\sigma}_t' + W_2^* \widetilde{\phi}_t' \gamma(u_t)\end{array} \tag{2.52}$$

where

$$\widetilde{W}_{1,t} = W_{1,t} - W_1^*, \quad \widetilde{W}_{2,t} = W_{2,t} - W_2^*$$

Define Lyapunov function candidate as

$$V_t := \Delta_t^T P \Delta_t + \tfrac{1}{2} tr \left[\widetilde{W}_{1,t}^T K_1^{-1} \widetilde{W}_{1,t} \right] + \tfrac{1}{2} tr \left[\widetilde{W}_{2,t}^T K_2^{-1} \widetilde{W}_{2,t} \right] + \tfrac{1}{2} tr \left[\widetilde{V}_{1,t}^T K_3^{-1} \widetilde{V}_{1,t} \right] + \tfrac{1}{2} tr \left[\widetilde{V}_{2,t}^T K_4^{-1} \widetilde{V}_{2,t} \right]$$

Calculating its derivative and using **A1** implies

$$\begin{aligned} \dot{V}_t \le 2\Delta_t^T P \dot{\Delta}_t + 2tr \left[\dot{\widetilde{W}}_{1,t}^T K_1^{-1} \widetilde{W}_{1,t} \right] \\ +2tr \left[\dot{\widetilde{W}}_{2,t}^T K_2^{-1} \widetilde{W}_{2,t} \right] + 2tr \left[\dot{\widetilde{V}}^T_{1,t} K_3^{-1} \widetilde{V}_{1,t} \right] + 2tr \left[\dot{\widetilde{V}}^T_{2,t} K_4^{-1} \widetilde{V}_{2,t} \right] \end{aligned} \tag{2.53}$$

Substitute (2.52) into (2.53), we get

$$\begin{aligned} 2\Delta_t^T P \dot{\Delta}_t = 2\Delta_t^T P A \Delta_t + 2\Delta_t^T P \left(W_1^* \widetilde{\sigma}_t + W_2^* \widetilde{\phi}_t \gamma(u_t) \right) \\ +2\Delta_t^T P \left(\widetilde{W}_{1,t} \sigma + \widetilde{W}_{2,t} \phi \gamma(u_t) \right) + 2\Delta_t^T P \left(W_1^* \widetilde{\sigma}_t' + W_2^* \widetilde{\phi}_t' \gamma(u_t) \right) \end{aligned} \tag{2.54}$$

In view of the matrix inequality (8.38) and using **A2.7,** the term $\Delta_t^T P \left(W_1^* \widetilde{\sigma}_t + W_2^* \widetilde{\phi}_t u_t \right)$ in (2.54) may be conclude as

$$\begin{aligned} 2\Delta_t^T P W_1^* \widetilde{\sigma}_t \le \Delta_t^T P W_1^* \Lambda_1^{-1} W_1^{*T} P \Delta_t + \widetilde{\sigma}_t^T \Lambda_1 \widetilde{\sigma}_t \le \Delta_t^T \left(P \overline{W}_1 P + \Lambda_\sigma \right) \Delta_t \\ 2\Delta_t^T P W_2^* \widetilde{\phi}_t \gamma(u_t) \le \Delta_t^T \left(P \overline{W}_2 P + \overline{u}^2 \Lambda_\phi \right) \Delta_t \end{aligned} \tag{2.55}$$

Using **A2.7,** the last term in (2.54) can be rewritten as

$$\begin{aligned} 2\Delta_t^T P W_1^* \widetilde{\sigma}_t' = 2\Delta_t^T P W_{1,t} D_\sigma \widetilde{V}_{1,t} \widehat{x}_t - 2\Delta_t^T P \widetilde{W}_{1,t} D_\sigma \widetilde{V}_{1,t} \widehat{x}_t + 2\Delta_t^T P W_1^* \nu_\sigma \\ 2\Delta_t^T P W_2^* \widetilde{\phi}_t' \gamma(u_t) = 2\Delta_t^T P W_{2,t} \sum_{i=1}^{q} \left[D_{i\phi} \widetilde{V}_{2,t} \widehat{x}_t \right] \gamma_i(u_t) \\ -2\Delta_t^T P \widetilde{W}_{2,t} \sum_{i=1}^{q} \left[D_{i\phi} \widetilde{V}_{2,t} \widehat{x}_t \right] \gamma_i(u_t) + 2\Delta_t^T P W_2^* \sum_{i=1}^{q} \nu_{i\phi} \gamma_i(u_t) \end{aligned} \tag{2.56}$$

The term $2\Delta_t^T P W_1^* \nu_\sigma$ in (2.56) may be estimated as

$$2\Delta_t^T P W_1^* \nu_\sigma \le \Delta_t^T P W_1^{*T} \Lambda_1^{-1} W_1^* P \Delta_t + l_1 \left\| \widetilde{V}_{1,t} \widehat{x}_t \right\|_{\Lambda_1}^2 \tag{2.57}$$

as well as the term $2\Delta_t^T P W_2^* \sum_{i=1}^{q} \nu_{i\phi}\gamma_i(u_t)$ in (2.56) may be estimated as

$$\begin{array}{l} 2\Delta_t^T P W_2^* \sum\limits_{i=1}^{q} \nu_{i\phi}\gamma_i(u_t) \leq \Delta_t^T P W_2^{*T} \Lambda_2^{-1} W_2^* P \Delta_t \\ \qquad + q \sum\limits_{i=1}^{q} \gamma_i^T(u_t) \nu_{i\phi}^T \Lambda_2 \nu_{i\phi}\gamma_i(u_t) \\ \leq \Delta_t^T P \overline{W}_2 P \Delta_t + q l_2 \bar{u} \left\| \widetilde{V}_{2,t}\widehat{x}_t \right\|_{\Lambda_2}^2 \end{array} \tag{2.58}$$

Substitute the estimations into (2.54)

$$\dot{V}_t \leq \Delta_t^T L \Delta_t + L_{w1} + L_{w2} + L_{v1} + L_{v2} - \Delta_t^T Q \Delta_t \tag{2.59}$$

where

$$\begin{array}{l} L = PA + A^T P + PRP + Q \\ L_{w1} = 2tr\left[\dot{\widetilde{W}}_{1,t}^T K_1^{-1} \widetilde{W}_{1,t} \right] + 2\Delta_t^T P \widetilde{W}_{1,t}\sigma - 2\Delta_t^T P \widetilde{W}_{1,t} D_\sigma \widetilde{V}_{1,t}\widehat{x}_t \\ L_{w2} = 2tr\left[\dot{\widetilde{W}}_{2,t}^T K_2^{-1} \widetilde{W}_{2,t} \right] + 2\Delta_t^T P \widetilde{W}_{2,t}\phi\gamma(u_t) \\ \qquad -2tr\left[\sum\limits_{i=1}^{q} \left(D_{i\phi}\gamma_i(u_t)\right) \widetilde{V}_{2,t}\widehat{x}_t \Delta_t^T P \tilde{W}_{2,t} \right] \\ L_{v1} = 2tr\left[\left(\dot{\widetilde{V}}_{1,t} \right)^T K_3^{-1} \widetilde{V}_{1,t} \right] + 2\Delta_t^T P W_{1,t} D_\sigma \widetilde{V}_{1,t}\widehat{x}_t + l_1 \left\| \widetilde{V}_{1,t}\widehat{x}_t \right\|_{\Lambda_1}^2 \\ L_{v2} = 2tr\left[\left(\dot{\widetilde{V}}_{2,t} \right)^T K_4^{-1} \widetilde{V}_{2,t} \right] + 2tr\left[\widehat{x}_t \Delta_t^T P W_{2,t} \sum\limits_{i=1}^{q} \left(D_{i\phi}\gamma_i(u_t)\right) \widetilde{V}_{2,t} \right] \\ \qquad + q l_2 \bar{u} \left\| \widetilde{V}_{2,t}\widehat{x}_t \right\|_{\Lambda_2}^2 \end{array}$$

Using **A2.8,** we can conclude that

$$\dot{V}_t \leq -\Delta_t^T Q \Delta_t \tag{2.60}$$

where $Q > 0$. So, integrating both sides of this inequality from $t = 0$ up to $t = T$, finally, it follows

$$\int_{t=0}^{T} \|\Delta_t\|_Q^2 \, dt \leq V_0 - V_T \leq V_0 < \infty$$

and, hence, we get

$$\int_{t=0}^{\infty} \|\Delta_t\|_Q^2 \, dt \le V_0 < \infty$$

So, the process $\{\Delta_t\}$ is quadratically integrable ($\Delta_t \in L_2$) and bounded ($\Delta_t \in L_\infty$) (because $\{V_t\}$ is bounded process too), that is,

$$\Delta_t \in L_2 \cap L_\infty$$

In view of this fact, from the error equation (2.60) we also conclude that its derivative is also bounded, that is,

$$\dot{\Delta}_t \in L_\infty$$

Using Barlalat's Lemma (Appendix A) it follows (2.51). As the signals u_t, $\sigma(\widehat{x}_t)$, $\phi(\widehat{x}_t)$ and P are bounded, so we can conclude (2.50). Theorem is proved. ■

Remark 2.2 *One can see that the learning law (2.49) of the multilayer dynamic neural network (2.41) has the similar structure with the backpropagation of the multilayer perceptrons (see [7]). If we consider $K_i P$ as an updating rate, the first terms of the differential equations in (2.49) exactly correspond to the backpropagation scheme. The second terms are absolutely new ones and are used here to assure the stable learning. A big learning rates $\overline{K}_i := K_i P$ can be achieved by a special selection of the gain matrices K_i.*

Remark 2.3 *Even the proposed learning law looks like the backpropagation algorithms with an additional term, due to the fact that it is derived using the Lyapunov approach, the global asymptotic error stability is guaranteed. So, the locally minimal convergence problem, which is the major concern in static neural networks learning, does not exist in this situation.*

2.3.3 Unmodeled Dynamics Presence

In this paragraph, the more realistic case is considered, when the dynamic neural networks does not match exactly the nonlinear system and, as a result, we deal with system description including known structure and, obligatory, unmodeled dynamic part. So, to justify the implementation of any learning scheme for applied neural networks, its robustness property with respect to unmodeled dynamic incorporated in to a real dynamics should be proven. As a fact, it is not permitted to be large enough to keep stability as well as good behavior of dynamic neural networks.

A2.9: *There exists a bounded control* u_t $\left(\|\gamma\ (u_t)\|^2 \leq \overline{u}\right)$, *such that the closed-loop system is quadratic stable, that is, there exists a Lyapunov function* $V^0 > 0$ *and a positive constant* λ *such that*

$$\frac{\partial V^0}{\partial x} f\left(x_t, u_t, t\right) \leq -\lambda \left\|x_t\right\|^2, \quad \lambda > 0 \tag{2.61}$$

Remark 2.4 *The condition of Assumption **A2.9** is a special case of Assumption **A2.2**, these two assumptions are coincide.*

Let us fixe some weight matrices W_1^*, W_2^*, V_1^*, V_2^* and a stable matrix A which can be selected below. In view of (2.41), we define the modelling error $\widetilde{f}_t$ as

$$\widetilde{f}_t := f(x_t, u_t, t) - [A\widehat{x}_t + W_1^* \sigma(V_1^* \widehat{x}_t) + W_2^* \phi(V_2^* \widehat{x}_t)\gamma(u_t)] \tag{2.62}$$

So, the original nonlinear system (2.1) can be represented in the following form:

$$\boxed{\dot{x}_t = Ax_t + W_1^* \sigma(V_1^* x_t) + W_2^* \phi(V_2^* x_t)\gamma(u_t) + \widetilde{f}_t} \tag{2.63}$$

In view of the fact

$$\left\|f\left(x_t, u_t, t\right)\right\|^2 \leq C_1 + C_2 \left\|x_t\right\|^2$$

(C_1 and C_2 are positive constants) which is needed to guarantee the global existence of the function x_t (see Theorem 12.1 of Appendix B), and taking into account that the sigmoid functions σ and ϕ, participating in the neural networks structure with hidden layer (2.2), are bounded, we can conclude that the unmodeled dynamics $\widetilde{f}_t$ verifies following properties:

A2.10: *For any normalizing matrix* Λ_f *there exist positive constants* $\overline{\eta}$ and $\overline{\eta}_1$ *such that*

$$\left\|\widetilde{f}_t\right\|_{\Lambda_f}^2 \leq \overline{\eta} + \overline{\eta}_1 \left\|x_t\right\|_{\Lambda_f}^2, \quad \Lambda_f = \Lambda_f^T > 0$$

Similar with **A2.8** we can also select Q_1 to satisfy following assumption.

A2.11: *For a given stable matrix* A *there exits a strictly positive defined matrix* Q_0 *such that the matrix Riccati equation (2.47) with*

$$R = 2\overline{W}_1 + 2\overline{W}_2 + \Lambda_1^{-1}, \quad Q = Q_1 + \Lambda_\sigma + \Lambda_\phi \overline{u} \tag{2.64}$$

has a positive solution. Here $\overline{u}$ is defined by **A2.2**.

The following theorem states the robust and stable learning law when the modelling error takes place.

Theorem 2.5 *Let us consider the unknown nonlinear system (2.1) and parallel neural network (2.41) with modelling error as in (2.62) whose weights are adjusted as*

$$\begin{array}{c}
\dot{W}_{1,t} = -s_t K_1 P \Delta_t \sigma^T + s_t K_1 P \Delta_t \widehat{x}_t^T \left(V_{1,t} - V_1^*\right)^T D_\sigma \\
\dot{W}_{2,t} = -s_t K_2 P \Delta_t \left(\phi\gamma(u_t)\right)^T + s_t K_2 P \Delta_t \widehat{x}_t^T \left(V_{2,t} - V_2^*\right)^T \sum_{i=1}^{q} \left(\gamma_i(u_t) D_{i\phi}^T\right) \\
\dot{V}_{1,t} = -s_t K_3 D_\sigma^T W_{1,t}^T P \Delta_t \widehat{x}_t^T - s_t \frac{l_1}{2} K_3 \Lambda_1 \left(V_{1,t} - V_1^*\right) \widehat{x}_t \widehat{x}_t^T \\
\dot{V}_{2,t} = -s_t K_4 \sum_{i=1}^{q} \left(\gamma_i(u_t) D_{i\phi}^T\right) W_{2,t}^T P \Delta_t \widehat{x}_t^T \\
-s_t \frac{q l_2 \overline{u}}{2} K_4 \Lambda_2 \left(V_{2,t} - V_2^*\right) \widehat{x}_t \widehat{x}_t^T, \quad V_1^* = V_{1,0}, V_2^* = V_{2,0}
\end{array} \tag{2.65}$$

where ***dead-zone function*** s_t *is defined as*

$$s_t := \left[1 - \frac{\mu_1}{\left\|P^{1/2}\Delta\right\|}\right]_+, \quad [z]_+ = \begin{cases} z & z \geq 0 \\ 0 & z < 0 \end{cases} \tag{2.66}$$

$$\mu_1 = \overline{\eta} / \lambda_{\min}\left(P^{-1/} Q_0 P^{-1/2}\right) \tag{2.67}$$

Assuming also that ***A2.2, A2.7, A2.9-A2.11*** *are verified, the following facts hold:*

a)

$$\Delta_t, W_{1,t}, W_{2,t} \in L_\infty \tag{2.68}$$

b) *the identification error* Δ_t *satisfies the following tracking performance*

$$\limsup_{T\to\infty} T^{-1} \int_0^T \Delta_t^T Q_0 \Delta_t dt \le \overline{\eta} \tag{2.69}$$

Proof. Because

$$\begin{array}{c}\dot{\Delta}_t = A\Delta_t + \widetilde{W}_{1,t}\sigma + \widetilde{W}_{2,t}\phi\gamma(u_t) + W_1^*\widetilde{\sigma}_t \\ +W_2^*\widetilde{\phi}_t\gamma(u_t) + W_1^*\widetilde{\sigma}_t' + W_2^*\widetilde{\phi}_t'\gamma(u_t) + \widetilde{f}_t\end{array} \tag{2.70}$$

Defining Lyapunov function candidate as

$$\begin{array}{c}V_t := V^0 + \left[\left\|P^{1/2}\Delta\right\| - \mu\right]_+^2 + tr\left[\widetilde{W}_{1,t}^T K_1^{-1}\widetilde{W}_{1,t}\right] + tr\left[\widetilde{W}_{2,t}^T K_2^{-1}\widetilde{W}_{2,t}\right] \\ +tr\left[\widetilde{V}_{1,t}^T K_3^{-1}\widetilde{V}_{1,t}\right] + tr\left[\widetilde{V}_{2,t}^T K_4^{-1}\widetilde{V}_{2,t}\right]\end{array}$$

where $P = P^T > 0$, then, in view of Lemma 11.6 in Appendix A, we derive:

$$\begin{array}{c}\dot{V}_t \le -\lambda\left\|x_t\right\|^2 + 2\left[\left\|P^{1/2}\Delta\right\| - \mu\right]_+ \frac{\left(P^{1/2}\Delta\right)^T}{\left\|P^{1/2}\Delta\right\|} P^{1/2}\dot{\Delta} \\ +2tr\left[\dot{\widetilde{W}}_{1,t}^T K_1^{-1}\widetilde{W}_{1,t}\right] + 2tr\left[\dot{\widetilde{W}}_{2,t}^T K_2^{-1}\widetilde{W}_{2,t}\right] \\ +2tr\left[\dot{\widetilde{V}}^T{}_{1,t} K_3^{-1}\widetilde{V}_{1,t}\right] + 2tr\left[\dot{\widetilde{V}}^T{}_{2,t} K_4^{-1}\widetilde{V}_{2,t}\right] \\ = 2\left[1 - \mu\left\|P^{1/2}\Delta\right\|^{-1}\right]_+ \Delta P\dot{\Delta} \\ +2tr\left[\dot{\widetilde{W}}_{1,t}^T K_1^{-1}\widetilde{W}_{1,t}\right] + 2tr\left[\dot{\widetilde{W}}_{2,t}^T K_2^{-1}\widetilde{W}_{2,t}\right] \\ +2tr\left[\dot{\widetilde{V}}^T{}_{1,t} K_3^{-1}\widetilde{V}_{1,t}^T\right] + 2tr\left[\dot{\widetilde{V}}^T{}_{2,t} K_4^{-1}\widetilde{V}_{2,t}^T\right]\end{array} \tag{2.71}$$

If we define s_t as in (2.66), then (2.71) becomes

$$\begin{array}{l}\dot{V}_t \le -\lambda\left\|x_t\right\|^2 + s_t 2\Delta^T P\dot{\Delta} \\ +2tr\left[\dot{\widetilde{W}}_{1,t}^T K_1^{-1}\widetilde{W}_{1,t}\right] + 2tr\left[\dot{\widetilde{W}}_{2,t}^T K_2^{-1}\widetilde{W}_{2,t}\right] \\ +2tr\left[\dot{\widetilde{V}}_{1,t} K_3^{-1}\widetilde{V}_{1,t}^T\right] + 2tr\left[\dot{\widetilde{V}}_{2,t} K_4^{-1}\widetilde{V}_{2,t}^T\right]\end{array}$$

From (2.70) we have

$$\begin{aligned} 2\Delta_t^T P \dot{\Delta}_t &= 2\Delta_t^T P A \Delta_t + 2\Delta_t^T P \left(W_1^* \widetilde{\sigma}_t + W_2^* \widetilde{\phi}_t \gamma(u_t) \right) \\ &+ 2\Delta_t^T P \left(\widetilde{W}_{1,t} \sigma + \widetilde{W}_{2,t} \phi \gamma(u_t) \right) + 2\Delta_t^T P \left(W_1^* \widetilde{\sigma}_t' + W_2^* \widetilde{\phi}_t' \gamma(u_t) \right) \end{aligned} \tag{2.72}$$

In view of the matrix inequality

$$X^T Y + \left(X^T Y \right)^T \leq X^T \Lambda^{-1} X + Y^T \Lambda Y$$

which is valid for any $X, Y \in \Re^{n \times k}$ and for any positive defined matrix $0 < \Lambda = \Lambda^T \in \Re^{n \times n}$, and using **A2.2**, **A2.7** the term

$$\Delta_t^T P \left(W_1^* \widetilde{\sigma}_t + W_2^* \widetilde{\phi}_t \gamma(u_t) \right)$$

in (2.72) can be estimated as

$$\begin{aligned} 2\Delta_t^T P W_1^* \widetilde{\sigma}_t \leq \Delta_t^T P W_1^* \Lambda_1^{-1} W_1^{*T} P \Delta_t + \widetilde{\sigma}_t^T \Lambda_1 \widetilde{\sigma}_t \leq \Delta_t^T \left(P \overline{W}_1 P + \Lambda_\sigma \right) \Delta_t \\ 2\Delta_t^T P W_2^* \widetilde{\phi}_t \gamma(u_t) \leq \Delta_t^T \left(P \overline{W}_2 P + \overline{u}^2 \Lambda_\phi \right) \Delta_t \end{aligned}$$

Using **A2.7,** the last term in (2.72) can be rewritten as

$$\begin{aligned} 2\Delta_t^T P W_1^* \widetilde{\sigma}_t' &= 2\Delta_t^T P W_{1,t} D_\sigma \widetilde{V}_{1,t} \widehat{x}_t - 2\Delta_t^T P \widetilde{W}_{1,t} D_\sigma \widetilde{V}_{1,t} \widehat{x}_t + 2\Delta_t^T P W_1^* \nu_\sigma \\ 2\Delta_t^T P W_2^* \widetilde{\phi}_t' \gamma(u_t) &= 2\Delta_t^T P W_{2,t} \sum_{i=1}^{q} \left[D_{i\phi} \widetilde{V}_{2,t} \widehat{x}_t \right] \gamma_i(u_t) \\ &- 2\Delta_t^T P \tilde{W}_{2,t} \sum_{i=1}^{q} \left[D_{i\phi} \widetilde{V}_{2,t} \widehat{x}_t \right] \gamma_i(u_t) + 2\Delta_t^T P W_2^* \sum_{i=1}^{q} \nu_{i\phi} \gamma_i(u_t) \end{aligned} \tag{2.73}$$

The term $2\Delta_t^T P W_1^* \nu_\sigma$ in (2.73) may be also estimated from below as

$$\begin{aligned} 2\Delta_t^T P W_1^* \nu_\sigma &\leq \Delta_t^T P W_1^{*T} \Lambda_1^{-1} W_1^* P \Delta_t + \nu_\sigma^T \Lambda_1 \nu_\sigma \\ &\leq \Delta_t^T P \overline{W}_1 P \Delta_t + l_1 \left\| \widetilde{V}_{1,t} \widehat{x}_t \right\|_{\Lambda_1}^2 \end{aligned}$$

as well as the term $2\Delta_t^T P W_2^* \sum_{i=1}^{q} \nu_{i\phi} \gamma_i(u_t)$ in (2.73) may be estimated as

$$2\Delta_t^T P W_2^* \sum_{i=1}^{q} \nu_{i\phi} \gamma_i(u_t) \leq \Delta_t^T P \overline{W}_2 P \Delta_t + q l_2 \left\| \gamma(u_t) \right\|^2 \left\| \widetilde{V}_{2,t} \widehat{x}_t \right\|_{\Lambda_2}^2$$

By the analogous way, in view of **A2.10** the term $2\Delta_t^T P\widetilde{f}_t$ can be estimated as

$$2\Delta_t^T P\widetilde{f}_t \leq \Delta_t^T P\Lambda_f^{-1}P\Delta_t + \widetilde{f}_t^T\Lambda_f\widetilde{f}_t \leq \Delta_t^T P\Lambda_f^{-1}P\Delta_t + s_t\overline{\eta} + \overline{\eta}_1\left\|x_t\right\|_{\Lambda_f}^2$$

Using all these upper estimates, (2.71) can be rewritten as

$$\begin{array}{c}\dot{V}_t \leq s_t\Delta_t^T L\Delta_t + L_{w1} + L_{w2} + L_{v1} + L_{v2} - s_t\Delta_t^T Q\Delta_t \\ -\lambda\left\|x_t\right\|^2 + \overline{\eta}_1\left\|x_t\right\|_{\Lambda_f}^2 + s_t\overline{\eta}\end{array} \tag{2.74}$$

where

$$L := PA + A^T P + PRP + Q$$

$$L_{w1} := 2tr\left[\dot{\widetilde{W}^T}_{1,t}\ K_1^{-1}\widetilde{W}_{1,t}\right] + 2s_t\Delta_t^T P\widetilde{W}_{1,t}\sigma - 2s_t\Delta_t^T P\widetilde{W}_{1,t}D_\sigma\widetilde{V}_{1,t}\widehat{x}_t$$

$$\begin{array}{c}L_{w2} := 2tr\left[\dot{\widetilde{W}^T}_{2,t}\ K_2^{-1}\widetilde{W}_{2,t}\right] + 2s_t\Delta_t^T P\widetilde{W}_{2,t}\phi\gamma(u_t) \\ -2tr\left[s_t\sum\limits_{i=1}^{q}\left(D_{i\phi}\gamma_i(u_t)\right)\widetilde{V}_{2,t}\widehat{x}_t\Delta_t^T P\tilde{W}_{2,t}\right]\end{array}$$

$$L_{v1} := 2tr\left[\dot{\widetilde{V}^T}_{1,t}\ K_3^{-1}\widetilde{V}_{1,t}\right] + 2s_t\Delta_t^T PW_{1,t}D_\sigma\widetilde{V}_{1,t}\widehat{x}_t + s_t l_1\left\|\widetilde{V}_{1,t}\widehat{x}_t\right\|_{\Lambda_1}^2$$

$$\begin{array}{c}L_{v2} = 2tr\left[\dot{\widetilde{V}^T}_{2,t}\ K_4^{-1}\widetilde{V}_{2,t}\right] + 2tr\left[s_t\widehat{x}_t\Delta_t^T PW_{2,t}\sum\limits_{i=1}^{q}\left(D_{i\phi}\gamma_i(u_t)\right)\widetilde{V}_{2,t}\right] \\ +s_t q l_2\left\|\gamma\left(u_t\right)\right\|^2\left\|\widetilde{V}_{2,t}\widehat{x}_t\right\|_{\Lambda_2}^2\end{array}$$

R and Q are defined in (2.64). Because

$$\overline{\eta}_1\left\|x_t\right\|_{\Lambda_f}^2 \leq \overline{\eta}_1\left\|\Lambda_f\right\|\left\|x_t\right\|^2$$

if we select

$$\left\|\Lambda_f\right\| \leq \frac{\lambda}{\overline{\eta}_1}$$

and using **A2.11,** we obtain

$$\dot{V}_t \leq L_{w1} + L_{w1} + L_{v1} + L_{v2} - s_t\Delta_t^T Q\Delta_t + \overline{\eta}$$

Using the updating law as in (2.65), and in view of the following properties

$$\begin{array}{c}\dot{\widetilde{W}}_{1,t} = \dot{W}_{1,t},\ \dot{\widetilde{W}}_{2,t} = \dot{W}_{2,t} \\ \dot{\widetilde{V}}_{1,t} = \dot{V}_{1,t},\ \dot{\widetilde{V}}_{2,t} = \dot{V}_{2,t}\end{array}$$

we get

$$L_{w1} = 0,\ L_{w2} = 0,\ L_{v1} = 0,\ L_{v2} = 0$$

Finally, the expression (2.74) get the form

$$\dot{V}_t \leq -s_t \left[\Delta_t^T Q_0 \Delta_t - \overline{\eta}\right]$$

The right-hand side of the last inequality can be estimated by the following way:

$$\dot{V}_t \leq -s_t \lambda_{\min}\left(P^{-1/} Q_0 P^{-1/2}\right)\left(\left\|P^{1/2}\Delta\right\|^2 - \mu\right) \leq 0 \tag{2.75}$$

where μ is defined as in (2.67). So, V_t is bounded. The statement a) (2.68) is proved. Since

$$0 \leq s_t \leq 1$$

from (2.75) we have

$$\dot{V}_t \leq -\Delta_t^T Q_0 \Delta_t s_t + \overline{\eta} s_t \leq -\Delta_t^T Q_0 \Delta_t s_t + \overline{\eta}$$

Integrating (2.75) from 0 up to T yields

$$V_T - V_0 \leq -\int_0^T \Delta^T Q_0 \Delta s_t dt + \overline{\eta}$$

So,

$$\int_0^T \Delta^T Q_0 \Delta s_t dt \leq V_0 - V_T + \overline{\eta} T \leq V_0 + \overline{\eta} T$$

Because $W_{1,0} = W_1^*$ and $W_{2,0} = W_2^*$, and V_0 is bounded, (2.69) is obtained and b) is proved. ■

Remark 2.5 *If we have no any unmodeled dynamic ($\widetilde{f} = 0$), we obtain $\overline{\eta} = \overline{\eta}_1 = 0$ and, hence, from (2.69) we conclude that the asymptotic stability is guaranteed, i.e.,*

$$\limsup_{t\to\infty} \frac{1}{T}\int_0^T \|\Delta\|_{Q_0}^2 s_t dt = 0$$

from which we directly obtain

$$\limsup_{t\to\infty} s_t \|\Delta_t\| = 0$$

2.4 Illustrating Examples

To illustrate the applicability of the suggested the approach, we would like to consider the following two numerical examples.

Example 2.1 *Consider* ***the nonlinear system*** *defined by*

$$\begin{aligned}\dot{x}_1 &= -5x_1 + 3\ sign\,(x_2) + u_1\\ \dot{x}_2 &= -10x_2 + 2\ sign\,(x_1) + u_2\end{aligned}$$

with the initial conditions

$$x_1(0) = 10, x_2(0) = -10$$

This nonlinear system even simple, is interesting enough, because it has multiple isolated equilibrium. We will compare the suggested algorithm (6.45) with single-layer dynamic neural network as in [18] and [13].

1) Single Layer Network

Let select the neural network as $(\gamma(u_t) = u_t)$

$$\boxed{\dot{\widehat{x}}_t = A\widehat{x}_t + W_{1,t}\sigma(\widehat{x}_t) + W_{2,t}\phi(\widehat{x}_t)u_t}$$

We select

$$W_{1,t}, W_{2,t} \in R^{2\times 2}$$

and suppose that the sigmoid functions are

$$\begin{aligned}\sigma\,(x_i) &= 2/(1+e^{-2x_i})2 - 0.5\\ \phi\,(x_i) &= 0.2/(1+e^{-0.2x_i}) - 0.05\end{aligned}$$

We also select

$$A = \begin{bmatrix} -15 & 0 \\ 0 & -10 \end{bmatrix}, \qquad \widehat{x}_0 = [-5,-5]^T$$
$$W_{1,0} = \begin{bmatrix} 1 & 10 \\ 10 & 1 \end{bmatrix} W_{2,0} = \begin{bmatrix} .1 & 0 \\ 0 & .1 \end{bmatrix}$$

We use the learning law

$$
\begin{array}{|l|}\hline
\dot{W}_{1,t} = -s_t K_1 P \Delta_t \sigma(\widehat{x}_t)^T \\
\dot{W}_{2,t} = -s_t K_2 P \Delta_t u_t{}^T \phi(\widehat{x}_t)^T \\ \hline
\end{array}
$$

where ***dead-zone function*** s_t *is defined as*

$$
s_t := \left[1 - \frac{\mu}{\left\|P^{1/2}\Delta\right\|}\right]_+, \quad [z]_+ = \begin{cases} z & z \geq 0 \\ 0 & z < 0 \end{cases}
$$

$$
\mu = \left(\overline{\eta}_\sigma + \overline{\eta}_\phi \overline{u}\right) / \lambda_{\min}\left(P^{-1/} Q_0 P^{-1/2}\right)
$$

Input signals u_1, u_2 *are chosen as sine wave and saw-tooth function with the gain equals to 1, so*

$$
\bar{u} = 1, \qquad \overline{\eta}_\phi = \overline{\eta}_\sigma = 2, \qquad K_1 = K_2 = 10I
$$

If we take

$$
Q_0 = I, R = \begin{bmatrix} 8 & 2 \\ 2 & 8 \end{bmatrix}, Q = \begin{bmatrix} 2 & 1 \\ 1 & 2 \end{bmatrix}
$$

the solution of the Riccati equation (2.47) is equal to

$$
P = \begin{bmatrix} 0.06 & 0.04 \\ 0.04 & 0.106 \end{bmatrix}
$$

The identification results are shown in Figures 2.3- 2.4.

2) Multilayer Network

Here we will use the dynamic multilayer neural network as

$$
\boxed{\dot{\widehat{x}}_t = A\widehat{x}_t + W_{1,t}\sigma(V_{1,t}\widehat{x}_t) + W_{2,t}\phi(V_{2,t}\widehat{x}_t)u_t}
$$

The vectors $\widehat{x}_0$, u_1, u_2, σ *and* ϕ *are the same as before, but now*

$$
W_{1,t} \text{ and } W_{2,t} \in R^{2\times 3}
$$

and

$$
V_{1,t} \text{ and } V_{2,t} \in R^{3\times 2}
$$

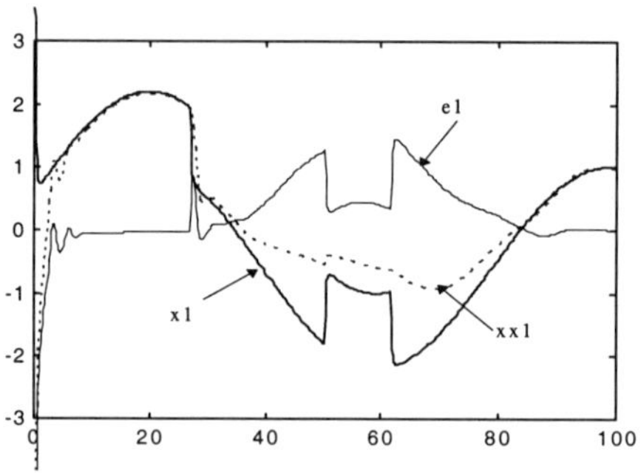

FIGURE 2.3. Identification result for x_1 (without hidden layer).

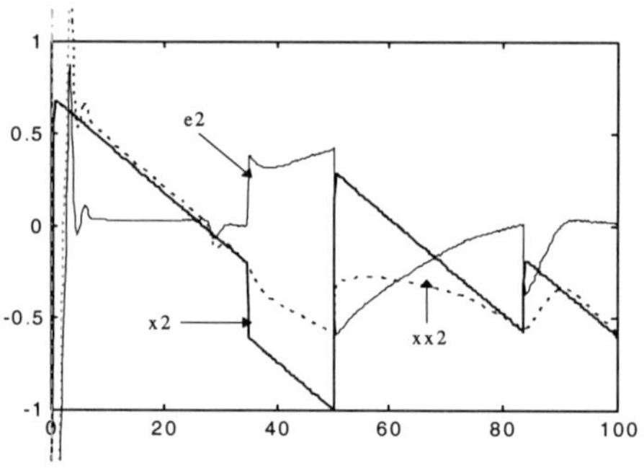

FIGURE 2.4. Identification result for x_2 (without hidden layer).

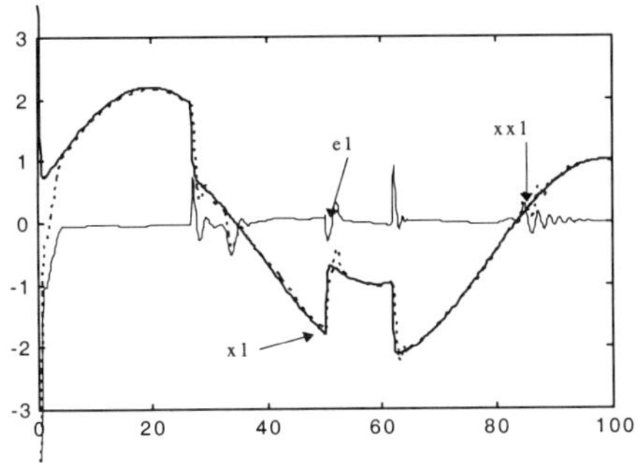

FIGURE 2.5. Identification result for x_1 (with hidden layer).

We select

$$W_{1,0}=W_{2,0}=V_{10}^T=V_{2,0}^T=\begin{bmatrix}1&1&2\\1&2&1\end{bmatrix}$$

The updating law is

$$\boxed{\begin{array}{c}\dot{W}_{1,t}=-s_tK_1P\Delta_t\sigma^T+s_tK_1P\Delta_t\widehat{x}_t^T\left(V_{1,t}-V_1^0\right)^TD_\sigma\\ \dot{W}_{2,t}=-s_tK_2P\Delta_t\left(\phi\gamma(u_t)\right)^T+s_tK_2P\Delta_t\widehat{x}_t^T\left(V_{2,t}-V_2^0\right)^T\sum_{i=1}^{q}\left(\gamma_i(u_t)D_{i\phi}^T\right)\\ \dot{V}_{1,t}=-s_tK_3D_\sigma^TW_{1,t}^TP\Delta_t\widehat{x}_t^T-s_t\frac{l_1}{2}K_3\Lambda_1\left(V_{1,t}-V_1^0\right)\widehat{x}_t\widehat{x}_t^T\\ \dot{V}_{2,t}=-s_tK_4\sum_{i=1}^{q}\left(\gamma_i(u_t)D_{i\phi}^T\right)W_{2,t}^TP\Delta_t\widehat{x}_t^T-s_t\frac{ql_2\bar{u}}{2}K_4\Lambda_2\left(V_{2,t}-V_2^0\right)\widehat{x}_t\widehat{x}_t^T\end{array}}$$

The constants are same as before. Identification results are shown at Figures 2.5-2.6. One can see that the multilayer dynamic neural network turns out to be more effective.

Example 2.2 *Let consider the* ***Van der Pol oscillator*** *given by*

$$\begin{bmatrix}\dot{x}_1\\ \dot{x}_2\end{bmatrix}=\begin{bmatrix}0&1\\0&0\end{bmatrix}\begin{bmatrix}x_1\\x_2\end{bmatrix}+\begin{bmatrix}0\\1.5\end{bmatrix}\left[\left(1-x_1^2\right)x_2-x_1\right] \tag{2.76}$$

Because this nonlinear system has no any control input, the dynamic neural net can be selected in more simple manner:

$$\boxed{\dot{\widehat{x}}_t=A\widehat{x}_t+W_{1,t}\sigma(V_{1,t}\widehat{x}_t)}$$

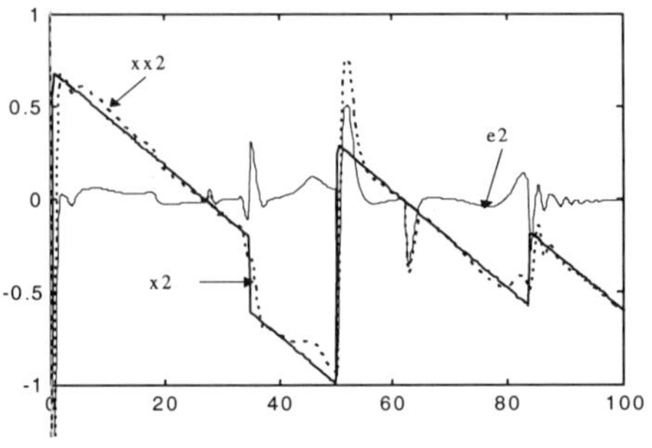

FIGURE 2.6. Identification result for x_2 (with hidden layer).

We use the same algorithm (with the same parameters) as in Example 1, but with

$$W_{2,t} = 0$$

It means

$$\boxed{\begin{array}{l} \dot{W}_{1,t} = -s_t K_1 P \Delta_t \sigma^T + s_t K_1 P \Delta_t \widehat{x}_t^T \left(V_{1,t} - V_1^0\right)^T D_\sigma \\ \dot{V}_{1,t} = -s_t K_3 D_\sigma^T W_{1,t}^T P \Delta_t \widehat{x}_t^T - s_t \frac{l_1}{2} K_3 \Lambda_1 \left(V_{1,t} - V_1^0\right) \widehat{x}_t \widehat{x}_t^T \end{array}}$$

with the same initial conditions. The corresponding results are shown in Figure 2.7 - 2.10. In this case the single-layer dynamic neural network cannot successfully follow the given trajectory, but the multilayer dynamic neural network can do its job well.

Example 2.3 *Dynamic Neural Networks identifier for* ***Vehicle Idle Speed Control***

The engine idle speed control is a challenging problem. The engine operation at idle is a nonlinear process that is far from its optimal range. Because it does not require any large degree of instrumentation or external sensing capabilities, the idle speed control is also accessible and can be formatted as a benchmark problem for control society. The aim of idle speed control is to regulate the engine speed at a desire level in the presence of torque disturbance which are often unmeasured. A challenge of the idle speed control is to use information that is available to a vehicle's power-train control module (PCM) to coordinate effectively the two available

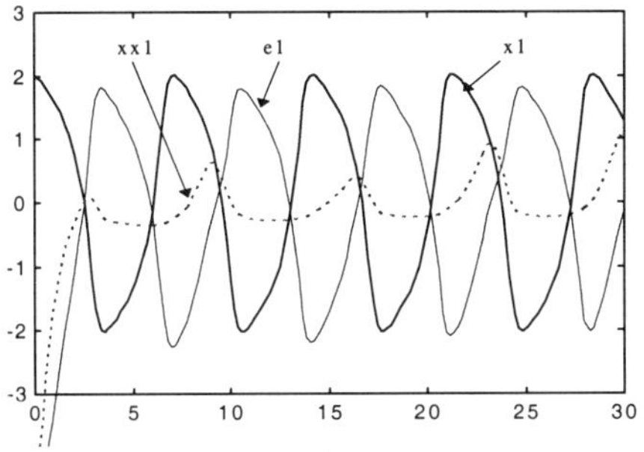

FIGURE 2.7. Identification for x_1 (without hidden layer).

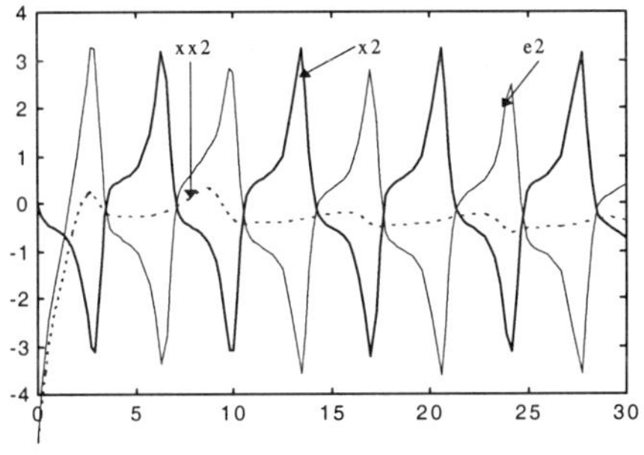

FIGURE 2.8. Identification for x_2 (without hidden layer).

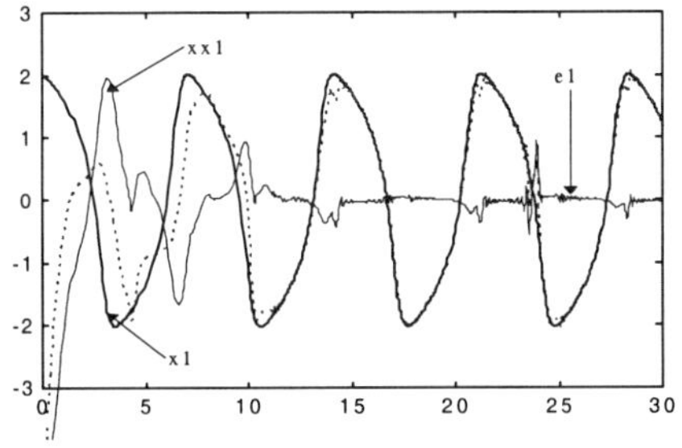

FIGURE 2.9. Identification for x_1 (with hidden layer).

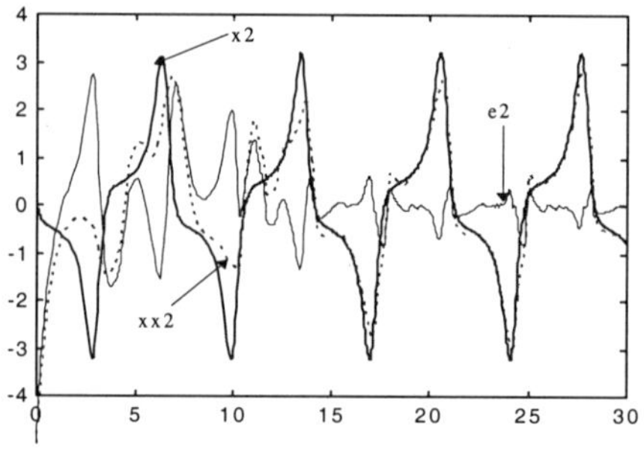

FIGURE 2.10. Identification for x_2 (with hidden layer).

controls: bypass air and spark advance. The two controls have different dynamic effects on the engine. The passby air command, which is a signal between zero and unity that determines the duty cycle for a solenoid valve, regulates the amount of air allowed into the intake manifold of an engine under conditions of closed throttle. The control range of the passby air signal is large, but its effect is delay by a time inversely proportional to engine speed. The spark advance command has an immediate effect on engine speed, but over a small range.

In a typical production vehicle, the engine idle speed control strategy runs as a background PCM process that is interrupted by a foreground process which performs operations required before each cylinder fires. The background process thus executes asynchronously with the application of controls that occur at every engine event, and the time required to execute a complete background process increases with engine speed. For the vehicle considered in this chapter, the background process of the PCM executes in approximately 30ms. The control commands for bypass air and spark advance are computed within the background process as a function of measured PCM variables, such as engine speed and mass air flow. The occurrence of torque disturbance is asynchronous with the computation and application of controls.

In this chapter the robust tracking problem of a engine idle speed with disturbance and unknown parameter is considered. *The main result* consists in the proposition of a robust nonlinear controller which can guarantee a certain accuracy of a tracking process.

The process of engine at idle has time delays that vary inversely with engine speed and is time-varying due to aging of components and environmental changes such as engine warm-up after a cold start. The measurement of system outputs occurs asynchronously with the calculation of control signals. We assume that the occurrence of plant disturbances, such as engagement of air conditioner compressor, shift from neutral to drive in automatic transmissions, application and release of electric loads, and power steering lock-up, are not directly measured.

The dynamic engine model employed in this chapter was derived from steady-state engine map data and empirical information [30], with revisions as described in [33]. The engine model parameters are for a 1.6 liter, 4-cylinder fuel injected engine. The model is a two inputs and two outputs system [31]:

$$\begin{aligned} \dot{P} &= k_P\left(\dot{m}_{ai} - \dot{m}_{ao}\right) \\ \dot{N} &= k_N\left(T_i - T_L\right) \\ \dot{m}_{ai} &= \left(1 + k_{m1}\theta + k_{m2}\theta^2\right) g(P) \\ \dot{m}_{ao} &= -k_{m3}N - k_{m4}P + k_{m5}NP + k_{m6}NP^2 \end{aligned} \tag{2.77}$$

where

$$g(P) = \begin{cases} 1 & P < 50.6625 \\ 0.0197\sqrt{101.325P - P^2} & P \geq 50.6625 \end{cases}$$

$$\begin{aligned} T_i &= -39.22 + 325024m_{ao} - 0.0112\delta^2 + 0.635\delta \\ &+ \tfrac{2\pi}{60}(0.0216 + 0.000675\delta)N - \left(\tfrac{2\pi}{60}\right)^2 0.000102N^2 \\ T_L &= \left(\tfrac{N}{263.17}\right)^2 + T_d \\ m_{ao} &= \dot{m}_{ao}(t-\tau)/(120N) \end{aligned}$$

the parameters are

$$\begin{aligned} k_P = 42.40, &\quad k_N = 54.26 \\ k_{m1} = 0.907, &\quad k_{m2} = 0.0998 \\ k_{m3} = 0.0005968, \quad k_{m4} &= 0.0005341, \quad k_{m5} = 0.000001757 \\ \tau &- 45/N \end{aligned}$$

The system outputs are manifold press P (kPa) and engine speed N (rpm). The control inputs are thottle angle θ (degree) and the spark advance δ (degree). Disturbances act to the engine in the form of unmeasured accessory torque T_d (N-m). The variable $\dot{m}_{ai}$ and $\dot{m}_{ao}$ refer to the mass air flow into and out of the manifold. m_{ao} is the air mass in the cylinder. The parameter τ is a dynamic transport time delay. The function $g(P)$ is a manifold pressure influence function. T_i is the engine's internally developed torque, T_L is the load torque.

Let us now represent this system in the standard form which will be in force throughout of this chapter. To do this, we introduces the extended vector

$$x = (P,\ N)^T, \quad u = (\theta,\ \delta)^T$$

and in view of this definition we can rewrite the dynamic equation (2.77) as follows:

$$\dot{x} = \begin{pmatrix} \dot{x}_1 \\ \dot{x}_2 \end{pmatrix} = \begin{pmatrix} f_1(x,u) \\ f_2(x,u) \end{pmatrix}.$$

f_1 and f_2 are assumed to be unknown and only x and u are measurable. The identification results for manifold press and engine speed are shown in Figure 2.11 and Figure 2.12. One can see that the dynamic neural networks is a good identifier for idle speed.

2.5 Conclusion

In this chapter we propose a new adaptive neural identifier, which is applied to two types of nonlinear plants:

- with exact Neural Network Modelling;
- with non exact Neural Network Modelling.

We establish bounds for both type of the identification errors. It is worth mentioning that the given systems is assumed to be nonlinear one, belonging to a wide enough class. So, the developed structure allows the implementation of many applications.

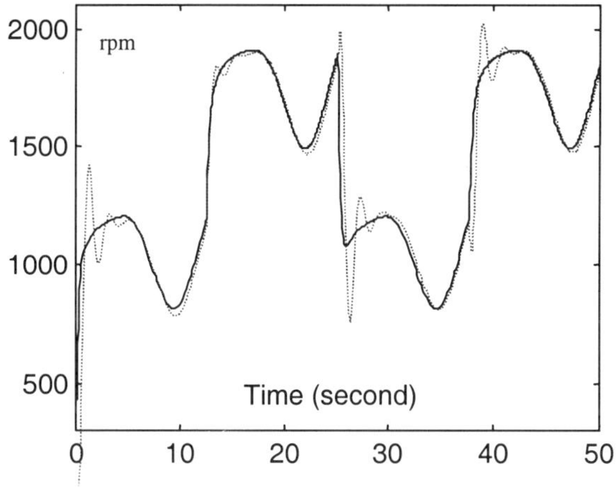

FIGURE 2.11. Identification for engine speed

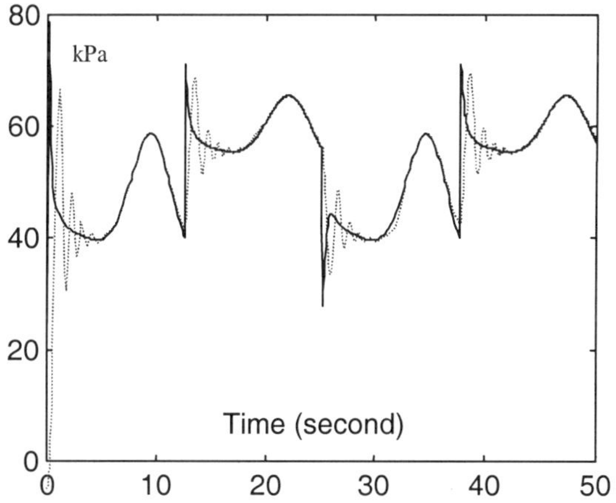

FIGURE 2.12. Identification for manifold press

We also have proposed a new stable learning law for a multilayer dynamic neural network and demonstrate its application to nonlinear system on-line identification. By means of a Lyapunov-like analysis we determine stability conditions for the hidden layer and output layer weights. An algebraic Riccati equation is used to give a bound for the identification error. The new learning procedure looks like the backpropagation of multilayer perceptrons containing an additional correcting term. With this updating law we can assure the learning procedure is global stability.

We illustrate the applicability of these results by two examples: one of them deals with the system which has multiple equilibrium and is described by the vector field which is not differentiable, the second one deals with the identification of a Van der Pol oscillator. The obtained results seem to be very promising. In the future chapters (4 and 5) we intend to extend this approach for nonlinear system tracking as well as state space observation problems.

In this chapter we have suggested **a new differential learning law** to tune the corresponding weight matrices. The question:

" Why differential learning? May be something more simple is also applicable!?"

seams to be reasonable. In the next chapter we show that *Sliding Mode approach* to be applied to Dynamic Neural Networks adaptation leads to **a non-differential (algebraic type) learning law** which may have some advantages as well as disadvantages with respect to the differential one considered in this chapter.

2.6 References

[1] E.Alcorta and E.N.Sanchez, "Nonlinear identification via neural networks", in Proc.4th I*FAC, Intl. Symp. on Adapt. Syst.. in Cont. and Sign. Process.*, pp 675-679, June 1992.

[2] C.A.Desoer ann M.Vidyasagar, *Feedback Systems: Input-Output Properties*, New York: Academic, 1975

[3] A.Delgado, C.Kambahmpati, and K.Warwick, "Dynamic recurrent neural network for systems identification and control", *IEE Proc.-Cont. Theo. Appl.*, Vol. 142, No 4, July 1995.

[4] K.Funahashi, and Y.Nakamura, "Approximation of dynamical systems by continuous time recurrent neural networks", Neural Networks, Vol. 6, pp 801-806, 1993.

[5] S.Haykin, *Neural Networks- A comprehensive Foundation*, Macmillan College Publ. Co., New York, 1994.

[6] J.J.Hopfield, "Neurons with grade response have collective computational propierties like those of a two-state neurons", *Proc. of the National Academy of Science, USA*, vol. 81, 3088-3092, 1984.

[7] K.J.Hunt, D.Sbarbaro, R.Zbikowski and P.J.Gawthrop, "Neural Networks for Control Systems- A Survey", *Automatica*, vol. 28, pp. 1083-1112, 1992.

[8] P.A.Ioannou and J.Sun, *Robust Adaptive Control*, Prentice-Hall, Inc, Upper Saddle River: NJ, 1996

[9] A.Isidori, *Nonlinear control systems*, 2nd edition, Ed. Springer Verlag, 1989.

[10] E.B.Kosmatopoulos, M.M.Polycarpou, M.A.Christodoulou and P.A.Ioannpu, "High-Order Neural Network Structures for Identification of Dynamical Systems", *IEEE Trans. on Neural Networks*, Vol.6, No.2, 442-431, 1995.

[11] E.B.Kosmatopoulos, M.A.Christodoulou and P.A.Ioannpu, "Dynamical Neural Networks that Ensure Exponential Identification Errror Convergence", *Neural Networks*, Vol.10, No.2, 299-314, 1997ol.1, 4-27.

[12] F.L.Lewis, A.Yesildirek and K.Liu, Multilayer Neural-Net Robot Controller with Guaranteed Tracking Performance, *IEEE Trans. on Neural Network*, Vol. 7, No 2, 388-398, 1996.

[13] R.R.Selmic and F.L.Lewis, Neurocontrol for Compensation of Actuator Nonlinearities, *10th Yale Workshop on Adaptive and Learning System,* 83-92, 1998.

[14] W.T.Miller, S.A.Sutton and P.J.Werbos, *Neural Networks for Control*, MIT Press, Cambridge, MA, 1990

[15] K.S.Narendra and K.Parthasarathy, "Identification and Control of Dynamical Systems Using Neural Networks", *IEEE Trans. on Neural Networks*, Vol.1, 4-27 for invited session, *Proc.34th IEEE CDC*, February 1995.

[16] M.Nikolaou, and V.Hanagandi, "Control of nonlinear dynamical systems modeled by recurrent neural networks", *AIChE Journal*, Vol. 39, No 11, pp 1890-1894, November 1993.

[17] V.M.Popov, *Hyperstability of Control Systems*, Springer-Verlag, New York, 1973

[18] A.S. Poznyak, Learning for Dynamic Neural Networks, *10th Yale Workshop on Adaptive and Learning System,* 38-47, 1998.

[19] A.S. Poznyak, W.Yu , E.N. Sanchez and Jose P. Perez, Stability Analysis of Dynamic Neural Control, *Expert System with Applications*, Vol.14, No.1, 227-236, 1998.

[20] G.A.Rovithakis and M.A.Christodoulou, " Adaptive Control of Unknown Plants Using Dynamical Neural Networks", *IEEE Trans. on Syst., Man and Cybern.*, Vol. 24, 400-412, 1994.

[21] G.A.Rovithakis and M.A.Christodoulou, "Direct Adaptive Regulation of Unknown Nonlinear Dynamical System via Dynamical Neural Networks", *IEEE Trans. on Syst., Man and Cybern.*, Vol. 25, 1578-1594, 1994.

[22] E.Sontag, "Neural nets as systems models and controllers", in *Proc. 7th Yale Workshop on Adaptive and Learning Systems*, pp 73-79, Yale University, 1992.

[23] I.W. Sandberg, "Approximation theorems for discrete-time systems", *IEEE Trans. on Circ. and Syst.*, Vol. 38, pp 564-566, 1991.

[24] I.W.Sandberg, "Uniform approximation and the circle criterion", *IEEE Trans. on Automatic Control*, Vol. 38, pp 1450-1458., 1992.

[25] E.N.Sanchez, "Dynamic neural network for nonlinear systems identification", in *Proc.33rd IEEE CDC,* pp 2480-2481, December 1994.

[26] K.Wang, and A.N.Mitchel, "Robustness and perturbation analysis of a class of nonlinear systems with applications to neural networks", *IEEE Trans. on Circ. and Syst.*, Part 1, Vol. 41, No 1, pp 24-32, January 1994.

[27] K.Wang, and A.N.Mitchel, "Robustness and perturbation analysis of a class of artificial neural networks", *Neural Networks*, Vol. 7, No 2, pp 251-257, 1994

[28] J.C.Willems, "Least squares optimal control and algebraic Riccati equations", *IEEE Trans. on Automatic Control*, Vol. 16, No 6, pp 621-634, 1971.

[29] H.K.Wimmer, Monotonicity of Maximal Solutions of Algebraic Riccati Equations, *System and Control Letters*, Vol.5, pp317-319, 1985

[30] B.K.Powell and J.A.Cook, Nonlinear low frequency phenomenological engine modeling and analysis, *Proc. of the 1987 American Control Conference*, Vol.1 336-340, 1987

[31] G.V.Puskorius and L.A.Feldkamp, Neurocontrol of nonlinear dynamical systems with Kalman filter trained recurrent networks, *IEEE Trans. on Neural Networks*, Vol.5, No.2, 279-297, 1994

[32] G.V.Puskorius, L.A.Feldkamp and L.I.Davis, Dynamic neural network methods applied to on-vehicle idle speed control, *Procedings of the IEEE*, Vol.84, No.10, 1407-1420, 1996

[33] G.Vachtsevanos, S.S.Farinwata and D.K.Pirovolou, Fuzzy logic control of an automotive engine, *IEEE Control Systems Magine*, Vol.13, No.3, 62-68, 1993

3

Sliding Mode Identification: Algebraic Learning

In this chapter, the problem of identification of continuous, uncertain nonlinear systems in presence of bounded disturbances, is implemented using dynamic (differential) neural networks. The neural net weights are updated on-line by a learning algorithm based on the sliding mode technique. Even in the presence of any bounded external perturbations, affected to unknown nonlinear plant, the proposed neural identifier guarantees that the state estimation error tends to zero if the gain matrix, participating in the sliding mode control, is big enough. Numerical simulations illustrate its effectiveness, even for highly nonlinear systems in presence of important disturbances.

3.1 Introduction

Sliding modes is a high speed switching strategy, which provides a robust mean for controlling nonlinear plants. Essentially, it utilizes a switching control law to drive the plant state trajectory onto a perspectively sliding surface. This surface is also called the *switching surface* because the controller has a switching gain which provides the dynamic trajectories tendency to this surface. The plant dynamics tending to this surfaces constitutes the controlled system behavior. By proper design of the sliding surface, it is possible to attain the following control goals for nonlinear systems as [4]

- stabilization
- regulation
- ideal trajectory tracking.

Initially, the sliding mode control technique was mainly developed in the former Soviet Union [18]. Due to its robust properties, it is quite attractive for nonlinear system control and optimization[5],[16] and [19].

Recently, it has been proposed to implement sliding mode control for nonlinear systems represented as neural networks. This implementation is mostly done as follows:

- a neural network is adapted on-line in order to minimize the error between its output and the nonlinear systems one; so, the neural network reproduces the dynamic behavior of the system.
- then, based on this neural network model, a sliding mode controller is synthesized.

Initially, such applications were based on radial basis Gaussian networks [13], [17]. Recent publications consider other type of neural networks such as single layer perceptrons [2], or multi-layer perceptrons for robot control [12]. For all these applications, stability is established by means of the Lyapunov approach.

Contrasting with neural control applications, the sliding mode technique has almost not been applied to neural network adaptive learning. The first related publication [14] presents a class of adaptive learning algorithms, based on the theory of quasi-sliding modes in discrete time dynamic systems, for both single and multilayer perceptrons; convergence is assured trough the existence of a quasi-sliding mode on the zero learning error. These algorithms are at the basis of recently proposed identification and control schemes [3],[6]. In [15], the design of learning strategies in adaptive perceptrons, from the viewing point of sliding modes in continuous time, is addressed. The unique feature of the sliding mode approach lies in the enhanced insensitivity of the proposed adaptive learning algorithm with respect to bounded external perturbation signals and measurements noises; again, convergence is guaranteed by the existence of a zero sliding mode on the zero learning error.

In this chapter we present the application of the sliding mode technique to the adaptive learning of dynamic neural networks, in order to minimize the error between the system to be identified and a neural identifier of the simplest structure (without any hidden layers). Here we follow the approach given in [10]. The global stability property of this error is analyzed by means of a Lyapunov approach. The structure of the identifier is taken from a previous publication of our research group [9]. The obtain learning law turns out to be a non-differentiable (algebraic) type procedure. The applicability of the proposed scheme is illustrated via simulations dealing with the same nonlinear systems as in the previous chapter to make possible the comparison of the obtained results.

3.2 Sliding Mode Technique: Basic Principles

In recent years, increasing attention has been given to systems where control action are discontinuous [18]. By intelligent selection of control actions, the state trajectories may be changed correspondingly to give the desired properties to the processes in the system under control. The design control problem in such systems with discontinuous control actions can be reduced to the convergence problem to a special surface in the corresponding phase space. When certain relations are valid, a special kind of motion - the so called *sliding mode-* may arise [19]. Sliding modes have a number of attractive features, and so have long since been in use in solving various control problems. The basic idea behind design of system with sliding mode is in following two steps:

- first, a sliding motion in a certain sense is obtained by an appropriate choice of discontinuity surfaces;
- second, a control is chosen so that the sliding modes on the intersection of those discontinuity surface would be stable.

Formally, such discontinuous dynamic system may be described by the following equation

$$\dot{x} = f\left(x, t, u_t\right) \tag{3.1}$$

where $x_t \in \Re^n$ is the system state vector and $u_t \in \Re^m$ is a discontinuous control input, $t \in \Re^+$. A general class of discontinuous control is defined by the following relationships:

$$u_{t,i} = \begin{cases} u_{t,i}^{+}(x,t) & \text{with } s_i(x) > 0 \\ u_{t,i}^{-}(x,t) & \text{with } s_i(x) < 0 \end{cases} \qquad i = 1, 2 \ldots, m \tag{3.2}$$

where $u_t = [u_{t,1}, u_{t,2}, \ldots u_{t,m}]^T$ and all functions $u_{t,i}^{+}(x,t)$ and $u_{t,i}^{-}(x,t)$ are continuous. The function $s_i(x)$ is discontinuity surface (subspace) defined by

$$s_i(x) = 0, \quad s_i(x) \in \Re^1, \quad i = 1, 2 \ldots, m$$

More commonly used limiting surface is as follows:

$$s_i(x) \mid_{x = x_{t,i} - x_{t,i}^*} = k_i x, \quad k_i > 0$$

where $x_{t,i}^*$ is the i-th component of a desired trajectory.

Some switching strategy of the continuous control $u_i^+(x,t)$ and $u_i^-(x,t)$ may lead to a non-regular (discontinue) behavior. As a result, the corresponding trajectories may be singular (unbounded). An accepted term on discontinuity surface is sliding mode. A sliding mode does exist on a discontinuity surface when ever the distances, to this surface and the velocity of its change $\dot{s}$ are of opposite signs, i.e.,

$$\lim_{s \to -0} \dot{s} > 0 \text{ and } \lim_{s \to +0} \dot{s} < 0 \tag{3.3}$$

or

$$\dot{s}_i \, s_i < 0$$

Condition (3.3) is useful for determination of $u_i^+(x,t)$ and $u_i^-(x,t)$. In the case of zero input, the solution of (3.1) is known to exist and be unique if a Lipschitz constant L may be found such that for any two vector x_1 and x_2 the following inequality holds:

$$\|f(x_1,t) - f(x_2,t)\| \le L \|x_1 - x_2\|$$

But, in the neighborhood of discontinuous surfaces, this Lipschitz condition is obligatory violated. So, some additional effort is need to find a solution at an occurrence of a sliding mode. This problem of finding the sliding domain is reducible to the specific problem of stability of a nonlinear system.

Consider the positive definite quadratic forms

$$V_t = \frac{1}{2} s_t^T P s_t, \quad P = P^T > 0 \tag{3.4}$$

To find the sliding domains with the control given by (3.2) we have to solve the equivalent stability problem with the equilibrium position

$$s = 0$$

Let us define the motion projection on subspace s as follows:

$$\dot{s} = -D \, sign\,(s) \tag{3.5}$$

where

$$sign\,(s) := [sign\,(s_1)\,, \ldots sign\,(s_m)]^T$$

and

$$D \in \Re^{m \times m}$$

Finding the time derivative of function (3.4) on the trajectory of system (3.5), we derive:

$$\dot{V} = -s^T L \, sign\,(s) \tag{3.6}$$

where

$$L = PD$$

The right-side hand of (3.6) may be expressed as

$$\dot{V} = -\sum_{k=1}^{m} |s_k| \left(l_{kk} + \sum_{i=1, i \neq k}^{m} l_{ki} sign\,(s_k s_i) \right) \tag{3.7}$$

where l_{ij} are the elements of the matrix L, i.e.,

$$L = \{l_{ij}\}$$

In view of the following inequality which is assumed to be valid

$$l_{kk} > \sum_{i=1, i \neq k}^{m} |l_{ki}|, \quad k = 1, \ldots m, \tag{3.8}$$

then we conclude that

$$\dot{V} < 0$$

and

$$\sup_{\|s\|=R} \dot{V} = - \|s\| \Delta_{l_r}$$

where R is the entire sphere of $\|s\|$, Δ_{l_r} is the minimal number of all $\Delta_{l_{rk}}$ defined by

$$\Delta_{l_{rk}} = l_{kk} - \sum_{i=1, i \neq k}^{m} |l_{ki}| \tag{3.9}$$

So, generally if the inequality (3.8) does hold, the upper bound of the derivative $\dot{V}$ will be consequently negative. Hence, the entire manifold $s = 0$ becomes the sliding domain and the sliding modes (3.5) become stable in the large (see Appendix B).

Now we consider the affine nonlinear system

$$\dot{x} = f(x, t) + g(x, t)u \tag{3.10}$$

where f and g are all argument continuous vector and matrix of dimension $n \times 1$ and $n \times m$, respectively. Assume there exists a symmetric positive definite matrix P such that

$$L = -PGg, \quad G = \left\{ \frac{\partial s}{\partial x} \right\}$$

satisfy the condition (3.8). Select the control u as

$$u = \alpha F(x,t) sign\,(s) \tag{3.11}$$

where $F(x,t)$ is an upper estimate for any component of the equivalent control $u_{eq}(x,t)$

$$\begin{array}{c} F(x,t) > |u_{eq}(x,t)| \\ u_{eq}(x,t) := -\left[G\;g\right]^{-1} G\;f \end{array} \tag{3.12}$$

Let us define the following dynamics

$$\dot{s} = Gf + Ggu \tag{3.13}$$

Taking into account (3.12) we derive

$$\dot{s} = Gg\,(u - u_{eq})$$

If in (3.5) we put

$$\begin{array}{c} D = -GB \\ \|GB\| < M \end{array}$$

where M is a positive constant and then find the time derivative of the function (3.4) on the trajectory of (3.13), we obtain:

$$\dot{V} = \alpha F(x,t) s^T \left(sign\,(s) - \frac{1}{\alpha}\mu(x,t) \right), \quad \alpha > 0$$

where

$$\mu(x,t) = u_{eq}/F(x,t)$$

From (3.12) we conclude

$$\|\mu(x,t)\| \leq \sqrt{m}$$

According to (3.7) and (3.9) we can get the following relation:

$$\dot{V} \leq \alpha F(x,t)\,\|s\| \left(\Delta l_r - \frac{1}{\alpha}\,\|P\|\,M\sqrt{m} \right)$$

If we choose

$$\alpha \geq \|P\| \, M\sqrt{m} \tag{3.14}$$

the function $\dot{V}$ be negative, so with the sliding mode control (3.11), the entire manifold $s = 0$ turns out to be stable in the large (see Appendix B).

In the trajectory tracking case, we want to design a sliding mode control to force the system (3.10) to the piecewise continuous trajectories

$$\frac{d}{dt}x_t^* = \varphi(x_t^*)$$

Let

$$s := Ke_t, K = K^T > 0$$
$$e = x_t - x_t^*$$

So,

$$\dot{s} = K\,\dot{e} + \dot{x}_t^* = Kf + Kgu - \varphi(x_t^*) = Kf' + Kgu$$

where

$$f' = f - K^{-1}\varphi(x_t^*)$$

Compare with the sliding surface (3.13), if we use the control as in (3.11), satisfying the condition (3.12), and (3.14), we can conclude that the surface

$$s = 0$$

is stable, so

$$\lim_{t\to\infty} e_t = 0$$

In the next section, based on these concepts, we will derive the learning procedure for on-line adaptation of Dynamic Neural Network weights to obtain globally stable identification process.

3.3 Sliding Model Learning

We consider nonlinear systems given as:

$$\boxed{\dot{x}_t = f(x_t, u_t, t) + \xi_t} \tag{3.15}$$

where

$x_t \in \Re^n$ is the system state vector at time $t \in R^+ := \{t : t \geq 0\}$;

$u_t \in \Re^q$ is a given control action;

$f(\cdot) : \Re^n \times \Re^q \times [0, \infty) \to \Re^n$ is a unknown nonlinear function describing the dynamics of the system;

ξ_t is a vector valued function representing external disturbances, which satisfies the following assumption.

A3.1: ξ_t is a Riemann integrable with bounded norm, i.e.

$$\limsup_{t\to\infty} \|\xi_t\| = \Upsilon < \infty \tag{3.16}$$

So, hereafter we will consider **external bounded disturbances**.
We consider the following neural networks as Chapter 2

$$\boxed{\dot{\widehat{x}}_t = A\widehat{x}_t + W_{1,t}\sigma(\widehat{x}_t) + W_{2,t}\phi(\widehat{x}_t)\gamma(u_t)} \tag{3.17}$$

The identification error we define as

$$\Delta_t := \widehat{x}_t - x_t \tag{3.18}$$

According to the sliding mode technique, discussed in the subsection 3.2 , we would like to obtain the following dynamic behavior

$$\boxed{\dot{\Delta}_t - -\mathit{\Gamma} sign\,(\Delta_t) + \nu_t} \tag{3.19}$$

where P is a positive diagonal matrix

$$P = diag[P_1...P_n]$$

and the vector function $sign\,(\Delta_t)$ is defined as follows:

$$sign\,(\Delta_t) := (\text{sign}\,(\Delta_{1,t})\,, ..., \text{sign}\,(\Delta_{n,t}))^T$$

ν_t is an unmodeled dynamic part which can be evaluated using a priory information about the class of uncertainties and on the class of nonlinear system being considered.

From (3.15) and (3.17) we derive:

$$\dot{\Delta}_t = A\widehat{x}_t + W_{1,t}\sigma(\widehat{x}_t) + W_{2,t}\phi(\widehat{x}_t)\gamma(u_t) - f(x_t, u_t, t) - \xi_t \tag{3.20}$$

Because $f(x_t, u_t, t)$ is unknown, we will use the following approximation

$$f(x_t, u_t, t) = \frac{x_t - x_{t-\tau}}{\tau} + \delta_t \tag{3.21}$$

for a small enough $\tau \in R^+$. The vector δ_t is the approximation error at time t. In view of (3.15), its norm can be estimated as

$$\begin{array}{c} \|\delta_t\| = \left\|\tau^{-1}\left(x_t - x_{t-\tau}\right) - f(x_t, u_t, t)\right\| \\ = \left\|\tau^{-1} \int_{t-\tau}^{t} \dot{x}_s\, ds - f(x_t, u_t, t)\right\| \\ = \left\|\tau^{-1}\left[\int_{t-\tau}^{t} f(x_s, u_s, s) - f(x_t, u_t, t)\right] ds - \tau^{-1} \int_{t-\tau}^{t} \xi_s ds\right\| \\ \leq \tau^{-1} \int_{t-\tau}^{t} \|f(x_s, u_s, s) - f(x_t, u_t, t)\|\, ds + \sup_t \|\xi_t\| \end{array} \tag{3.22}$$

Let also assume

A3.2: the vector field $f(x_t, u_t, t)$ satisfies the following condition of **"bounded rate variations"**:

$$\boxed{\begin{array}{c} \|\dot{x}_s - \dot{x}_t\| = \\ \|f(x_s, u_s, s) - f(x_t, u_t, t)\| \leq C_\tau + D_\tau\, |s - t| \end{array}} \tag{3.23}$$

that is valid for any $s, t \in R^+$ and for any x_s, u_s, x_t, u_t satisfying (3.15) (C_τ and D_τ are known nonnegative constants). This condition can be applied to wide class of nonlinear functions, including continuous functions as well as discontinuous ones with bounded variations, i.e.,

$$f(x_t, u_t, t) = f_0(x_t, t) + f_1(x_t, t) \cdot sign\,(u_t)$$

where $f_0(x_t,t)$, $f_1(x_t,t)$ are assumed to be continuous. So, this assumption is absolutely not restrictive. In general, C_τ is an upper bound estimation for local variations (for example, in the case of $sign(u_t)$ we have $C_\tau = 2$). As for D_τ, we can consider it as an upper bound of "the cone-condition" (as in Popov's criterion for absolute stability of closed-loop systems [8]) valid for the function $f(x_t,u_t,t)$. So, taking into account the bounds (3.16) and (3.23) we can obtain directly from (3.22) that

$$\|\delta_t\| \leq C_\tau + \tau D_\tau + \Upsilon \tag{3.24}$$

After substituting(3.21) into (3.20), we conclude that in order to guarantee the sliding mode behavior (3.19), the following relation has to be verified:

$$-P\,\mathrm{sign}\,(\Delta_t) = A\widehat{x}_t + \begin{bmatrix} W_{1,t} & W_{2,t} \end{bmatrix} \begin{bmatrix} \sigma(\widehat{x}_t) \\ \phi(\widehat{x}_t)\gamma(u_t) \end{bmatrix} - \frac{x_t - x_{t-\tau}}{\tau} \tag{3.25}$$

As a result, we get

$$\nu_t = \xi_t + \delta_t \tag{3.26}$$

Selecting weights $\begin{bmatrix} W_{1,t} & W_{2,t} \end{bmatrix}$ to fulfill the relation (3.25), we can satisfy the property (3.19).

This is the algebraic (not differential) linear matrix equation!

One possible selection is the least square estimate [1] given by

$$\boxed{\begin{bmatrix} \widehat{W}_{1,t} & \widehat{W}_{2,t} \end{bmatrix} = [\tau^{-1}(x_t - x_{t-\tau}) - A\widehat{x}_t - P\mathrm{sign}\,(\Delta_t)] \begin{bmatrix} \sigma(\widehat{x}_t) \\ \phi(\widehat{x}_t)\gamma(u_t) \end{bmatrix}^+} \tag{3.27}$$

where $[\bullet]^+$ stands for the pseudoinverse matrix in Moor-Penrose sense [1].

Remark 3.1 *This learning law is just an algebraic relative depending on Δ_t which can be evaluated directly without the implementation of any integrating procedure.*

Taking into account that (see [1])

$$x^+ = \frac{x^T}{\|x\|^2},\ 0^+ = 0$$

the last formula (3.27) can be rewritten as follows:

$$\boxed{\left[\begin{array}{cc}\widehat{W}_{1,t} & \widehat{W}_{2,t}\end{array}\right]=\frac{\left[\tau^{-1}(x_t-x_{t-\tau})-A\widehat{x}_t-P\text{sign}(\Delta_t)\right]}{\|\sigma(\widehat{x}_t)\|^2+\|\phi(\widehat{x}_t)\gamma(u_t)\|^2}\left[\begin{array}{c}\sigma(\widehat{x}_t)\\ \phi(\widehat{x}_t)\gamma(u_t)\end{array}\right]^T} \tag{3.28}$$

Remark 3.2 *Notice that we do not need any persistent excitation condition, which is common for identification of constant parameters [7], because the suggested sliding mode algorithm (3.28) does not require the convergence of the parameters* $\widehat{W}_{1,t}, \widehat{W}_{2,t}$.

To analyze the equation (3.19), it is enough to consider the simplest Lyapunov function:

$$V_t = \frac{1}{2}\|\Delta_t\|^2$$

By calculating its derivative along the trajectories of the differential equation (3.19), we derive:

$$\begin{array}{l}\dot{V}_t = \Delta_t^T\dot{\Delta}_t\\ = \Delta_t^T(-P\text{sign}(\Delta_t)+\nu_t)\\ = -\sum_{i=1}^n P_i|\Delta_i| + \Delta_t^T\nu_t\\ \leq -\min_i P_i\|\Delta_t\| + \|\Delta_t\|\cdot\|\nu_t\|\end{array}$$

Using (3.26) and applying the bounds, given by (3.24), we obtain:

$$\|\nu_t\| \leq \|\xi_t\| + \|\delta_t\| \leq \Upsilon + C_\tau + \tau D_\tau$$

and, hence,

$$\dot{V}_t \leq -\|\Delta_t\|\left[\min_i P_i - (\Upsilon + C_\tau + \tau D_\tau)\right]$$

Selecting

$$\min_i P_i > \Upsilon + C_\tau + \tau D_\tau$$

we can guarantee the property $\Delta_t \to 0$.

Finally, we formulate our main result.

Theorem 3.1 *If under the assumptions **A3.1** and **A3.2**, the gain diagonal matrix coefficient P, in the learning procedure (3.28), is selected such a way that*

$$\min_i P_i > \Upsilon + C_\tau + \tau D_\tau \tag{3.29}$$

*then the identification error vector is **globally asymptotically stable**, i.e.,*

$$\Delta_t \to 0$$

Remark 3.3 *In order to guarantee the stability condition (3.29), it is desirable to select τ as small as possible.*

3.4 Simulations

In this section we present simulation results, which illustrate the applicability of the theoretical study given above. We consider two illustrative examples:

1. The first one, we consider a nonlinear system with signum type elements.
2. The second one, we applied the proposed scheme to the Van der Pole oscillator.

Example 3.1 *Consider the nonlinear system as*

$$\boxed{\begin{aligned} \dot{x}_1 &= -5x_1 + 3\ sign\,(x_2) + u_1 \\ \dot{x}_2 &= -10x_2 + 2\ sign\,(x_1) + u_2 \end{aligned}}$$

with the initial conditions

$$x_1(0) = 10, x_2(0) = -10$$

Let select the neural network as

$$\boxed{\dot{\widehat{x}}_t = A\widehat{x}_t + W_{1,t}\sigma(\widehat{x}_t) + W_{2,t}\phi(\widehat{x}_t)u_t}$$

We select

$$W_{1,t}, W_{2,t} \in R^{2\times 2}$$

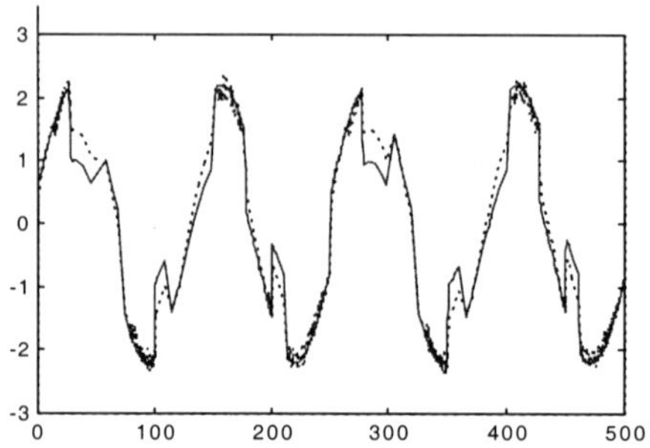

FIGURE 3.1. State 1 time evolution.

and suppose that the sigmoid functions are

$$\sigma(x_i) = 2/(1+e^{-2x_i}) - 0.5$$
$$\phi(x_i) = 0.2/(1+e^{-0.2x_i}) - 0.05$$

We also select $a_1 = -1, a_2 = -2, \widehat{x}_1(0) = 0, \widehat{x}_2(0) = 0$

$$A = \begin{bmatrix} -1 & 0 \\ 0 & -2 \end{bmatrix}, \qquad \widehat{x}_0 = [0,0]^T$$

The positive diagonal matrix P *is selected as*

$$P = \begin{bmatrix} 2 & 0 \\ 0 & 1 \end{bmatrix}$$

The input signal are a sine wave and a saw-tooth function. $\tau = 0.1$, *to adapt on-line the dynamic neural network weights we use the learning algorithm as*

$$\boxed{\begin{bmatrix} \widehat{W}_{1,t} & \widehat{W}_{2,t} \end{bmatrix} = \frac{\left[\tau^{-1}(x_t - x_{t-\tau}) - A\widehat{x}_t - Psign(\Delta_t)\right]}{\|\sigma(\widehat{x}_t)\|^2 + \|\phi(\widehat{x}_t)u_t\|^2} \begin{bmatrix} \sigma(\widehat{x}_t) \\ \phi(\widehat{x}_t)u_t \end{bmatrix}^T}$$

The corresponding results are shown in Figure 3.1, Figure 3.2 and Figure 3.3. The solid lines correspond to nonlinear system state trajectories, and the dashed line to neural network ones.

The time evolution for the weights of neural network is shown at Figure 3.4.

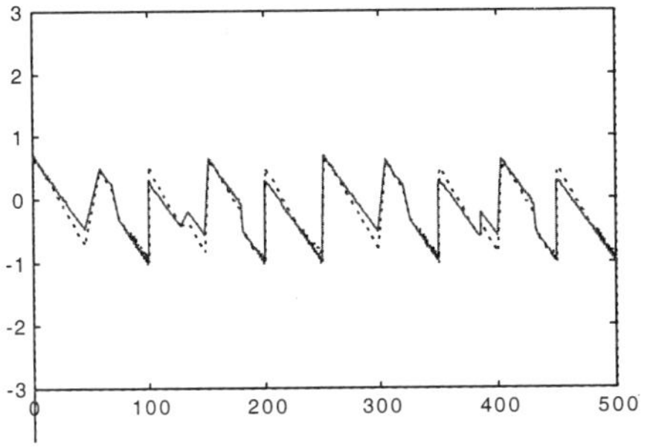

FIGURE 3.2. State 2 time evolution.

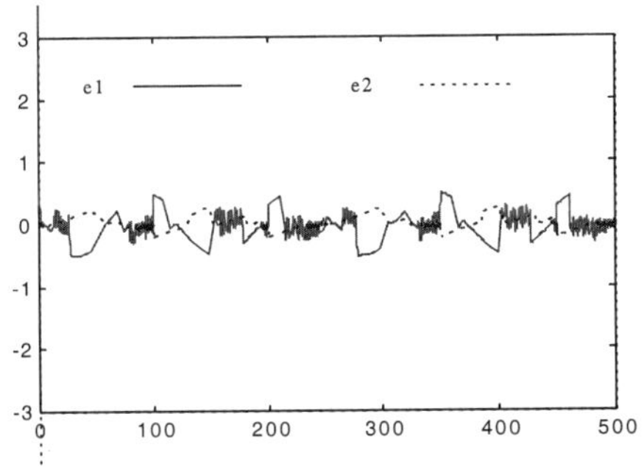

FIGURE 3.3. Identification errors.

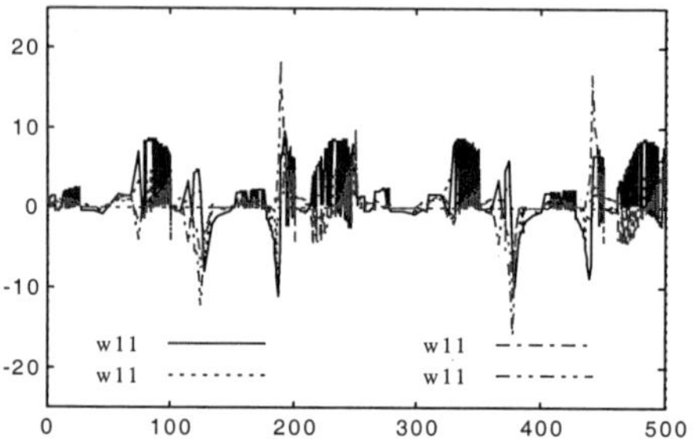

FIGURE 3.4. Weights time evolution.

Example 3.2 *Let consider the following Van der Pol oscillator:*

$$\begin{bmatrix} \dot{x}_1 \\ \dot{x}_2 \end{bmatrix} = \begin{bmatrix} 0 & 1 \\ 0 & 0 \end{bmatrix} \begin{bmatrix} x_1 \\ x_2 \end{bmatrix} + \begin{bmatrix} 0 \\ 1.5 \end{bmatrix} \left[\left(1 - x_1^2\right) x_2 - x_1 \right] \tag{3.30}$$

The neural network is the same as

$$\boxed{\dot{\widehat{x}}_t = \begin{bmatrix} -1 & 0 \\ 0 & -2 \end{bmatrix} \widehat{x}_t + W_{1,t}\sigma\left(\widehat{x}_t\right)}$$

where $\sigma\left(x_i\right) = \frac{2}{1+e^{-2x_i}} - 0.5$. *Now we select a bigger* P *as*

$$P = \begin{bmatrix} 30 & 0 \\ 0 & 20 \end{bmatrix}$$

If we choose $\tau = 0.1$ *and on-line the learning algorithm as*

$$\boxed{\widehat{W}_{1,t} = \left[\tau^{-1}\left(x_t - x_{t-\tau}\right) - A\widehat{x}_t - Psign\left(\Delta_t\right)\right] \sigma(\widehat{x}_t)^T / (\|\sigma(\widehat{x}_t)\|^2)}$$

The corresponding results are shown at Figure 3.5 and Figure 3.6 . The solid lines correspond to nonlinear system state trajectories, and the dashed line to neural network ones.

The time evolution for the weights of the corresponding neural network is shown at Figure 3.7. The limit circles $((x_1, x_2)$ *and* $(\widehat{x}_1, \widehat{x}_2))$ *are shown in Figure 3.8*

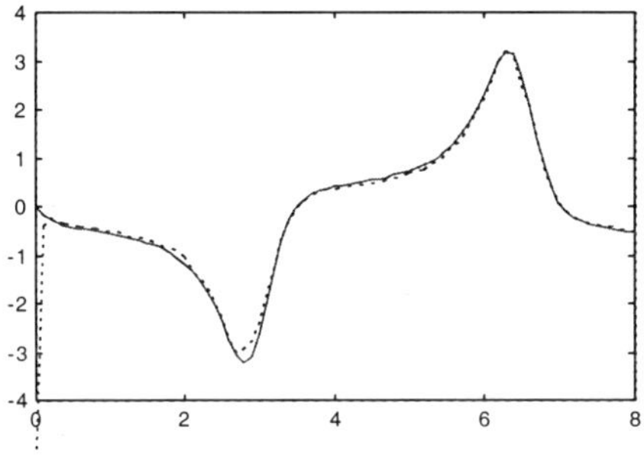

FIGURE 3.5. State 1 time evolution.

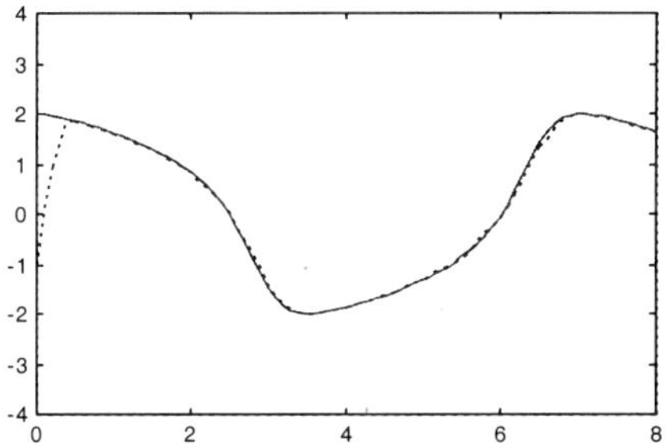

FIGURE 3.6. State 2 time evolution.

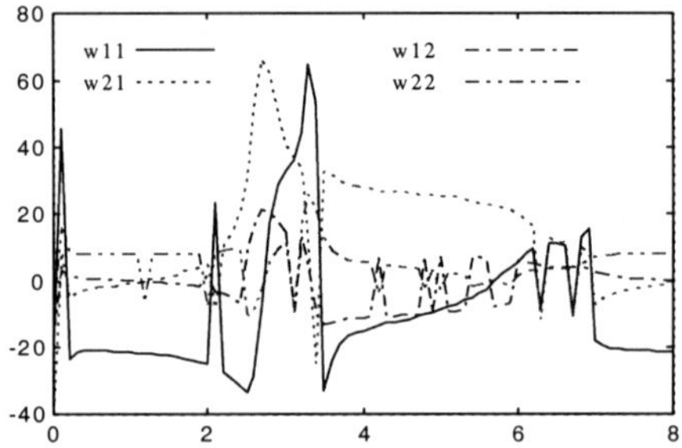

FIGURE 3.7. Weights time evolution.

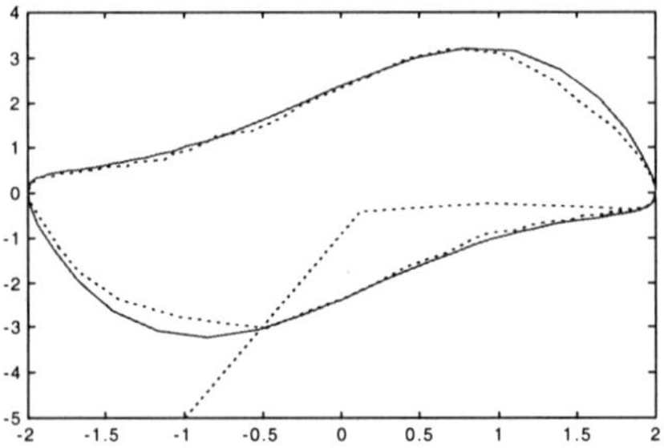

FIGURE 3.8. Limit circles.

3.5 Conclusion

In this chapter we have discussed the application of the sliding mode techniques to learning of dynamic neural networks. The suggested algorithms are utilized to implement a neural identifier. Global convergence of the identification error to zero is established via the Lyapunov approach.

In order to guarantee the existence of the sliding mode regime we suggest a new learning law to adapt on line the weights of the neural network identifier. This law belongs to the class of analytical functions given by the special algebraic linear matrix equations. So, we do not need any pre-integrating contour to realize this learning law! This is a big advantages of this algorithm with respect to the differential learning law described in the previous chapter.

We would like to emphasize that the approach, dealing with sliding mode technique, stops to work if we have any unmeasured noises presented in the available state observations. The next chapter is especially devoted to this problem and deals with the neuro state observers which are intended to construct on-line state estimates of an unobserved state vector even in the presence of a bound uncertainty (noise) in the output of a nonlinear system.

3.6 References

[1] A.Albert, "Regression and the Moore-Penrose Pseudoinverse", Academic Press, 1972.

[2] Y.J.Cao, S.J.Cheng, and Q.H.Wu, "Sliding mode control of nonlinear systems using neural network", *Proc. International Conference on Control '94*, pp. 855-859,1994.

[3] E.Colina-Morles, and N.Mort, "Neural network-based adaptive control design", *J. Systems Eng.*, vol.1, pp. 9-14, 1993.

[4] R.A.DeCarlo, S.H.Zak, and G.P.Matthews, "Variable structure control of nonlinear multivariable systems: A tutorial", *Proc. of IEEE*, vol.76, pp. 212-232, 1988.

[5] H.K.Khalil, *Nonlinear Systems*, 2nd Edition, Prentice Hall, 1996.

[6] J.G.Kuschewski, S.Hui, and S.H.Zak, "Application of feedforward networks to dynamical systems identification and control", *IEEE Trans. Contr. Syst. Techno.*, vol.1, pp. 37-49, 1993.

[7] L.Ljung, *System Identification: Theory for the User*, Prentice-Hall Inc., Englewood Cliff, NJ:07632, 1987

[8] V.M.Popov, *Hyperstability of Automatic Control System*, Springer-Verlag, New York, 1973

[9] A.S.Poznyak, and E.N. Sanchez, "Nonlinear system identification and trajectory tracking using dynamic neural networks", *Proc. 35th Conf. on Dec. and Contr.*, pp. 955-960, 1996.

[10] A.S.Poznyak, Wen Yu , Hebertt Sira Ramirez and Edgar N. Sanchez, Robust Identification by Dynamic Neural Networks Using Sliding Mode Learning, *Applied Mathematics and Computer Sciences*, Vol.8, No.1, 101-110, 1998

[11] G.A.Rovithakis, and M.A.Christodoulou, "Adaptive control of unknown plants using dynamical neural networks", *IEEE Trans. Syst., Man and Cybern.*, vol. 24, pp. 400-412, 1994.

[12] R.Safaric, K. Jezernik, A. Sabanovic, and S.Uran, "Sliding mode neural network robot controller", *Proc. of AMC 96-MIE,* pp. 395-400, 1996.

[13] R.M.Sanner, Stable adaptive control and recursive identification of nonlinear systems using radial Gaussian networks, *Ph. D. Dissertation*, MIT, Dept. of Aero. and Astron., May 1993.

[14] H.Sira-Ramirez, and S.H.Zak, "The adaptation of perceptrons with applications to inverse dynamics identification of unknown dynamic systems", *IEEE Trans. Syst., Man and Cybern.*, vol. 21, pp. 634-643, 1991.

[15] H.Sira-Ramirez, and E.Colina-Morles, "Asliding mode strategy for adaptive learning in adalines", *IEEE Trans. Circ. and Syst.-1*, vol. 42, pp. 101-1012, 1995.

[16] J.J.E.Slotine, and W.Li, *Applied Nonlinear Control*, Prentice-Hall, 1991.

[17] E.Tzirkel-Hancock, and F. Fallside, Stable neural control of multiple input-output systems, *CUED Report, TR.90*, Cambridge, England, January, 1992.

[18] V.I.Utkin, *Sliding Modes and Their Application in Variable Structure Systems*, MIR Publishers, Moscow, Soviet Union, 1978.

[19] V.I.Utkin, *Sliding Modes in Optimization and Control*, Springer-Verlag, 1992.

4

Neural State Estimation

The state observation using dynamic recurrent neural network, for continuous, uncertain nonlinear systems, subject to external and internal disturbances of bounded power, is discussed. The design of a suboptimal neuro-observer is proposed to achieve a prespecified accuracy of the estimation error which is defined as the weighted squares of its semi-norm. This error turns out to be a linear combination of the power levels of the external disturbances and internal uncertainties. The proposed robust neuro-observer has an extended Luneburger structure with weights learned on-line by a new adaptive gradient-like technique. The gain matrix is calculated by solving a matrix optimization problem and an inverted solution of a differential matrix Riccati equation. The numerical simulations of the proposed robust observer illustrate its effectiveness even in the case of a highly nonlinear systems in presence of unmodeled uncertainties.

4.1 Nonlinear Systems and Nonlinear Observers

4.1.1 The Nonlinear State Observation Problem

Nonlinear control is the hot topic of recent interest [19]. Most of the publications use the assumption on the complete accessibility of states of the systems to be controlled. But in reality this is not always true. That is why the solution of the nonlinear state observation problem seems to be very important.

In general, this sufficiently difficult problem has received much attention by many authors who obtained a lot of important and perspective results in different directions. The *Lie-algebraic method* is used widely [17] to construct the nonlinear observer based on the error linearization technique. The *adaptive observer* for nonlinear system [16] is suggested with the use of a special nonlinear transformation of the states of the given system into a nonlinear one given by the so-called canonical form. The *Lyapunov-like observers* are commonly used by many authors (see, for example, [25]). *The Linearization approach* can be considered as a relatively simple method to construct nonlinear observers. For example, in [33, 3] the systems are linearized close to the constant operating points but, unfortunately, this technique does not guarantee the global stability of the corresponding errors. Other kinds of nonlinear observer techniques are also known, for example, the *high-gain approach* [10, 27, 7], the *optimization-based observer* [15], the *reduced-order nonlinear observer* [7]. The observer proposed in [10] does not require a preliminary nonlinear change of coordinates.

All approaches mentioned above deal with the situation when the system model has no any uncertainties within its given description. In practice, we are working in the presence of external (noises or disturbances of observation) and internal (unmodeled dynamics) uncertainties of different physical nature. Less attention has been focused on the nonlinear observers with model and disturbance uncertainties. Such observers keeping a good performance within a class of uncertain models are called to be *the robust observers*. Some results have been achieved using the *variable-structure theory* [28, 32] that provided the property of the robust stability, but the design processes was too complex. The $H^{\infty}-$ approach was applied to construct the robust observer for the class of linear stationary systems with external perturbations of the L_2-bounded norm (in some sense, such disturbances tend to zero) [2]. The robust controllers of robots using a direct linear control and an *estimated linear state feedback* were suggested in [24] and [4].

In this chapter we consider the class of nonlinear SISO systems with the observer form as in [7], where both of the above features are presented: there are external nonrandom disturbances of a bounded power as well as some uncertainties in the model description. So, we deal with the so-called *mixed uncertainties.* Assuming that the Lipschitz and the uniformly observability properties for the right-hand side of the nonlinear nominal operator are valid, we suggest to apply a Luneburger-like observer with the gain matrix which is specially selected to guarantee the property of robustness within the given class of uncertain systems. To compute this matrix gain we use a differential matrix Riccati equation and a matrix optimization calculus. This approach enables us to specify the class of uncertain systems for which one can guarantee the existence of a robust observer. We also proof that there are conditions when the corresponding differential matrix Riccati equation with time varying parameters has a positive solution.

4.1.2 Observers for Autonomous Nonlinear System with Complete Information

When we talk about the *observability* of the given system we mean that the data of the output $y(t)$ within the time interval $[t_0, t_1]$, $t_1 > t_0$ completely determine the initial state $x(t_0)$ of the system.

Consider a single-output autonomous nonlinear system given by

$$\boxed{\begin{aligned} \dot{x}_t &= f(x_t, t) \\ y_t &= h(x_t, t) \end{aligned}} \tag{4.1}$$

where $x_t \in \Re^n$ is the state vector of this system at time $t \in R^+ := \{t : t \geq 0\}$, the given nonlinear function $f(\cdot,\cdot)$ describing the dynamic operator of this system is defined as

$$f(x_t, t) : \Re^n \times [0, \infty) \to \Re^n$$

and the function

$$h(x_t, t) : \Re^n \times [0, \infty) \to \Re$$

is an output scalar-valued function which is assumed to be completely known and measurable at each time t.

Let start with some mathematical preliminaries and definitions (see [11]) which are used throughout this chapter.

Definition 1 *Given f a C^∞ (infinitely times differentiable) vector field on $\Re^n$ and h a C^∞ scalar field on $\Re^n$, then the* Lie derivative *of h with respect to f is defined as*

$$L_f h := < dh, f >$$

where $< .,. >$ *denotes the dual product of* dh *and* f, *i.e.,*

$$< dh, f > := \sum_{i=1}^{n} \frac{\partial h}{\partial x_i} f_i$$

It is easy to see that this Lie derivative is also a C^{∞} scalar field on $\Re^n$. Thus, one can inductively define higher order of the Lie derivatives as follows

$$L_f^k(h) = L_f\left(L_f^{k-1}(h)\right) = < dL_f^{k-1}(h), f >, \; k \geq 2$$

Definition 2 *Given* $f, g \in C^{\infty}$ *vector fields on* $\Re^n$, *the* ***Lie Bracket*** $[f, g]$ *is the vector field defined by*

$$[f, g] := \frac{\partial g}{\partial x} f - \frac{\partial f}{\partial x} g$$

where $\partial g/\partial x$ *and* $\partial f/\partial x$ *are the Jacobians. The* Lie Bracket $[f, g]$ *is also a* C^{∞} *vector field on* $\Re^n$ *and one can define successive the Lie Brackets*

$$[f, [f, g]], [f, [f, [f, g]]]$$

etc.

The observability of the nonlinear system (4.1) at time t is equivalent to the following requirement [19]:

the set of functions (called *observability space*)

$$\Theta = \begin{bmatrix} h(x_t, t) \\ L_f h(x_t, t) \\ \vdots \\ L_f^{n-1} h(x_t, t) \end{bmatrix}$$

should "separate" the points from any physical subset $\Omega \in \Re^n$, that is, for any x_1, $x_2 \in \Omega$, there exists the index $i \in (0, ..., n-1)$ such that

$$L_f^i h(x_1) \neq L_f^i h(x_2)$$

In other words, the states of the given system are observable (see, [19]), if the corresponding *observability matrix* defined by

$$Q_t := \frac{d\Theta}{dx_t}$$

is non-singular for any $x_t \in \Re^n$ and any $t \in R^+$, that is, the given system (4.1) satisfies the following *observability rank condition:*

$$\boxed{\begin{array}{c} \operatorname{rank}\{Q_t\} = n \\ \forall x_t \in \Re^n, t \in R^+ \end{array}}$$

Definition 3 *Any nonlinear dynamic system defined by*

$$\dot{\hat{x}}_t = \mathcal{F}(\hat{x}_t, y_t, t) \tag{4.2}$$

we will call ***an observer*** *for the given dynamics (4.1) if*

1. *the dimension of* $\hat{x}_t$ *coincides with the dimension of* x_t *:*

$$x_t, \hat{x}_t \in \Re^n$$

2. *the upper bound of the difference* $(x_t - \hat{x}_t)$ *is finite, i.e.,*

$$\sup_{t \in R^+} \|(x_t - \hat{x}_t)\| \leq c < \infty$$

The value c of this upper bound defines the quality of the corresponding observer: the bound smaller, the quality of the given observation process is higher.

Following to the traditional technique of the differential geometry [11], we can consider (see [8, 13, 14]) *"the nonlinear high-gain observer"* given by

$$\boxed{\dot{\hat{x}}_t = f(\hat{x}_t, t) + K_t\left[y_t - h(\hat{x}_t, t)\right]} \tag{4.3}$$

as a possible candidate for the dynamic observer. Here the gain matrix is defined as

$$K_t := -S_t^{-1}\left[\frac{\partial}{\partial \hat{x}_t} h(\hat{x}_t, t)\right]^T \tag{4.4}$$

where

$$S_t^{-1} = \begin{pmatrix} 2\theta I_{2x2} & \theta^2 I_{2x2} \\ \theta^2 I_{2x2} & \theta^3 I_{2x2} \end{pmatrix}$$

The positive parameter θ determines the desired convergence rate. Moreover, S_t should be a positive solution of the algebraic equation

$$E_t^T \left(S_t + \frac{\theta}{2} I \right) + \left(S_t + \frac{\theta}{2} I \right) E_t = \left[\frac{\partial}{\partial \hat{x}_t} h(\hat{x}_t, t) \right]^T \left[\frac{\partial}{\partial \hat{x}_t} h(\hat{x}_t, t) \right] \tag{4.5}$$

with E_t defined as

$$E_t := \frac{\partial}{\partial \hat{x}_t} f(\hat{x}_t, t)$$

As it is shown in [8, 13, 14] that under certain technical assumptions (Lipschitz conditions for the given nonlinearities, etc.) this nonlinear observer has arbitrary exponential decay $\alpha\theta$, that is,

$$\|\hat{x}_t - x_t\| \leq C \exp\{-\alpha\theta t\}$$

4.1.3 Observers for Controlled Nonlinear Systems

Consider now a SISO (Single Input - Single Output) forced nonlinear system given by

$$\begin{aligned} \dot{x}_t &= f(x_t, t) + g(x_t, t) u_t \\ y_t &= h(x_t, t) \end{aligned} \tag{4.6}$$

where $u_t \in \Re$ is a control action at time t. Notice that the right-hand of the first equation in the system (4.6) is a linear function with respect to the control action u_t. Throughout of this book we will use basically such sort of nonlinear systems with dynamics linear with respect to control. So below, to simplify the presentations and discussions, we will discuss only this class of nonlinear systems.

Definition 4 *This system is said to be* ***observable*** *within a segment* $[0,t]$ *if for any pair of initial state* x_0 *and* x_1*, there exists a control* u_s $(s \in [0,t])$ *such that* x_0 *and* x_1 *are distinguishable by the observation of the corresponding outputs* $y_{0,s}$ *and* $y_{1,s}$.

So, **the observability property depends on the input** u_t **:**
one control could provide observability property and other one not!
Next definition specifies the observability property for a forced system subject to any bounded control.

Definition 5 *If on any finite time interval* $[0,T]$ *for any measurable bounded input* u_t*, defined on* $[0,T]$*, the initial state can be uniquely determined on the basis of the outputs* $\{y_t\}_{t\in[0,T]}$ *and the inputs* $\{u_t\}_{t\in[0,T]}$*, we say that this nonlinear system is* ***uniformly observable*** *for any input (UOAI).*

As it shown in [7], the controlled system (4.6) is UOAI at time t if and only if it can be rewritten in the following *canonical (Brunovskii) form*

$$\dot{x}_t = \begin{bmatrix} x_2 \\ x_3 \\ \vdots \\ x_n \\ \varphi(x) \end{bmatrix} + \begin{bmatrix} \psi(x_1) \\ \psi(x_1, x_2) \\ \vdots \\ . \\ \psi(x_1, x_2 \ldots, x_n) \end{bmatrix} u \qquad (4.7)$$

$$y_t = x_1 = Cx$$

where

$$C = [1, 0 \ldots 0]$$

and the function $\varphi(x)$ is globally Lipschitz on $\Re^n$. This property can be expressed as follows:

the *observability space* defined by the vector field

$$\Theta_t = \begin{bmatrix} h(x_t, t) \\ g(x_t, t) \\ L_{[ad^1 f,g]} h(x_t, t) \\ L_{[ad^1 g,f]} h(x_t, t) \\ L_{[ad^2 f,g]} h(x_t, t) \\ \vdots \end{bmatrix}$$

where

$$\begin{aligned} [ad^1 f, g] &:= [f, g] := \tfrac{\partial f}{\partial x} g - \tfrac{\partial g}{\partial x} f \\ \left[ad^k f, g\right] &:= \left[f, \left(ad^{k-1} f, g\right)\right] \end{aligned}$$

generates the observability matrix

$$Q_t = \frac{d\Theta_t}{dx_t}$$

which is non-singular for any $x_t \in \Re^n$ and any $t \in \Re^+$.

The corresponding state space observer also has the Luneburger-like structure [7] coinciding with (4.3):

$$\boxed{\dot{\hat{x}}_t = f(\hat{x}_t, t) + g(\hat{x}_t, t) u_t + K_t [y_t - h(\hat{x}_t, t)]} \tag{4.8}$$

4.2 Robust Nonlinear Observer

4.2.1 System Description

We start with the description of the class of SISO nonlinear uncertain systems given by

$$\boxed{\begin{aligned} \dot{x}_t &= f(x_t) + g(x_t) u_t + \xi_{1,t} \\ y_t &= Cx + \xi_{2,t} \end{aligned}} \tag{4.9}$$

where

$$f(x) = \begin{pmatrix} x_2 \\ \vdots \\ x_n \\ \varphi_0(x) + \Delta\varphi(x) \end{pmatrix}$$

$$g(x) = \begin{pmatrix} g_1(x_1) + \Delta g_1(x_1) \\ g_2(x_1, x_2) + \Delta g_2(x_1, x_2) \\ \vdots \\ g_n(x_1 \cdots x_n) + \Delta g_n(x_1 \cdots x_n) \end{pmatrix}$$

and

$$C = [1, 0 \cdots 0]$$

The vector fields $\Delta\varphi(x)$ and $\Delta g_i(x)$ define the *unmodeled dynamics* or, in other words, the *internal uncertainty.* The vector-functions $\xi_{1,t}$, $\xi_{2,t}$ present the external state and output perturbations and satisfies the assumption on *"bounded power"*:

A4.1:

$$\limsup_{T\to\infty} \frac{1}{T} \int_0^T \left\| \xi_{i,t} \right\|_{\Lambda_{\xi_i}}^2 dt = \Upsilon_i < \infty,\ 0 < \Lambda_{\xi_i} = \Lambda_{\xi_i}^T,\ i = 1, 2$$

The matrices Λ_{ξ_i} is assumed to be a priori given.

Notice that the plant (4.9) is already given in the observable form (4.7). So, the observer described before (with Luneburger-like structure) (4.3) can be applied in the case of no uncertainties in the given plant description.

In this section we show that in the presence of any uncertainty, satisfying A.4.1, the observer of the same structure but with another matrix gain is robust, it means that it guaranties a bound for average values of quadratic estimation error for the whole class of nonlinear systems containing uncertainties and, in the same time, this bound turns out to be *"tight" or "sharp"* (equal to zero if no any uncertainties at all).

So, let us consider a class of observable nonlinear systems which can be written as a nominal and an unmodeled dynamics parts, that is,

$$\begin{aligned} f(x_t,t) &= F(x,t)+\Delta f(x,t) \\ g(x_t,t) &= G(x,t)+\Delta g(x,t) \end{aligned} \tag{4.10}$$

where the nominal part is defined by

$$F(x,t)=[x_2,\ldots x_n,\varphi_0(x)]^T$$

$$G(x,t)=[g_1(x_1),\ldots g_n(x_1\cdots x_n)]^T$$

and the unmodeled dynamics is given as

$$\Delta f(x,t)=[0,\ldots 0,\Delta\varphi(x)]^T$$

and

$$\Delta g(x,t)=[\Delta g_1(x_1),\ldots \Delta g_n(x_1\cdots x_n)]^T$$

We only assume the nominal parts $F(x,t)$ and $G(x,t)$ are known.

Following to [7], we assume

A4.2: $F(x,t)$ and $G(x,t)$ are Lipschitz in a convex set $X\subset\Re^n$ (X may consider with $\Re^n$) uniformly in R^+:

$$\begin{aligned} \left\|F^T(x_1,t)-F^T(x_2,t)\right\|_{\Lambda_f} \le \|x_1-x_2\|_{\Lambda_{fx}}, \quad t\in R^+, \\ \left\|G^T(x_1,t)-G^T(x_2,t)\right\|_{\Lambda_g} \le \|x_1-x_2\|_{\Lambda_{gx}}, t\in R^+ \end{aligned} \tag{4.11}$$

where

$$\Lambda_f\in\Re^{n\times n},\ \Lambda_{fx}\in\Re^{n\times n},\ \Lambda_g\in\Re^{m\times m},\ \Lambda_{gx}\in\Re^{n\times n}$$

are the given strictly positive definite normalizing matrices (the Lipschitz constant matrices).

For the unmodeled dynamic mappings Δf and Δg we assume that they satisfy the "*strip-bound condition* [21]"

A4.3:

$$\begin{array}{c}\Delta f^T(x_t,t)\Lambda_{\Delta f}\Delta f(x_t,t) \leq C_{\Delta f} + x_t^T D_{\Delta f} x_t,\ t \in R^+\\ \Delta g^T(x_t,t)\Lambda_{\Delta g}\Delta g(x_t,t) \leq C_{\Delta g} + x_t^T D_{\Delta g} x_t,\ t \in R^+\end{array} \tag{4.12}$$

where

$$0 < \Lambda_{\Delta f}^T = \Lambda_{\Delta f} \in \Re^{n\times n}, 0 < D_{\Delta f}^T = D_{\Delta f} \in \Re^{n\times n}, 0 < D_{\Delta g}^T = D_{\Delta g} \in \Re^{n\times n}$$

are the known constant matrices, and $C_{\Delta f}$ and $C_{\Delta g}$ are the known positive constants characterizing the behavior of the corresponding unmodeled dynamics mappings in the point $x = 0$.

If

$$D_{\Delta f} = D_{\Delta g} = 0$$

the unmodeled dynamics are bounded.

If

$$C_{\Delta f} = C_{\Delta g} = 0$$

the unmodeled dynamics belong to sector nonlinearities [3].

If unmodeled dynamics are absent at all, we have

$$\Delta f(0,t) = 0,\ \Delta g(0,t) = 0$$

Definition 6 *Denote by $\mathcal{H}$ the set of nonlinear dynamic systems given by (4.9) with the perturbations $\xi_{i,t}$ of "bounded power" (the assumption **A4.1**) and containing the nominal part F and G satisfying Lipschitz condition (the assumption **A4.2**), and strip bounded unmodeled dynamics $\Delta f(\cdot)$, $\Delta g(\cdot)$ and $\Delta h(\cdot)$ (the assumption **A4.3**).*

In the next subsection we will discuss the possible structure of the corresponding robust observer.

4.2.2 Nonlinear Observers and The Problem Setting

Select the observer structure written in Luneburger's form (4.7) [7]. It uses only the available information on the nominal nonlinear mappings $F(\cdot)$, $G(.)$ and C and the on-line measurable function y_t:

$$\boxed{\dot{\widehat{x}}_t = F(\widehat{x}_t, t) + G(\widehat{x}_t, t)u_t + K_t \left[y_t - C\widehat{x}_t\right]} \tag{4.13}$$

where $K_t \in \Re^{n \times m}$ is a matrix to be found to provide a *"good behavior"* of the generated estimates $\widehat{x}_t$. The initial conditions $\widehat{x}_0$ are assumed to be fixed. This observer presents a copy of the nominal system with the correction term

$$K_t \left[y_t - C\widehat{x}_t\right]$$

In special case, when there are no any uncertainties and disturbances, the gain matrix K_t can be selected as a *time-invariant*

$$K_t = K$$

and the following inequality of the estimating process can be proved (see [7]):

$$\|x_t - \widehat{x}_t\| \leq k(\theta)\exp\left(-\frac{\theta}{3}t\right)\|x_0 - \widehat{x}_0\| \tag{4.14}$$

where θ is big enough positive constants, $k(\theta)$ is a positive function related to the observer gain. It means that the observation error is asymptotically stable uniformly on initial conditions. Obviously, in the presence of perturbations and unmodeled dynamics we may loose this property. But we can try to guarantee the observation error to be bounded and small enough by choosing a suitable *time-variant* gain K_t.

Let us define the observation error as

$$\Delta_t := \widehat{x}_t - x_t \tag{4.15}$$

With each nonlinear system (4.9) and its observer (4.13), we associate the performance index

$$J(\{K_t\}_{t\geq 0}) := \sup_{\mathcal{H}} \limsup_{T\to\infty} \frac{1}{T}\int_0^T \Delta_t^T Q \Delta_t dt \tag{4.16}$$

which characterizes the quality of the nonlinear observer (4.13) from the class of nonlinear systems $\mathcal{H}$. The strictly positive constant matrix Q is suggested to be a known normalizing matrix which gives an opportunity to work with the error vector Δ_t having the components of a different physical nature. This performance index depends on the matrix function $\{K_t\}_{t\geq 0}$ which has to be selected to obtain a good quality of the estimating process. The formal statement of the robust observation problem is presented next.

Statement of the problem: *For the given class of nonlinear systems $\mathcal{H}$ and the gain matrix function $\{K_t\}_{t\geq 0}$ obtain an upper bound $J^+(\{K_t\}_{t\geq 0})$ of the performance index (4.16). The main objective is to minimize this upper bound J^+ with respect to the gain matrix $\{K_t\}_{t\geq 0}$, i.e.,*

$$J(\{K_t\}_{t\geq 0}) \leq J^+(\{K_t\}_{t\geq 0}) \to \inf_{\{K_t^*\}_{t\geq 0}} \tag{4.17}$$

The following definition of the robust observer is used subsequently.

Definition 7 *If, within the class of nonlinear systems $\mathcal{H}$, the gain matrix function $\{K_t\}_{t\geq 0}$ is the solution of (4.17) with finite upper bound (tolerance level) which is "tight" (equal to zero in the case of no any uncertainties), then the nonlinear observer (4.13) is said to be* ***the robust observer*** .

In next section we will proof that the robust observer guarantees the stability of the observer error and the "seminorm" of estimation error defined by (4.16) turns out to be bounded in the "average sense".

4.2.3 The Main Result on The Robust Observer

The theorem presented below formulates the main result on the robust observer synthesize in the presence of mixed uncertainties. Suppose that in the addition to **A4.1-A4.3** the following technical assumption concerning the differential Riccati equation is fulfilled:

A4.4: There exist a stable matrix A_0 and strictly positive definite matrices Q and Π such that the matrix differential Riccati equation

$$\boxed{-\dot{P}_t = P_t A_0 + A_0^T P_t + P_t R_t P_t + Q_0} \tag{4.18}$$

for any $t \in R^+$ has the strictly positive solution $P_t = P_t^T > 0$.

The functional matrices $R_t := R(t, \widehat{x}_t)$ and Q_0 are defined by

$$\begin{array}{c} R_t := R(t, \widehat{x}_t) \\ = R_0 + \beta_t \left(C^+ \Lambda (C^+)^T\right)^{\frac{1}{2}} \left(I + \Pi^{-1}\right) \left(C^+ \Lambda (C^+)^T\right)^{\frac{1}{2}} \beta_t^T \\ Q_0 := \left(2\Lambda_{fx} + K_1 \lambda_{\max}(\Lambda_{\partial f}) I\right) + \left(2\Lambda_{gx} + K_2 \lambda_{\max}(\Lambda_{\partial g}) I\right) + Q \end{array} \tag{4.19}$$

where

$$\begin{array}{c} R_0 := 2\Lambda_f^{-1} + \Lambda_{\xi_1}^{-1} + \Lambda_{\triangle f}^{-1} + \Lambda_{\triangle g}^{-1} + \Lambda_{\partial f}^{-1} + \Lambda_{\partial g}^{-1} \\ \Lambda := \Lambda_{\xi_2}^{-1} \end{array} \tag{4.20}$$

K_1, K_2 are positive constants, and $\Lambda_{\partial f}^{-1}$ and $\Lambda_{\partial g}^{-1}$ are positive definite matrices. The matrix $\beta_t \in \Re^{n \times n}$ is defined for any $t \in R^+$ as

$$\beta_t := \frac{\partial}{\partial x_t} F(\widehat{x}_t, t) + \frac{\partial}{\partial x_t} G(\widehat{x}_t, t) - A_0 \tag{4.21}$$

C^+ is the pseudoinverse matrix in the Moore-Penrose sense [1].

Remark 4.1 *In fact, this assumption is related to some properties of the nonlinear system (see the equations (4.11), (4.12) and (4.21)). If we know that the pair $(A_0, R_0^{1/2})$ is controllable and the pair $(Q_0^{1/2}, A_0)$ is observable, the differential Riccati equation with the constant parameters A_0, R_0 and Q_0*

$$-\dot{P}_t' = P_t' A_0 + A_0^T P_t' + P_t' R_0 P_t' + Q_0 \tag{4.22}$$

has a positive solution $P_t' \geq \bar{P} > 0$. According to the Appendix A, we can compare the differential Riccati equation containing time variant parameters with (4.22). If the condition

$$\begin{bmatrix} Q_0 & A_0^T \\ A_0 & R_t \end{bmatrix} \leq \begin{bmatrix} Q_0 & A_0^T \\ A_0 & R_0 \end{bmatrix}$$

is satisfied (i.e., the uncertainties are not large enough), we can conclude that for the differential equation (4.18) we can guarantee

$$P_c(t) = P_c^T(t) > P_c'(t) \geq \bar{P} > 0, \quad \forall t \geq 0 \tag{4.23}$$

The matrix inequality given above can be satisfied by the corresponding selection the Hurwitz matrix A_0 *and the matrix* Q *.*

The strictly positivity condition

$$P_c^{'}(t) \geq \bar{P} > 0$$

can also be expressed as a special local frequency condition (see Appendix A)

$$\frac{1}{4}\left(A_0^T R_0^{-1} - R_0^{-1} A_0\right) R_0 \left(A_0^T R_0^{-1} - R_0^{-1} A_0\right)^T \leq A_0^T R_0^{-1} A_0 - Q_0 \tag{4.24}$$

If uncertainties are "big enough" we will loose the property (4.24). The condition (4.24) states some sort of the trade-off between the admissable uncertainties and the dynamics of the nominal model.

Next theorem presents the main contribution of this section and deals with the upper bound for the estimation error performance index. Its dependence on the gain matrix K_t is stated.

Theorem 4.1 *For the given class of nonlinear systems satisfying* ***A4.1-A4.4*** *and for any matrix sequence*

$$\left\{K_t = \widehat{K}_t C C^+\right\}_{t\geq 0}$$

the following upper bound for the performance index (4.16) holds:

$$J(\{K_t\}_{t\geq 0}) \leq J^+(\{K_t\}_{t\geq 0}) = \overline{C} + D + \Upsilon_1 + \Upsilon_2 + \phi_\delta\left(\left\{\widehat{K}_t\right\}_{t\geq 0}\right) \tag{4.25}$$

where the constants Υ_1, Υ_2 *are defined by* ***A4.1*** *and*

$$\begin{array}{c} 0 \leq \overline{C} := C_{\triangle f} + C_{\triangle g} \\ D := \sup\limits_{\mathcal{H}} \limsup\limits_{T\to\infty} \frac{1}{T}\int\limits_0^T x_t^T \left(D_{\triangle f} + D_{\triangle g}\right) x_t dt \\ \phi_\delta\left(\left\{\widehat{K}_t\right\}_{t\geq 0}\right) := \sup\limits_{\mathcal{H}} \limsup\limits_{T\to\infty} \\ \frac{1}{T}\int\limits_0^T \Delta_t^T \left(X_t\widehat{\Omega}^{\frac{1}{2}} + \widehat{\Omega}^{-\frac{1}{2}}\right)\left(X_t\widehat{\Omega}^{\frac{1}{2}} + \widehat{\Omega}^{-\frac{1}{2}}\right)^T \Delta_t dt \end{array} \tag{4.26}$$

$$X_t := P_t\left(\beta_t - \widehat{K}_t C C^{+}\right)$$
$$\Omega := \Lambda^{\frac{1}{2}}\left(I + \Pi\right)\Lambda^{\frac{1}{2}} > 0 \tag{4.27}$$

The robust optimal gain matrix $\widehat{K}_t$ *verifies:*

$$\widehat{K}_t C C^{*} = P_t^{-1}\Omega^{-1} + \beta_t \tag{4.28}$$

then this the gain matrix provides the property

$$\phi_\delta\left(\left\{\widehat{K}_t\right\}_{t\geq 0}\right) = 0 \tag{4.29}$$

Proof. To start the proof of this theorem, we need to derive the differential equation for the error vector. Taking into account the relations (4.9) and (4.13), we can get:

$$\begin{array}{c}\dot{\Delta}_t = \dot{\widehat{x}}_t - \dot{x}_t = F\left(\widehat{x}_t, t\right) + G(\widehat{x}_t, t)u_t + K_t\left[y_t - C\widehat{x}_t\right]\\ -F(x_t, t) - G(x_t, t)u_t - \triangle f(x_t, t) - \triangle g(x_t, t) - \xi_{1,t}\end{array} \tag{4.30}$$

Denote

$$\begin{array}{c}F_t = F(x_t, \Delta_t, u_t, t \mid K_t) := F(x_t + \Delta_t, t) - F(x_t, t) + \\ G(x_t + \Delta_t, t)u_t - G(x_t, t)u_t - K_t C\Delta_t \\ \triangle H_t = \triangle H(\xi_{1,t}, \xi_{2,t}, \triangle f \mid K_t) := K_t\xi_{2,t} - \triangle f(\cdot) - \triangle g(\cdot) - \xi_{1,t}\end{array} \tag{4.31}$$

The vector function F_t describes the dynamic of the nominal model and the function $\triangle H_t$ corresponding to unmodeled dynamics and external disturbances. So, we can represent the differential equation for error vector as follows:

$$\dot{\Delta}_t = F_t + \triangle H_t$$

Calculating the derivative of the quadratic Lyapunov function

$$V_t := \Delta_t^T P_t \Delta_t, \qquad P_t^T = P_t > 0 \tag{4.32}$$

on the trajectories of the differential equation (4.30) we derive:

$$\frac{dV_t}{dt} = \frac{\partial V_t}{\partial P_t} + \left(\frac{\partial V_t}{\partial \Delta_t}, \dot{\Delta}_t\right) = \Delta_t^T \dot{P}_t \Delta_t + 2\Delta_t^T P_t\left[F_t + \triangle H_t\right] \tag{4.33}$$

Using (4.11) and Lemma 12.5 of Appendix A we obtain:

$$F(x_t+\Delta_t,t)-F(x_t,t) = \frac{\partial}{\partial x}F^T(x_t,t)\Delta_t + v_{f,t} \tag{4.34}$$
$$G(x_t+\Delta_t,t)-G(x_t,t) = \frac{\partial}{\partial x}G^T(x_t,t)\Delta_t + v_{g,t}$$

where

$$\|v_{f,t}\|_{\Lambda_f} \le 2\|\Delta\|_{\Lambda_{fx}}, \; \|v_{g,t}\|_{\Lambda_g} \le 2\|\Delta\|_{\Lambda_{gx}} \tag{4.35}$$

Substituting (4.34) in to (4.31), we conclude that the following presentation holds:

$$\begin{aligned} F_t &= \frac{\partial F^T(x_t,t)}{\partial x_t}\Delta_t + v_f + \frac{\partial G^T(x_t,t)}{\partial x_t}\Delta_t + v_g - K_t C\Delta_t \\ &= \left[\frac{\partial F^T(x_t,t)}{\partial x_t} + \frac{\partial G^T(x_t,t)}{\partial x_t} - K_t C\right]\Delta_t + v_{f,t} + v_{g,t} \end{aligned} \tag{4.36}$$

Because $\Delta_t^T P_t v_{f,t}$ is the scalar, applying (4.35) and in view of the matrix inequality

$$X^T Y + \left(X^T Y\right)^T \le X^T \Lambda^{-1} X + Y^T \Lambda Y \tag{4.37}$$

which is valid for any $X, Y \in \Re^{n\times k}$ and for any positive defined matrix $0 < \Lambda = \Lambda^T \in \Re^{n\times n}$, we obtain

$$\begin{aligned} 2\Delta_t^T P_t v_{f,t} \le \Delta_t^T \left(P_t \Lambda_f^{-1} P_t + 2\Lambda_{fx}\right)\Delta_t \\ 2\Delta_t^T P_t v_{g,t} \le \Delta_t^T \left(P_t \Lambda_g^{-1} P_t + 2\Lambda_{gx}\right)\Delta_t \end{aligned} \tag{4.38}$$

Using the assumptions **A4.1,** we also get

$$\begin{aligned} -2\Delta_t^T P_t \xi_{1,t} \le \Delta_t^T P_t \Lambda_{\xi_1}^{-1} P_t \Delta_t + \xi_{1,t}^T, \Lambda_{\xi_1}\xi_{1,t} \\ 2\Delta_t^T P_t K_t \xi_{2,t} \le \Delta_t^T P_t K_t \Lambda_{\xi 2}^{-1} K_t^T P_t \Delta_t + \xi_{2,t}^T, \Lambda_{\xi_2}\xi_{2,t} \end{aligned} \tag{4.39}$$

In view of the assumptions **A4.3,** it follows

$$\begin{aligned} -2\Delta_t^T P_t \triangle f \le \Delta_t^T P_t \Lambda_{\triangle f}^{-1} P_t \Delta_t + C_{\triangle f} + x_t^T D_{\triangle f} x_t \\ -2\Delta_t^T P_t \triangle g \le \Delta_t^T P_t \Lambda_{\triangle g}^{-1} P_t \Delta_t + C_{\triangle g} + x_t^T D_{\triangle g} x_t \end{aligned} \tag{4.40}$$

Denote by

$$A_t := \frac{\partial F^T(x_t,t)}{\partial x_t} + \frac{\partial G^T(x_t,t)}{\partial x_t} - K_t C \tag{4.41}$$

Using the identity

$$2\Delta_t^T A_t \Delta_t = \Delta_t^T A_t \Delta_t + \Delta_t^T A_t^T \Delta_t$$

and adding and substraction the term

$$\Delta_t^T Q \Delta_t \left(Q = Q^T > 0\right)$$

in the right-hand side of (4.33) we obtain:

$$\begin{array}{c} \frac{dV_t}{dt} \leq \Delta_t^T L_t \Delta_t + \left(C_{\triangle f} + C_{\triangle g}\right) \\ +\Upsilon_t + x_t^T \left(D_{\triangle f} + D_{\triangle g}\right) x_t - \Delta_t^T Q \Delta_t \end{array} \tag{4.42}$$

where the following notations are used:

$$\Upsilon_t := \xi_{1,t}^T \Lambda_{\xi_1} \xi_{1,t} + \xi_{2,t}^T \Lambda_{\xi_2} \xi_{2,t}$$

$$\begin{array}{c} L_t = \dot{P}_t + P_t A_t + A_t^T P_t \\ +P_t \left(\Lambda_f^{-1} + \Lambda_g^{-1} + \Lambda_{\xi_1}^{-1} + \Lambda_{\triangle g}^{-1}\right) P_t \\ +P_t K_t \Lambda_{\xi_2}^{-1} K_t^T P_t + 2\left(\Lambda_{fx} + \Lambda_{gx}\right) + Q \end{array} \tag{4.43}$$

Choosing a Hurwitz matrix A_0, we can rewrite (4.41) as follows:

$$\begin{array}{c} A_t = A_0 + \left(\frac{\partial}{\partial x_t} F^T(\widehat{x}_t, t) + \frac{\partial}{\partial x_t} G^T(\widehat{x}_t, t) - K_t C - A_0\right) + \\ \left(\frac{\partial}{\partial x} F^T(x_t, t) - \frac{\partial}{\partial x} F^T(\widehat{x}_t, t)\right) + \left(\frac{\partial}{\partial x} G^T(x_t, t) - \frac{\partial}{\partial x} G^T(\widehat{x}_t, t)\right) \end{array} \tag{4.44}$$

Denote

$$\begin{array}{r} \widehat{A}_t := \frac{\partial}{\partial x_t} F^T(\widehat{x}_t, t) + \frac{\partial}{\partial x_t} G^T(\widehat{x}_t, t) - K_t C - A_0 \\ \partial f_t := \frac{\partial}{\partial x} F^T(x_t, t) - \frac{\partial}{\partial x} F^T(\widehat{x}_t, t) \\ \partial g_t := \frac{\partial}{\partial x} G^T(x_t, t) - \frac{\partial}{\partial x} G^T(\widehat{x}_t, t) \end{array}$$

From (4.34) we have

$$\partial f_t = \left(F(x_t + \Delta_t, t) - F(x_t, t)\right) - \left(F(x_t + 2\Delta_t, t) - F(x_t + \Delta_t, t)\right)$$

Similarly to (4.39), we deduce:

$$
\begin{aligned}
2\Delta_t^T P_t \partial f_t &\leq \Delta_t^T \left(P_t \Lambda_{\partial f}^{-1} P_t + \partial f_t \Lambda_{\partial f} \partial f_t^T \right) \Delta_t \\
&\leq \Delta_t^T \left(P_t \Lambda_{\partial f}^{-1} P_t + K_1 \lambda_{\max}(\Lambda_{\partial f}) I \right) \Delta_t \\
2\Delta_t^T P_t \partial g_t &\leq \Delta_t^T \left(P_t \Lambda_{\partial g}^{-1} P_t + K_2 \lambda_{\max}(\Lambda_{\partial g}) I \right) \Delta_t
\end{aligned}
$$

where K_1 and K_2 are positive constants. Substituting these inequalities into (4.43) and using (4.44), we obtain:

$$
\begin{aligned}
L_t \leq & \left(\dot{P}_t + P_t A_0 + A_0^T P_t + P_t R_0 P_t + Q_0 \right) \\
& + \left(P_t K_t \Lambda K_t^T P_t + P_t \widehat{A}_t + \widehat{A}_t^T P_t \right)
\end{aligned} \tag{4.45}
$$

where the matrices R_0, Q_0 and Λ are defined by (4.20).

Based on the definition (4.21), in view of the pseudoinverse property

$$
CC^+C = C
$$

we can rewrite

$$
\begin{aligned}
& P_t K_t \Lambda K_t^T P_t + P_t \widehat{A}_t + \widehat{A}_t^T P_t \\
= & P_t \left(\beta_t - \widehat{K}_t C C^+ \right) + \left(\beta_t - \widehat{K}_t C C^+ \right)^T P_t + P_t \widehat{K}_t C G C^T \widehat{K}_t^T P_t
\end{aligned} \tag{4.46}
$$

where

$$
G := C^+ \Lambda (C^+)^T \tag{4.47}
$$

The definition (4.27) and (4.46) imply

$$
\begin{aligned}
\Delta_t^T P_t \widehat{K}_t C G C^T \widehat{K}_t^T P_t \Delta_t &= \left\| \Lambda^{\frac{1}{2}} \left(\widehat{K}_t C C^+ \right)^T P_t \Delta_t \right\|^2 \\
&= \left\| \Lambda^{\frac{1}{2}} X_t^T \Delta_t - \Lambda^{\frac{1}{2}} \beta_t P_t \Delta_t \right\|^2 \\
&\leq \Delta_t^T X_t \Lambda^{\frac{1}{2}} \left(I + \Pi \right) \Lambda^{\frac{1}{2}} X_t^T \Delta_t \\
&+ \Delta_t^T P_t \beta_t \Lambda^{\frac{1}{2}} \left(I + \Pi^{-1} \right) \Lambda^{\frac{1}{2}} \beta_t^T P_t \Delta_t
\end{aligned}
$$

where Π is any positive definite matrix. Defining

$$
\Phi_t := \beta_t \Lambda^{\frac{1}{2}} \left(I + \Pi^{-1} \right) \Lambda^{\frac{1}{2}} \beta_t^T
$$

and in view of (4.27), the first term in (4.46) can be estimated as

$$P_tK_t\Lambda K_t^TP_t \leq X_t\Omega X_t^T + P_t\Phi_tP_t$$

Taking in to account that $\Omega > 0$, the expression (4.46) can be rewritten as

$$\begin{aligned}
P_tK_t\Lambda K_t^TP_t + P_t\widehat{A}_t + \widehat{A}_t^TP_t &\leq X_t + X_t^T + X_t\Omega X_t^T + P_t\Phi_tP_t \\
&\leq \left(X_t\Omega^{\frac{1}{2}} + \Omega^{-\frac{1}{2}}\right)\left(X_t\Omega^{\frac{1}{2}} + \Omega^{-\frac{1}{2}}\right)^T \\
&\quad -\Omega^{-1} - \delta X_tX_t^T + P_t\Phi_tP_t \\
&\leq \left(X_t\Omega^{\frac{1}{2}} + \Omega^{-\frac{1}{2}}\right)\left(X_t\Omega^{\frac{1}{2}} + \Omega^{-\frac{1}{2}}\right)^T \\
&\quad +P_t\Phi_tP_t
\end{aligned}$$

If we define R_t as in (4.19), using definition of (4.20) and (4.26), we transform (4.42) to the follow form:

$$\frac{dV_t}{dt} \leq \Delta_t^TL_t'\Delta_t + \overline{C} + \Upsilon_t + D_t - \Delta_t^TQ\Delta_t + \Delta_t^T\widehat{\gamma}_t\Delta_t \tag{4.48}$$

where

$$\begin{aligned}
L_t' &:= \dot{P}_t + P_tA_0 + A_0^TP_t + P_tR_tP_t + Q_0 \\
\gamma_t &:= \left(X\Omega^{\frac{1}{2}} + \Omega^{-\frac{1}{2}}\right)\left(X_t\Omega^{\frac{1}{2}} + \Omega^{-\frac{1}{2}}\right)^T \\
D_t &= X_t^T\left(D_{\Delta f} + D_{\Delta g}\right)X_t
\end{aligned}$$

The assumption **A4.4** implies

$$L_t' = 0$$

Integrating (4.48) within the interval $t \in [0, T]$ and dividing both sides on T, we obtain:

$$\begin{aligned}
\frac{1}{T}\int_0^T \Delta_t^TQ\Delta_tdt &\leq \overline{C} + \frac{1}{T}\int_0^T \Upsilon_tdt + \frac{1}{T}\int_0^T D_tdt + \frac{1}{T}\int_0^T \Delta_t^T\gamma_t\Delta_t - \frac{1}{T}\left(V_T - V_0\right) \\
&\leq \overline{C} + \frac{1}{T}\int_0^T \Upsilon_tdt + +\frac{1}{T}\int_0^T D_tdt + \frac{1}{T}\int_0^T \Delta_t^T\gamma\Delta_t - \frac{1}{T}V_0
\end{aligned}$$

Taking limit on $T \to \infty$ we finally obtain (4.25).

To minimize the right side hand of the expression in (4.25), we must choose

$$X_t = -\Omega^{-1} \tag{4.49}$$

that leads to

$$P_t(\beta_t - \widehat{K}_t CC^+) = -\Omega^{-1}$$

or, in equivalent form,

$$\widehat{K}_t CC^+ = \beta_t + P_t^{-1}\Omega^{-1} \tag{4.50}$$

Theorem is proved. ■

Corollary 4.1 *The robust observer is defined by the following gain matrix:*

$$\begin{aligned} K_t^* &:= \widehat{K}_t CC^+ = \left[P_t^{-1}\Omega^{-1} + \beta_t\right] \\ &= P_t^{-1}\Lambda^{\frac{1}{2}}\left(I + \Pi\right)^{-1}\Lambda^{\frac{1}{2}} + \beta_t \end{aligned} \tag{4.51}$$

that guarantees that the upper bound (4.16) reaches minimum

$$J^+(\{K_t^*\}_{t\geq 0}) = \overline{C} + D + \Upsilon_1 + \Upsilon_2$$

Proof. It follows directly from (4.50) and (4.28). ■

Remark 4.2 *If there are no any unmodeled dynamics* $\left(\overline{C} = D = 0\right)$ *and no any external disturbances* $(\Upsilon_1 = \Upsilon_2 = 0)$, *the robust observer (4.13) with the optimal matrix gain given by (4.28) guarantees* ***"the stability in average"****:*

$$\sup_{\mathcal{H}} \lim_{t\to\infty} \frac{1}{T}\int_0^T \Delta_t^T Q \Delta_t dt = 0$$

that, in some sense, is equivalent to the fact

$$\lim_{t\to\infty} \Delta_t = 0$$

4.3 The Neuro-Observer for Unknown Nonlinear Systems

The approach presented in the last section assumes that the structure of the system at least partially known (it consists of a nominal model dynamic part plus unmodeled dynamics or perturbations). In this section we will show that even with an incomplete knowledge of a nominal model, the dynamic neural network technique can be successfully applied to provide a good enough state estimation process for such unknown systems.

Some authors have already discussed the application of neural networks techniques to construct state observers for nonlinear systems with incomplete description. In [20] a nonlinear observer, based on the ideas of [7], is combined with a feedforward neural network, which is used to solve a matrix equation. [11] uses a nonlinear observer to estimate the nonlinearities of an input signal. As far as we know, the first observer for nonlinear systems, using dynamic neural networks, has been presented in [10]. The stability of this observer with on-line updating of neural network weights is analyzed, but several restrictive assumptions are used: the nonlinear plant has to contain a known linear part and a strictly positive real (SPR) condition should be fulfilled to prove the stability of the estimation error .

In this section we consider more general class of nonlinear systems containing external nonrandom disturbances of bounded power as well as unmodeled dynamics. We apply a Luneburger-like observer with a gain matrix that is specifically constructed to guarantee the robustness property for a given class of uncertainties. To calculate this matrix gain we use a differential matrix Riccati equation with time varying parameters and the pseudoinverse operator technique. A new updating law for the neural network weights is used to guarantee their boundness and provide a high accuracy for the estimation error.

4.3.1 The Observer Structure and Uncertainties

We consider the class of nonlinear systems given by

$$\boxed{\begin{array}{c} \dot{x}_t = f(x_t, u_t, t) + \xi_{1,t} \\ y_t = Cx_t + \xi_{2,t} \end{array}} \tag{4.52}$$

where

$x_t \in \Re^n$ is the state vector of the system at time $t \in \Re^+ := \{t : t \geq 0\}$;

$u_t \in \Re^q$ is a given control action;

$y_t \in \Re^m$ is the output vector, which is measurable at each time t;

$f(\cdot) : \Re^n \rightarrow \Re^n$ is an unknown vector valued nonlinear function describing the system dynamics;

$C \in \Re^{m \times n}$ is a unknown output matrix;

$\xi_{1,t}$, $\xi_{2,t}$ are vector-functions representing external perturbations with the "bounded power" (**A4.1**),

C satisfies following assumptions:

A4.5:

$$C = C_0 + \Delta C$$

where C_0 is known, ΔC verifies a kind of "strip bounded condition"as **A2.4**

$$\Delta C^T \Lambda_{\Delta C} \Delta C \leq C_{\Delta C}, \quad \forall t \in R^+$$

Remark 4.3 *If a closed-loop system is exponentially stable,* $f(x_t, u_t, t)$ *does not depend on t and is the Lipschitz function with respect to both arguments, then the converse Lyapunov theorem implies **A4.5**. But the assumption **A4.5** is weaker and easy to be satisfied.*

Below the motivation of the observer structure selection is given. Following to the standard technique [18], in the case of the complete knowledge on a nonlinear system (without unmodeled dynamics and external perturbation terms) the structure of the corresponding nonlinear observer can be selected as follows:

$$\frac{d}{dt}\hat{x}_t = f(\hat{x}_t, u_t, t) + L_{1,t}\left[y_t - C\hat{x}_t\right]$$

The first term in the right-hand side of (4.52) repeats the known dynamics of the nonlinear system and the second one is intended to correct the estimated trajectory based on the current residual values. If $L_{1,t} = L_{1,t}(\hat{x}_t)$, it is named as **"differential algebra" type observer** (see [7] and [16]). In the case of $L_{1,t} = L_1 = const$, we name it as **"high-gain" type observer** which was studied in [21].

If apply such observers to a class of mechanical systems when only the position measurements are available (the velocities are not available), as a rule, the corresponding velocity estimates are not so good because of the following effects:

- The original dynamic mechanical system, in general, is given as

$$\ddot{z}_t = F(z_t, u_t, t)$$
$$y = z_t$$

or, in the equivalent standard Cauchy form,

$$\dot{x}_{1,t} = x_{1,t}$$
$$\dot{x}_{2,t} = \tilde{F}(x_t, u_t, t)$$
$$y = x_{1,t}$$

So, the corresponding nonlinear observer is

$$\frac{d}{dt}\begin{pmatrix} \hat{x}_{1,t} \\ \hat{x}_{2,t} \end{pmatrix} = \begin{pmatrix} \hat{x}_{2,t} \\ \tilde{F}(\hat{x}_t, u_t, t) \end{pmatrix} + \begin{pmatrix} L_{1,1,t} \\ L_{1,2,t} \end{pmatrix} [y_t - \hat{x}_{1,t}] \tag{4.53}$$

That means, the observable state components are estimated very well that leads to small value of the residual term $[y_t - \hat{x}_{1,t}]$. As a fact, it practically has no any effect in (6.32). But any current information containing the output $y_t(x_{1,t})$ has no any practical affect to the velocity estimate $\hat{x}_{2,t}$. So, this velocity estimate turns out to be extremely bad. One of possible overcomings of this problem consists in the addition a new term

$$L_{2,t}\left[h^{-1}(y_t - y_{t-h}) - Ch^{-1}(\hat{x}_t - \hat{x}_{t-h})\right]$$

which can be considered as a *"derivative estimation error"* and can be used for the adjustment of the velocity estimates. This new modified observer can be described as

$$\boxed{\begin{array}{c}\frac{d}{dt}\hat{x}_t = f(\hat{x}_t, u_t, t) + L_{1,t}\left[y_t - C\hat{x}_t\right] + \\ L_{2,t}h^{-1}\left[(y_t - y_{t-h}) - C\left(\hat{x}_t - \hat{x}_{t-h}\right)\right]\end{array}}$$

- If we have no a complete information on the nonlinear function $f(\hat{x}_t, u_t, t)$, it seems to be natural to construct its estimate $\hat{f}(\hat{x}_t, u_t, t \mid W_t)$ depending on parameters W_t which can be adjusted on-line to obtain the best nonlinear approximation of the unknown dynamic operator. That implies the following observation scheme:

$$\begin{array}{c}\frac{d}{dt}\hat{x}_t = \hat{f}(\hat{x}_t, u_t, t \mid W_t) + L_{1,t}\left[y_t - C\hat{x}_t\right] + \\ L_{2,t}h^{-1}\left[(y_t - y_{t-h}) - C\left(\hat{x}_t - \hat{x}_{t-h}\right)\right]\end{array}$$

  with a special *updating* (learning) law

$$\dot{W}_t = \Phi\left(W_t, \hat{x}_t, u_t, t, y_t\right)$$

  Such "robust adaptive observer" seems to be a more advanced device which provides a good estimation in the absence of a dynamic information and incomplete state measurements. Below we present the detail analysis of this estimator.

4.3.2 The Signal Layer Neuro Observer without A Delay Term

First, start with the simplest situation to understand better all arising problems: select the recurrent neural networks (2.2) with the only one added correction term that leads to the following Luneburger-like observer structure [10]:

$$\boxed{\begin{array}{c}\dot{\widehat{x}}_t = A\widehat{x}_t + W_{1,t}\sigma(\widehat{x}_t) + W_{2,t}\phi(\widehat{x}_t)\gamma(u_t) + K_t\left[y_t - \widehat{y}_t\right] \\ \widehat{y}_t = C_0\widehat{x}_t\end{array}} \tag{4.54}$$

where $K_t \in \Re^{n\times m}$ is the observer gain matrix, $\widehat{x}_t$ is the state of the neural network. The initial conditions $\widehat{x}_0$ is assumed to be fixed. We name the proposed scheme as a *neuro-observer*. The estimation error is defined as in (4.15).

From Chapter 2 it follows that the function $f(x_t, u_t, t)$ in (4.52) can be modelled by the dynamic neural network (2.2) with fixed W_1^* and W_2^* as in (2.7) plus a modeling error $\triangle f\ (\cdot)$:

$$\triangle f(x_t, u_t, t) = f(x_t, u_t, t) - [Ax_t + W_1^* \sigma(x_t) + W_2^* \phi(x_t)\gamma(u_t)] \quad (4.55)$$

The modeling error reflects the effect of unmodeled dynamics and satisfies

$$\begin{array}{c} \triangle f^T(x_t, u_t, t)\Lambda_{\triangle f} \triangle f(x_t, u_t, t) \le C_{\triangle f} + x_t^T D_{\triangle f} x_t \\ \forall u \in \Re^q, t \in R^+ \end{array} \quad (4.56)$$

Here

$$0 < \Lambda_{\triangle f}^T = \Lambda_{\triangle f} \in \Re^{n \times n},\ 0 < D_{\triangle f}^T = D_{\triangle f} \in \Re^{n \times n}$$

are the known constant matrices, and $C_{\triangle f}$ and $D_{\triangle f}$ are the known positive constants.

Throughout this chapter we will denote the class of the nonlinear systems, satisfying the assumptions **A4.1** and **A4.5,** by $\mathcal{H}$.

For each nonlinear system (4.52) and neuro-observer (4.54), we associate the performance index as in (4.16). The constant strictly positive definite matrix Q normalizes the components of the error vector Δ_t, which could have different physical meanings. This performance index depends on the matrix $\{K_t\}_{t \ge 0}$ and on the weights $\{W_{1,t}\}_{t \ge 0}$, $\{W_{2,t}\}_{t \ge 0}$ which must be selected to provide a good quality of the estimation process for the given class $\mathcal{H}$.

Statement of the problem: *For the given class $\mathcal{H}$ of nonlinear systems, for any given gain matrix function $\{K_t\}_{t \ge 0}$ and for any weight matrices $\{W_{1,t}\}_{t \ge 0}$, $\{W_{2,t}\}_{t \ge 0}$ obtain an upper bound*

$$J^+ = J^+(\{K_t\}_{t \ge 0}, \{W_{1,t}\}_{t \ge 0}, \{W_{2,t}\}_{t \ge 0})$$

for the performance index J ($J \le J^+$) and, then, minimize this bound with respect to the matrices $\{K_t\}_{t \ge 0}$ and $\{W_{1,t}\}_{t \ge 0}, \{W_{2,t}\}_{t \ge 0}$, i.e., realize

$$\inf_{\{K_t\}_{t \ge 0}, \{W_{1,t}\}_{t \ge 0}, \{W_{2,t}\}_{t \ge 0}} J^+ \quad (4.57)$$

Suppose that, in the addition to **A4.1** and **A4.5,** the following assumption is fulfilled:

A4.6: There exist a strictly positive defined matrix Q, a stable matrix A and a positive constant δ such that the matrix Riccati equation

$$L = PA + A^T P + PR_0 P + Q_0 = 0 \tag{4.58}$$

has a strictly positive solution $P = P^T > 0$. The matrices R_0, Q_0 are defined as follows:

$$R_0 := \overline{W}_1 + \overline{W}_2 + \Lambda_{\xi_1}^{-1} + \Lambda_{\triangle f}^{-1},$$
$$Q_0 := D_{\sigma} + D_{\phi}\overline{u} + Q + 2\delta, \quad \delta > 0$$

Define also

$$\begin{array}{c} \widetilde{y}_t := C_0 \widehat{x}_t - y_t = C_0 \Delta_t - \left(\triangle C x_t + \xi_{2,t} \right) \\ N := C_0^T \left(C_0^T \right)^+ \end{array} \tag{4.59}$$

Theorem 4.2 *Under assumption **A4.6,** for the given class $\mathcal{H}$ of nonlinear systems given by (4.52), and for any matrix sequences*

$$\{K_t\}_{t\geq 0}, \{W_{1,t}\}_{t\geq 0}, \{W_{2,t}\}_{t\geq 0}$$

the following upper bound for the performance index (4.16) of the neuro-observer holds:

$$J \leq J^+ = C_{\triangle f} + D + \Upsilon + \varphi\left(\{K_t\}_{t\geq 0}\right) + \psi\left(\{W_{1,t}\}_{t\geq 0}, \{W_{2,t}\}_{t\geq 0}\right) \tag{4.60}$$

where the constants D, Υ, φ and ψ are equal to

$$\begin{array}{c} D := \sup\limits_{\mathcal{H}} \overline{\lim\limits_{T\to\infty}} \frac{1}{T} \int\limits_0^T x_t^T \left[D_{\triangle f} + 3C_{\triangle C} \right] x_t dt \\ \Upsilon = \sup\limits_{\mathcal{H}} \overline{\lim\limits_{T\to\infty}} \frac{1}{T} \int\limits_0^T \left(\xi_{1,t}^T \Lambda_{\xi_1} \xi_{1,t} + 3\xi_{2,t}^T \Lambda_{\xi_2} \xi_{2,t} \right) dt \\ \varphi\left(\{K_t\}_{t\geq 0}\right) := \\ \sup\limits_{\mathcal{H}} \overline{\lim\limits_{T\to\infty}} \frac{1}{T} \int\limits_0^T \Delta_t^T \left(P_t K_t \Omega^{\frac{1}{2}} - C_0^T \Omega^{-\frac{1}{2}} \right) \left(P_t K_t \Omega^{\frac{1}{2}} - C_0^T \Omega^{-\frac{1}{2}} \right) \Delta_t dt \\ \psi\left(\{W_{1,t}\}_{t\geq 0}, \{W_{2,t}\}_{t\geq 0}\right) := \\ \sup\limits_{\mathcal{H}} \overline{\lim\limits_{T\to\infty}} \frac{1}{T} \int\limits_0^T \left\| tr\left(\widetilde{W}_{1,t}^T L_{w1,t} + \widetilde{W}_{2,t}^T L_{w1,t} \right) \right\| dt \end{array} \tag{4.61}$$

here

$$\Omega := \Lambda_{\triangle C}^{-1} + \Lambda_{\xi_2}^{-1}$$

the matrix functions $L_{w1,t}$ and $L_{w1,t}$ are defined by

$$\begin{array}{c} L_{w1,t} = \dot{\widetilde{W}}_{1,t} + M_{1,t} \\ + \left(\Gamma_{1,t} + \delta\Gamma_{2,t}\right) \widetilde{W}_{1,t}\sigma(\widehat{x}_t)\sigma(\widehat{x}_t)^T \\ L_{w2,t} = \dot{\widetilde{W}}_{2,t} + M_{2,t} \\ + \left\|\gamma(u)\right\|^2 \left(\Gamma_{1,t} + \delta\Gamma_{2,t}\right) \widetilde{W}_{2,t}\phi(\widehat{x}_t)\phi(\widehat{x}_t)^T \end{array} \tag{4.62}$$

where

$$\begin{array}{c} \Gamma_{1,t} = \left(PN^{-T}C_0^{+}\right) \left(\Lambda_{\triangle C}^{-1} + \Lambda_{\xi_2}^{-1}\right) \left(C_0^{+T}N^{-1}P\right) \\ \Gamma_{2,t} = PN^{-T}N^{-1}P \\ M_{1,t} = 2PN^{-T}\sigma(\widehat{x}_t)\widetilde{y}_t^T C_0^{+} \\ M_{2,t} = 2P_tN^{-T}\phi(\widehat{x})\gamma(u_t)\widetilde{y}_t^T C_0^{+} \end{array} \tag{4.63}$$

Proof. We start the proof with the deriving the differential equation for the error vector (4.15). Taking into account (4.52), (2.2) we obtain

$$\begin{array}{c} \dot{\Delta}_t = \dot{\widehat{x}}_t - \dot{x}_t \\ = A\widehat{x}_t + W_{1,t}\sigma(\widehat{x}_t) + W_{2,t}\phi(\widehat{x}_t)\gamma(u_t) \\ + K_t\left[y_t - C_0\widehat{x}_t\right] - f\left(x_t, u_t, t\right) - \xi_{1,t} \end{array} \tag{4.64}$$

The calculation the derivative of the Lyapunov function candidate

$$V_t := \Delta_t^T P \Delta_t + \frac{1}{2} tr\left[\widetilde{W}_{1,t}^T \widetilde{W}_{1,t}\right] + \frac{1}{2} tr\left[\widetilde{W}_{2,t}^T \widetilde{W}_{2,t}\right] \tag{4.65}$$

for $P^T = P > 0$ along the trajectories of the differential equation (4.64) leads to the following expression:

$$\begin{array}{c} \frac{d}{dt} V_t = 2\Delta_t^T P \dot{\Delta}_t \\ + tr\left[\widetilde{W}_{1,t}^T \dot{\widetilde{W}}_{1,t}\right] + tr\left[\widetilde{W}_{2,t}^T \dot{\widetilde{W}}_{2,t}\right] \end{array} \tag{4.66}$$

The use of (4.55) and (4.64) implies

$$
\begin{gathered}
2\Delta_t^T P \dot{\Delta}_t = 2\Delta_t^T P_t \\
\Big[(A - K_t C_0)\,\Delta_t + W_1^* \widetilde{\sigma}_t + W_2^* \widetilde{\phi}_t \gamma(u_t) \\
+ \widetilde{W}_{1,t}\sigma(\widehat{x}_t) + \widetilde{W}_{2,t}\phi(\widehat{x}_t)\gamma(u_t) \Big] \\
+2\Delta_t^T P \left[K_t \xi_{2,t} + K_t \Delta C x_t - \triangle f - \xi_{1,t} \right]
\end{gathered}
$$

Based on the assumptions **A4.1,** we derive the following inequalities:

$$
\begin{aligned}
2\Delta_t^T P W_1^* \widetilde{\sigma}_t &\leq \Delta_t^T \left(P\overline{W}_1 P + D_\sigma \right) \Delta_t \\
2\Delta_t^T P W_2^* \widetilde{\phi}_t \gamma(u_t) &\leq \Delta_t^T \left(P\overline{W}_2 P + D_\phi \overline{u} \right) \Delta_t
\end{aligned}
\tag{4.67a}
$$

The terms $2\Delta_t^T P_t \xi_{1,t}$ and $2\Delta_t^T P_t K_t \xi_{2,t}$ can be estimated as in (4.39):

$$
\begin{aligned}
2\Delta_t^T P \xi_{1,t} &\leq \Delta_t^T P \Lambda_{\xi_1}^{-1} P \Delta_t + \xi_{1,t}^T \Lambda_{\xi_1} \xi_{1,t} \\
-2\Delta_t^T P K_t \xi_{2,t} &\leq \Delta_t^T P K_t \Lambda_{\xi_2}^{-1} K_t^T P \Delta_t + \xi_{2,t}^T \Lambda_{\xi_2} \xi_{2,t}
\end{aligned}
\tag{4.68}
$$

Using (4.56) and **A4.5**, the terms $2\Delta_t^T P_t \triangle f$ and $2\Delta_t^T P_t K_t \triangle C x_t$ can estimated as in (4.68)

$$
\begin{aligned}
-2\Delta_t^T P \triangle f &\leq \Delta_t^T P \Lambda_{\triangle f}^{-1} P \Delta_t + C_{\triangle f} + x_t^T D_{\triangle f} x_t \\
2\Delta_t^T P K_t \triangle C x_t &\leq \Delta_t^T P K_t \Lambda_{\triangle C}^{-1} K_t^T P \Delta_t + x_t^T C_{\triangle C} x_t
\end{aligned}
\tag{4.69}
$$

The definition of (4.59) implies

$$
\begin{aligned}
\Delta_t^T &= \Delta_t^T N N^{-1} = \Delta_t^T \left(C_0 C_0^+ + \delta I \right) N^{-1} \\
&= \left[\left(\widetilde{y}_t^T + x_t^T \triangle C^T + \xi_{2,t}^T \right) C_0^{+T} + \delta \Delta_t^T \right] N^{-1}
\end{aligned}
$$

and

$$
\begin{gathered}
2\Delta_t^T P \widetilde{W}_{1,t}\sigma(\widehat{x}_t) = 2\widetilde{y}_t^T C_0^{+T} N^{-1} P \widetilde{W}_{1,t}\sigma(\widehat{x}_t) \\
+2\left(\triangle C x_t + \xi_{2,t} \right)^T C_0^{+T} N^{-1} P \widetilde{W}_{1,t}\sigma(\widehat{x}_t) + 2\delta \Delta_t^T N^{-1} P \widetilde{W}_{1,t}\sigma(\widehat{x}).
\end{gathered}
\tag{4.70}
$$

Since Δ_t is not measurable, we can not use it in the updating law as in Chapter 2.

But $\sigma^T(\widehat{x}_t)$ is bounded, so similar to (4.68), the second term of (4.70) can be estimated as follows:

$$\begin{array}{c} 2\left(\triangle Cx_t + \xi_{2,t}\right)^T C_0^{+T} N^{-1} P\widetilde{W}_{1,t}\sigma(\widehat{x}_t) \leq \\ \left(\sigma^T(\widehat{x}_t)\widetilde{W}_{1,t}^T\right)\Gamma_1\left(\widetilde{W}_{1,t}\sigma(\widehat{x}_t)\right) + \xi_{2,t}^T\Lambda_{\xi_2}\xi_{2,t} + x_t^T C_{\triangle C} x_t \end{array} \tag{4.71}$$

where Γ_1 is defined as in (4.63). The same relationship for $\Delta_t^T P_t \widetilde{W}_{2,t}\phi(\widehat{x})\gamma(u_t)$ is valid:

$$\begin{array}{c} 2\Delta_t^T P\widetilde{W}_{2,t}\phi(\widehat{x})\gamma(u_t) \leq 2\widetilde{y}_t^T C_0^{+T} N^{-1} P\widetilde{W}_{2,t}\phi(\widehat{x})\gamma(u_t) \\ + \left\|\gamma(u)\right\|^2 \left(\phi(\widehat{x})\widetilde{W}_{2,t}^T\right)\Gamma_1\left(\widetilde{W}_{2,t}\phi(\widehat{x})\right) \\ + \xi_{2,t}^T\Lambda_{\xi_2}\xi_{2,t} + x_t^T C_{\triangle C} x_t \\ + \delta\left\|\gamma(u)\right\|\phi(\widehat{x})^T\widetilde{W}_{2,t}^T\Gamma_2\widetilde{W}_{2,t}\phi(\widehat{x}) + \delta\Delta_t^T\Delta_t \end{array} \tag{4.72}$$

Since

$$\begin{array}{c} x_1^T A x_2 = tr(x_1^T A x_2) = tr(A x_1 x_2^T) \\ x_1, x_2 \in \Re^n,\, A \in \Re^{n\times n} \end{array}$$

$$2\widetilde{y}_t^T C_0^{+T} N^{-1} P\widetilde{W}_{1,t}\sigma(\widehat{x}_t) = 2tr\left[\widetilde{W}_{1,t}^T P N^{-T}\sigma(\widehat{x}_t)\widetilde{y}_t^T C_0^{+T}\right]$$

$$\left(\sigma^T(\widehat{x}_t)\widetilde{W}_{1,t}^T\right)\Gamma_1\left(\widetilde{W}_{1,t}\sigma(\widehat{x}_t)\right) = tr\left[\widetilde{W}_{1,t}^T\Gamma_1\widetilde{W}_{1,t}\sigma(\widehat{x}_t)\sigma(\widehat{x}_t)^T\right]$$

Using the identity

$$2\Delta_t^T A_t\Delta_t = \Delta_t^T A_t\Delta_t + \Delta_t^T A_t^T\Delta_t$$

and adding and substraction the term $\Delta_t^T Q\Delta_t$ $\left(Q = Q^T > 0\right)$ into the right-hand side of (4.65), and taking into account (4.67a), (4.68), (4.69), (4.71) and (4.72), we obtain:

$$\begin{array}{c} \frac{dV_t}{dt} \leq \Delta_t^T L\Delta_t + tr\left[\widetilde{W}_{1,t}^T L_{w1}\right] + tr\left[\widetilde{W}_{2,t}^T L_{w2}\right] \\ + \left(C_{\triangle f} + \Upsilon_t + x_t^T\left[D_{\triangle_f} + 3C_{\triangle C}\right]x_t, -\Delta_t^T\left(Q - 2\delta I\right)\Delta_t\right) \\ + \Delta_t^T\left[PK_t\left(\Lambda_{\triangle C}^{-1} + \Lambda_{\xi_2}^{-1}\right)K_t^T P - PK_tC_0 - C_0^T K_t^T P\right]\Delta_t \end{array} \tag{4.73}$$

where L_t is defined in (4.58), L_{w1}, and L_{w1} are defined in (4.62),

$$\Upsilon_t = \xi_{1,t}^T\Lambda_{\xi_1}\xi_{1,t} + 3\xi_{2,t}^T\Lambda_{\xi_2}\xi_{2,t}$$

Defining $\Omega = \left(\Lambda_{\triangle C}^{-1} + \Lambda_{\xi_2}^{-1}\right)$ the last term in (4.73) can be represented as follows:

$$\begin{array}{c} G_t := PK_t\left(\Lambda_{\triangle C}^{-1} + \Lambda_{\xi_2}^{-1}\right)K_t^T P - PK_t C_0 - C_0^T K_t^T P \\ = \left(PK_t\Omega^{\frac{1}{2}} - C_0^T\Omega^{-\frac{1}{2}}\right)\left(PK_t\Omega^{\frac{1}{2}} - C_0^T\Omega^{-\frac{1}{2}}\right)^T - C_0^T\Omega^{-1}C_0 \\ \leq \left(PK_t\Omega^{\frac{1}{2}} - C_0^T\Omega^{-\frac{1}{2}}\right)\left(PK_t\Omega^{\frac{1}{2}} - C_0^T\Omega^{-\frac{1}{2}}\right)^T \end{array}$$

The differential equations (4.62) can be solved for initial conditions $\widetilde{W}_{1,0} = 0$ and $\widetilde{W}_{2,0} = 0$. As the result, we obtain

$$L_{w1,t} = 0,\ L_{w2,t} = 0$$

Using assumption **A4.6** we verify $L_t = 0$ that leads to

$$\frac{dV_t}{dt} \leq C_{\triangle f} + \Delta_t^T G_t \Delta_t + \Upsilon_t + D_t - \Delta_t^T\left(Q - 2\delta I\right)\Delta_t \tag{4.74}$$

Integrating (4.74), within the interval $t \in [0, T]$, and dividing both sides by T, we obtain:

$$\begin{array}{c} \frac{1}{T}\int\limits_0^T \Delta_t^T\left(Q - 2\delta I\right)\Delta_t dt \leq \frac{1}{T}\int\limits_0^T \Delta_t^T G_t\Delta_t dt + \\ \frac{1}{T}\int\limits_0^T \Upsilon_t dt + \frac{1}{T}\int\limits_0^T \left(C_{\triangle f} + D_t\right)dt - \frac{1}{T}\left(V_T - V_0\right) \leq \\ \frac{1}{T}\int\limits_0^T \Delta_t^T G_t\Delta_t + \frac{1}{T}\int\limits_0^T \Upsilon_t dt + \frac{1}{T}\int\limits_0^T \left(C_t + D_t\right)dt - \frac{1}{T}V_0 \end{array} \tag{4.75}$$

Taking limit for $T \to \infty$ and using upper limits for both sides of (4.75), we finally obtain (4.60). ■

Corollary 4.2 *If the neuro-observer gain matrix K_t is equal to*

$$K_t = P_t^{-1}C_0^T\Omega^{-1} \tag{4.76}$$

and

$$L_{w1,t} = 0, \quad L_{w2,t} = 0 \tag{4.77}$$

then the neuro-observer guarantees the properties

$$\varphi\left(\{K_t\}_{t\geq 0}\right) = 0$$
$$\psi\left(\{W_{1,t}\}_{t\geq 0}, \{W_{2,t}\}_{t\geq 0}\right) = 0$$

The corresponding upper bound in (4.57) is equal to

$$J^{+} = C_{\triangle f} + D + \Upsilon \tag{4.78}$$

Remark 4.4 *The current weight $W_{1,t}, W_{2,t}$ are updated as*

$$\begin{aligned}
\dot{W}_{1,t} &= -\left(\Gamma_{1,t} + \delta\Gamma_{2,t}\right)\left[W_{1,t} - W_1^*\right]\sigma(\widehat{x}_t)\sigma(\widehat{x}_t)^T - M_{1,t} \\
\dot{W}_{2,t} &= -\left\|\gamma(u)\right\|^2\left(\Gamma_{1,t} + \delta\Gamma_{2,t}\right)\left[W_{2,t} - W_2^*\right]\phi(\widehat{x}_t)\phi(\widehat{x}_t)^T - M_{2,t} \\
W_{1,0} &= W_1^*, \quad W_{2,0} = W_2^*
\end{aligned} \tag{4.79}$$

that also guarantees (4.77).

The equation (4.79) represents the new differential learning law for the dynamic neural network weights.

Remark 4.5 - *The term $C_{\triangle f}$ in (4.61) is related to the assumptions **A4.5** , and is equal to 0 if the unmodeled dynamics is obtained:*

$$\Delta f(0, u, t) = 0$$

*-The term D in (4.61) is also related to the assumptions **A4.5** and (4.61), and is equal to zero if we deal with an bounded unmodeled dynamics that corresponds to the case:*

$$D_{\Delta f} = C_{\Delta C} = 0$$

-If $D \neq 0$, to guarantee that right-hand side of (4.61) would be bounded, we have to assume additionally that the class $\mathcal{H}$ contains only nonlinear systems which are "stable in average", i.e.,

$$\sup_{\mathcal{H}} \lim_{T\to\infty} \frac{1}{T}\int_0^T \left\|x_t\right\|^2 dt < \infty$$

-In the case of no unmodeled dynamics we have

$$C_{\triangle f} = D = 0$$

-If we deal with a nonlinear system without any unmodeled dynamics, that is, neural network matches the plant exactly ($C_{\triangle f} = D = 0$) *and without any external disturbances* ($\Upsilon_1 = \Upsilon_2 = 0$), *then the neuro-observer (4.54) with the matrix gain given by (4.76) guarantees "the stability in average", i.e.,*

$$\sup_{\mathcal{H}} \lim_{T\to\infty} \frac{1}{T} \int_0^T \Delta_t^T (Q - 2\delta I) \Delta_t dt = 0$$

Since the integrand is a positive definite quadratic term

$$(Q - 2\delta I) = (Q - 2\delta I)^T > 0$$

we can conclude that

$$\lim_{t\to\infty} \Delta_t \to 0$$

4.3.3 Multilayer Neuro Observer with Time-Delay Term

In this subsection we will consider the Luneburger-like "second order" neuro observer with the new additional time-delay term [21]. It has the following structure:

$$\boxed{\begin{array}{c} \dot{\widehat{x}}_t = A\widehat{x}_t + W_{1,t}\sigma(V_{1,t}\widehat{x}_t) + W_{2,t}\phi(V_{2,t}\widehat{x}_t)u_t \\ + L_1 [y_t - \widehat{y}_t] + L_2/h [(y_t - y_{t-h}) - (\widehat{y}_t - \widehat{y}_{t-h})] \\ \widehat{y}_t = C\widehat{x}_t \end{array}} \tag{4.80}$$

The vector $\widehat{x}_t \in \Re^n$ is the state of the neural network, $u_t \in \Re^q$ is its input. The matrix $A \in \Re^{n\times n}$ is a stable fixed matrix which will be specified below. The matrices $W_{1,t} \in \Re^{n\times m}$ and $W_{2,t} \in \Re^{n\times q}$ are the weights of the output layer. $V_1 \in \Re^{m\times n}$ and $V_2 \in \Re^{q\times n}$ are the weights of the hidden layer. $\sigma(\cdot) \in \Re^m$ is the sigmoidal vector functions, $\phi(\cdot)$ is $\Re^{q\times q}$ a diagonal matrix, i.e.,

$$\phi(\cdot) - \text{diag}\left[\phi_1(V_{2,t}\widehat{x}_t)_1 \cdots \phi_q(V_{2,t}\widehat{x}_t)_q\right]$$

$L_1 \in \Re^{n\times m}$ and $L_2 \in \Re^{n\times m}$ are first and second order gain matrices to be selected. The scalar h is assumed to be positive.

Remark 4.6 *The most simple structure without any hidden layers (containing only input and output layers) corresponds to the case when*

$$m = n, \ V_1 = V_2 = I, \quad L_2 = 0 \tag{4.81}$$

This single-layer dynamic neural networks with Luneburger-like observer was considered in [10].

Remark 4.7 *The structure of the observer (4.80) has three parts:*

- *the neural networks identifier*

$$A\widehat{x}_t + W_{1,t}\sigma(V_{1,t}\widehat{x}_t) + W_{2,t}\phi(V_{2,t}\widehat{x}_t)u_t$$

- *the Luneburger tuning term* $L_1 \left[y_t - \widehat{y}_t \right]$*;*
- *the additional time-delay term*

$$L_2 h^{-1} \left[(y_t - y_{t-h}) - (\widehat{y}_t - \widehat{y}_{t-h}) \right]$$

where $(y_t - y_{t-h})/h$ *and* $(\widehat{y}_t - \widehat{y}_{t-h})/h$ *are introduced to estimate the derivatives* $\dot{y}_t$*and* $\dot{\widehat{y}}_t$*, correspondingly.*

To simplify all mathematical calculus the assumptions of **A4.5** is changed a little bit, since now it is assumed that $\Delta C = 0$.

The nonlinear system satisfies the following assumption.

A4.7: *For a realized bounded nonlinear feedback control* $(\|u_t(x_t)\| \leq \overline{u})$, *the nominal (unperturbed) closed-loop nonlinear system is* ***quadratically stable****, that is, there exists a Lyapunov (may be, unknown) function* $\overline{V}_t \geq 0$ *satisfying*

$$\frac{\partial \overline{V}_t}{\partial x} f(x_t, u_t) \leq -\lambda_1 \|x_t\|^2, \quad \left\| \frac{\partial \overline{V}_t}{\partial x} \right\| \leq \lambda_2 \|x_t\|, \quad \lambda_1, \lambda_2 > 0$$

Let us define the estimation error at time t as

$$\Delta_t := \widehat{x}_t - x_t \tag{4.82}$$

Then, the output error is

$$e_t = \widehat{y}_t - y_t = C\Delta_t - \xi_{2,t}$$

that implies

$$\begin{gathered} C^T e_t = C^T \left(C\Delta_t - \xi_{2,t}\right) = \\ \left(C^T C + \delta I\right)\Delta_t - \delta I \Delta_t - C^T \xi_{2,t} \\ \Delta_t = C_\delta^+ e_t + \delta N_\delta \Delta_t + C_\delta^+ \xi_{2,t} \end{gathered} \tag{4.83}$$

where

$$\begin{gathered} C_\delta^+ = \left(C^T C + \delta I\right)^{-1} C^T \\ N_\delta = \left(C^T C + \delta I\right)^{-1} \end{gathered}$$

and δ is a small positive scalar.

It is clear that all sigmoidal functions, commonly used in neural networks, satisfy the following conditions (see Chapter 2 and Appendix A).

$$\begin{gathered} \widetilde{\sigma}_t' := \sigma(V_{1,t}\widehat{x}_t) - \sigma(V_1^* \widehat{x}_t) = D_\sigma \widetilde{V}_{1,t}\widehat{x}_t + \nu_\sigma \\ \widetilde{\phi}_t' u_t := \phi(V_{2,t}\widehat{x}_t)u_t - \phi(V_2^* \widehat{x}_t)u_t \\ = \sum_{i=1}^{q} \left[\phi_i(V_2^* \widehat{x}_t) - \phi_i(V_{2,t}\widehat{x}_t)\right] u_{i,t} = \sum_{i=1}^{q} \left[D_{i\phi}\widetilde{V}_{2,t}\widehat{x}_t + \nu_{i\phi}\right] u_{i,t} \\ u_{i,t} \text{ is a scalar } (i\text{-th component of } u_t \\ D_\sigma = \frac{\partial \sigma(Z)}{\partial Z} |_{Z=V_{1,t}\widehat{x}_t} \in \Re^{m\times m}, \quad \|\nu_\sigma\|_{\Lambda_1}^2 \le l_1 \left\|\widetilde{V}_{1,t}\widehat{x}_t\right\|_{\Lambda_1}^2, \quad l_1 > 0 \\ D_{i\phi} = \frac{\partial \phi_i(Z)}{\partial Z} |_{Z=V_{2,t}\widehat{x}_t} \in \Re^{m\times m}, \quad \|\nu_\phi\|_{\Lambda_2}^2 \le l_2 \left\|\widetilde{V}_{2,t}\widehat{x}_t\right\|_{\Lambda_2}^2, \quad l_2 > 0 \\ \widetilde{V}_{1,t} = V_{1,t} - V_1^*, \quad \widetilde{V}_{2,t} = V_{2,t} - V_2^* \end{gathered} \tag{4.84}$$

where

$$\begin{gathered} \widetilde{\sigma}_t := \sigma(V_1^* x_t) - \sigma(V_1^* \widehat{x}_t), \quad \widetilde{\phi}_t := \phi(V_2^* x_t) - \phi(V_2^* \widehat{x}_t) \\ \widetilde{\sigma}_t' := \sigma(V_1^* \widehat{x}_t) - \sigma(V_{1,t}\widehat{x}_t), \quad \widetilde{\phi}_t' u_t := \phi(V_2^* \widehat{x}_t)u_t - \phi(V_{2,t}\widehat{x}_t)u_t \\ D_\sigma = \frac{\partial \sigma^T(Z)}{\partial Z} |_{Z=V_{1,t}\widehat{x}_t} \in R^{m\times m} \\ \|\nu_\sigma\|_{\Lambda_1}^2 \le l_1 \left\|\widetilde{V}_{1,t}\widehat{x}_t\right\|_{\Lambda_1}^2, \quad l_1 > 0 \\ D_\phi = \frac{\partial (\phi_i u_i)^T(Z)}{\partial Z} |_{Z=V_{2,t}\widehat{x}_t} \in R^{m\times m} \\ \|\nu_\phi\|_{\Lambda_2}^2 \le l_2 \left\|\widetilde{V}_{2,t}\widehat{x}_t\right\|_{\Lambda_2}^2, \quad l_2 > 0 \end{gathered}$$

Define also

$$\widetilde{W}_{1,t} := W_{1,t} - W_1^*, \quad \widetilde{W}_{2,t} := W_{2,t} - W_2^*$$

In general case, when the neural network

$$\dot{\widehat{x}}_t = A\widehat{x}_t + W_{1,t}\sigma(V_{1,t}\widehat{x}_t) + W_{2,t}\phi(V_{2,t}\widehat{x}_t)u_t$$

can not exactly match the given nonlinear system (4.52), the plant can be represented as

$$\dot{x}_t = Ax_t + W_1^*\sigma(V_1^*x_t) + W_2^*\phi(V_2^*x_t)u_t + \widetilde{f}_t \tag{4.85}$$

where $\widetilde{f}_t$ is unmodeled dynamic term and W_1^*, W_2^*, V_1^* and V_2^* are any known matrices which are selected below as initial for the designed differential learning law.

To guarantee the global existence of the solution for (4.52), the following condition should be satisfied

$$\left\| f\left(x_t, u_t, t\right)\right\|^2 \leq C_1 + C_2 \left\|x_t\right\|^2$$

C_1 and C_2 are positive constants [4]. In view of this fact and taking into account that the sigmoid functions σ and ϕ are uniformly bounded, the following assumption, concerning the unmodeled dynamics $\widetilde{f}_t$, seems also to be natural:

A4.9: *There exist positive constants* $\overline{\eta}$ and $\overline{\eta}_1$ *such that*

$$\left\|\widetilde{f}_t\right\|_{\Lambda_f}^2 \leq \overline{\eta} + \overline{\eta}_1 \left\|x_t\right\|_{\Lambda_f}^2, \quad \Lambda_f = \Lambda_f^T > 0$$

The next construction plays the key role in this study. It is well known [31] that if a matrix A is stable, the pair $(A, R^{1/2})$ is controllable, the pair $(Q^{1/2}, A)$ is observable, and the special *local frequency condition* or its "matrix equivalent"

$$A^T R^{-1} A - Q \geq \frac{1}{4}\left[A^T R^{-1} - R^{-1}A\right] R \left[A^T R^{-1} - R^{-1}A\right]^T \tag{4.86}$$

is fulfilled (see Appendix A), then the matrix Riccati equation

$$A^T P + PA + PRP + Q = 0 \tag{4.87}$$

has a positive solution. In view of this fact we will demand the following additional assumption.

A4.8: *There exist a stable matrix A and a positive parameter δ such that the matrix Riccati equation (4.87) with*

$$\begin{array}{c} R = 2\overline{W}_1 + 2\overline{W}_2 + \Lambda_f^{-1} + \Lambda_{\xi_1}^{-1} + \delta R_1 \\ Q = \Lambda_\sigma + \overline{u}^2 \Lambda_\phi, + P_1 + Q_1 - 2C^T \Lambda_\xi C \\ \overline{W}_1 := W_1^{*T} \Lambda_1^{-1} W_1^*, \quad \overline{W}_2 := W_2^{*T} \Lambda_2^{-1} W_2^* \end{array} \tag{4.88}$$

has a positive solution P. Here Q_1 is a positive defined matrix

$$\begin{array}{c} R_1 = 2N_\delta K_1^T \Lambda_\sigma^{-1} K_1 N_\delta^T + 2N_\delta K_1^T \Lambda_\phi^{-1} K_1 N_\delta^T + \\ N_\delta K_3^T \Lambda_\sigma^{-1} K_3 N_\delta^T + N_\delta K_4^T \Lambda_\phi^{-1} K_4^T N_\delta^T \end{array}$$

This conditions can be easily verified if we select A as a stable diagonal matrix. Denote by $\mathcal{H}$ the class of unknown nonlinear systems satisfying **A4.7-A4.9.**

Consider the new differential learning law given by the following system of matrix differential equations:

$$\boxed{\begin{array}{l} \dot{W}_{1,t} = -K_1 P C_\delta^+ e_t \sigma^T - (1+\delta) W_{\delta 1} + \\ \qquad K_1 P C_\delta^+ e_t \widehat{x}_t^T \widetilde{V}_{1,t}^T D_\sigma \\ \dot{W}_{2,t} = -K_2 P C_\delta^+ e_t (\phi u_t)^T - (1+\delta) W_{\delta 2} + \\ \qquad K_2 P C_\delta^+ e_t \widehat{x}_t^T (V_{2,t} - V_2^*)^T \sum\limits_{i=1}^{q} \left(u_{i,t} D_{i\phi}^T\right) \\ \dot{V}_{1,t} = -K_3 P W_{1,t} D_\sigma C_\delta^+ e_t \widehat{x}_t^T - (1+\delta) V_{\delta 1} - \\ \qquad \frac{l_1}{2} K_3 \Lambda_1 \widetilde{V}_{1,t} \widehat{x}_t \widehat{x}_t^T \\ \dot{V}_{2,t} = -K_4 P W_{2,t} D_\phi C_\delta^+ e_t \widehat{x}_t^T - (1+\delta) V_{\delta 2} - \\ \qquad \frac{q l_2 \bar{u}}{2} K_4 \Lambda_2 (V_{2,t} - V_2^*) \widehat{x}_t \widehat{x}_t^T \end{array}} \tag{4.89}$$

where

$$\begin{aligned} W_{\delta 1} &:= \sigma^T \Lambda_\sigma^{-1} \widetilde{W}_{1,t} \sigma + \widehat{x}_t^T \widetilde{V}_{1,t}^T D_\sigma \Lambda_\sigma^{-1} D_\sigma \widetilde{V}_{1,t} \widehat{x}_t \\ W_{\delta 2} &:= (\phi u_t)^T \Lambda_\phi^{-1} \widetilde{W}_{2,t} (\phi u_t) + \widehat{x}_t^T \widetilde{V}_{2,t}^T D_\phi \Lambda_\phi^{-1} D_\phi \widetilde{V}_{2,t} \widehat{x}_t \\ V_{\delta 1} &:= \widehat{x}_t^T \widetilde{W}_{1,t}^T D_\sigma \Lambda_\sigma^{-1} D_\sigma \widetilde{W}_{1,t} \widehat{x}_t \\ V_{\delta 2} &:= \widehat{x}_t^T \widetilde{W}_{2,t}^T D_\phi \Lambda_\phi^{-1} D_\phi \widetilde{W}_{2,t} \widehat{x}_t \end{aligned} \tag{4.90}$$

$K_i \in R^{n \times n}$ $(i = 1 \cdots 4)$ are positive defined matrices, P is the solution of the matrix Riccati equation given by (4.87). The initial weight are

$$W_{1,0} = W_1^*, W_{2,0} = W_2^*, V_{1,0} = V_1^*, V_{2,0} = V_2^*$$

Remark 4.8 *One can see that the learning law (4.89) of the neuro observer (4.80) consists of three parts:*

- the first term $(K_1 P) C_\delta^+ e_t \sigma^T$ exactly corresponds to the backpropagation scheme as in multilayer networks [19];

- the second term $K_1 P C_\delta^+ e_t \widehat{x}_t^T \widetilde{V}_{1,t}^T D_\sigma$ is intended to assure the robust stable learning;

- the last term is caused by pseudoinverse analogue of output error.

Even though the proposed learning law looks like the backpropagation algorithm, the global asymptotic error stability is guaranteed because of the fact that it is derived based on the Lyapunov approach (see the next Theorem). Hence, the global convergence problem (which is a major concern in static NN learning) does not arise in this case.

Theorem 4.3 *If the gain matrices are selected as*

$$L_1 = P^{-1} C^T \Lambda_{\xi_2}, \quad L_2 = h P^{-1} C^T \Lambda_\xi \tag{4.91}$$

and the weights are adjusted as in (4.89), for a given class $\mathcal{H}$ of nonlinear systems (4.52), the following properties hold:

(a) the weight matrices remain bounded during all learning period, that is,

$$W_{1,t} \in L^\infty, \quad W_{2,t} \in L^\infty, \quad V_{1,t} \in L^\infty, \quad V_{2,t} \in L^\infty \tag{4.92}$$

(b) the identification error Δ_t satisfies the following tracking performance

$$\limsup_{T\to\infty} T^{-1}\int_0^T \Delta_t^T Q_1 \Delta_t dt \le \overline{d} \tag{4.93}$$

where

$$\mu_1 := \overline{d}/\lambda_{\min}\left(P^{-1/}Q_1 P^{-1/2}\right), \quad \overline{d} := \overline{\eta} + \Upsilon_1 + 10\Upsilon_2 \tag{4.94}$$

Proof. Taking into account (4.15), (4.85) and (4.13), we obtain:

$$\begin{array}{c} \dot{\Delta}_t = A\Delta_t + \widetilde{W}_{1,t}\sigma + \widetilde{W}_{2,t}\phi u_t \\ +W_1^*\widetilde{\sigma}_t + W_2^*\widetilde{\phi}_t u_t + W_1^*\widetilde{\sigma}_t' + W_2^*\widetilde{\phi}_t' u_t + \widetilde{f}_t + \xi_{1,t} \\ -L_1\left[y_t - \widehat{y}_t\right] - L_2/h\left[(y_t - y_{t-h}) - (\widehat{y}_t - \widehat{y}_{t-h})\right] \end{array} \tag{4.95}$$

Let us define the Lyapunov-Krasovskii function as

$$\begin{array}{c} V_t := \overline{V}_t + \Delta^T P\Delta + tr\left[\widetilde{W}_{1,t}^T K_1^{-1}\widetilde{W}_{1,t}\right] + tr\left[\widetilde{W}_{2,t}^T K_2^{-1}\widetilde{W}_{2,t}\right] \\ +tr\left[\widetilde{V}_{1,t}K_3^{-1}\widetilde{V}_{1,t}^T\right] + tr\left[\widetilde{V}_{2,t}K_4^{-1}\widetilde{V}_{2,t}^T\right] + \int\limits_{t-h}^{t} s_t\Delta_\tau^T P_1\Delta_\tau d\tau \end{array} \tag{4.96}$$

where P and P_1 are positive defined matrices. Calculating its derivative and using **A4.7,** we obtain

$$\begin{array}{c} \dot{V}_t \le -\left(\lambda_1 - \lambda_2\left\|\Lambda_{\xi_2}\right\|\right)\left\|x_t\right\|^2 \\ +2\Delta^T P\,\dot{\Delta} \\ +2tr\left[\dot{\widetilde{W}}_{1,t}^T K_1^{-1}\widetilde{W}_{1,t}\right] + 2tr\left[\dot{\widetilde{W}}_{2,t}^T K_2^{-1}\widetilde{W}_{2,t}\right] \\ +2tr\left[\dot{\widetilde{V}}_{1,t} K_3^{-1}\widetilde{V}_{1,t}^T\right] + 2tr\left[\dot{\widetilde{V}}_{2,t} K_4^{-1}\widetilde{V}_{2,t}^T\right] \\ +s_t\Delta_t^T P_1\Delta_t - s_t\Delta_{t-h}^T P_1\Delta_{t-h} + \Upsilon_2 \end{array} \tag{4.97}$$

Substituting (4.95) into (4.96), we derive

$$\begin{array}{c} 2\Delta_t^T P\,\dot{\Delta}_t = 2\Delta_t^T PA\Delta_t + 2\Delta_t^T P\left(W_1^*\widetilde{\sigma}_t + W_2^*\widetilde{\phi}_t u_t\right) \\ +2\Delta_t^T P\left(\widetilde{W}_{1,t}\sigma + \widetilde{W}_{2,t}\phi u_t\right) + 2\Delta_t^T P\left(W_1^*\widetilde{\sigma}_t' + W_2^*\widetilde{\phi}_t' u_t\right) \\ +2\Delta_t^T P\widetilde{f}_t + 2\Delta_t^T P\xi_{1,t} \\ -2\Delta_t^T P\left\{L_1\left[y_t - \widehat{y}_t\right] + L_2/h\left[(y_t - y_{t-h}) - (\widehat{y}_t - \widehat{y}_{t-h})\right]\right\} \end{array} \tag{4.98}$$

Using (4.37) and **A4.3,** the terms in (4.98) can be estimated as follows

$$\begin{aligned} 2\Delta_t^T P W_1^* \widetilde{\sigma}_t &\le \Delta_t^T P W_1^* \Lambda_1^{-1} W_1^{*T} P \Delta_t + \widetilde{\sigma}_t^T \Lambda_1 \widetilde{\sigma}_t \\ &\le \Delta_t^T \left(P \overline{W}_1 P + \Lambda_\sigma \right) \Delta_t \\ 2\Delta_t^T P W_2^* \widetilde{\phi}_t u_t &\le \Delta_t^T \left(P \overline{W}_2 P + \overline{u}^2 \Lambda_\phi \right) \Delta_t \end{aligned} \tag{4.99}$$

In view of (4.84), by the analogous way, we obtain

$$\begin{gathered} 2\Delta_t^T P W_1^* \widetilde{\sigma}_t' = 2\Delta_t^T P W_{1,t} D_\sigma \widetilde{V}_{1,t} \widehat{x}_t + 2\Delta_t^T P \widetilde{W}_{1,t} D_\sigma \widetilde{V}_{1,t} \widehat{x}_t + 2\Delta_t^T P W_1^* \nu_\sigma \\ 2\Delta_t^T P W_2^* \widetilde{\phi}_t' u_t = 2\Delta_t^T P W_{2,t} \sum_{i=1}^{q} \left[D_{i\phi} \widetilde{V}_{2,t} \widehat{x}_t \right] u_{i,t} \\ -2\Delta_t^T P \widetilde{W}_{2,t} \sum_{i=1}^{q} \left[D_{i\phi} \widetilde{V}_{2,t} \widehat{x}_t \right] u_{i,t} + 2\Delta_t^T P W_2^* \sum_{i=1}^{q} \nu_{i\phi} u_{i,t} \end{gathered} \tag{4.100}$$

The term $2\Delta_t^T P W_2^* \sum_{i=1}^{q} \nu_{i\phi} u_{i,t}$ in (4.100) may be also estimated as

$$\begin{aligned} 2\Delta_t^T P W_2^* \sum_{i=1}^{q} \nu_{i\phi} u_{i,t} &\le \Delta_t^T P W_2^{*T} \Lambda_2^{-1} W_2^* P \Delta_t + q \sum_{i=1}^{q} u_{i,t}^T \nu_{i\phi}^T \Lambda_2 \nu_{i\phi} u_{i,t} \\ &\le \Delta_t^T P \overline{W}_2 P \Delta_t + q l_2 \bar{u} \left\| \widetilde{V}_{2,t} \widehat{x}_t \right\|_{\Lambda_2}^2 \end{aligned}$$

as well as $2\Delta_t^T P W_1^* \nu_\sigma$ in (4.100):

$$2\Delta_t^T P W_1^* \nu_\sigma \le \Delta_t^T P \overline{W}_1 P \Delta_t + l l_1 \left\| \widetilde{V}_{1,t} \widehat{x}_t \right\|_{\Lambda_1}^2 \tag{4.101}$$

The term $2\Delta_t^T P \widetilde{f}_t$ can be estimated as

$$\begin{aligned} 2\Delta_t^T P \widetilde{f}_t &\le \Delta_t^T P \Lambda_f^{-1} P \Delta_t + \widetilde{f}_t^T \Lambda_f \widetilde{f}_t \\ &\le \Delta_t^T P \Lambda_f^{-1} P \Delta_t + \overline{\eta} + \overline{\eta}_1 \left\| x_t \right\|_{\Lambda_f}^2 \end{aligned} \tag{4.102}$$

For the term $2\Delta_t^T P \xi_{1,t}$ in (4.98) we have

$$2\Delta_t^T P \xi_{1,t} \le \Delta_t^T P \Lambda_{\xi_1}^{-1} P \Delta_t + \xi_{1,t}^T \Lambda_{\xi_1} \xi_{1,t} \le \Delta_t^T P \Lambda_{\xi_1}^{-1} P \Delta_t + \Upsilon_1 \tag{4.103}$$

The last term in (4.98) is equal

$$\begin{gathered} -\Delta_t^T P \left\{ L_1 \left[y_t - \widehat{y}_t \right] + L_2/h \left[(y_t - \widehat{y}_t) - (y_{t-h} - \widehat{y}_{t-h}) \right] \right\} \\ = \Delta_t^T P L_1 C \Delta_t - \Delta_t^T P L_1 \xi_{2,t} \\ + \Delta_t^T P L_2/h C \Delta_t - \Delta_t^T P L_2/h \xi_{2,t} \\ - \Delta_t^T P L_2/h C \Delta_{t-h} + \Delta_t^T P L_2/h \xi_{2,t-h} \end{gathered} \tag{4.104}$$

Similar with (4.103), $\Delta_t^T P\left(L_1 + L_2/h\right)\xi_{2,t}$, $\Delta_{t-h}^T P L_2/h\xi_{2,t-h}$ and $\Delta_t^T P L_2/hC\Delta_{t-h}$ in (4.104) can be estimated as

$$\begin{array}{c} -2\Delta_t^T P L_1 \xi_{2,t} \leq \Delta_t^T P L_1 \Lambda_{\xi_2}^{-1} L_1^T P \Delta_t + \Upsilon_2 \\ -2\Delta_t^T P L_2/h\xi_{2,t} \leq 1/h^2 \Delta_t^T P L_2 \Lambda_{\xi_3}^{-1} L_2^T P \Delta_t + \Upsilon_2 \\ 2\Delta_t^T P L_2/h\xi_{2,t-h} \leq 1/h^2 \Delta_t^T P L_2 \Lambda_{\xi_4}^{-1} L_2^T P \Delta_t + \Upsilon_2 \\ 2\Delta_t^T P L_2/hC\Delta_{t-h} \leq 1/h^2 \Delta_t^T P L_2 \Lambda_{\xi_5}^{-1} L_2^T P \Delta_t + \Delta_{t-h}^T \Lambda_{\xi_5} \Delta_{t-h} \end{array} \tag{4.105}$$

So, finally we have

$$\begin{array}{c} \dot{V}_t \leq \Delta_t^T L \Delta_t + L_{w1} + L_{w2} + L_{v1} + L_{v2} \\ -\Delta_t^T Q \Delta_t + \Omega - \left(\lambda_1 - \lambda_2 \left\|\Lambda_{\xi_2}\right\|\right) \left\|x_t\right\|^2 \\ +\overline{\eta}_1 \left\|x_t\right\|_{\Lambda_f}^2 + \Delta_{t-h}^T \Lambda_{\xi_5} \Delta_{t-h} - \Delta_{t-h}^T P_1 \Delta_{t-h} \\ +\Delta_t^T \left(P L_1 \Lambda_{\xi_2}^{-1} L_1^T P - 2 P L_1 C\right) \Delta_t \\ +\Delta_t^T \left(1/h^2 P L_2 \Lambda_{\xi}^{-1} L_2^T P - 2/h P L_2 C\right) \Delta_t \end{array} \tag{4.106}$$

where

$$\Lambda_{\xi}^{-1} := \Lambda_{\xi_3}^{-1} + \Lambda_{\xi_4}^{-1} + \Lambda_{\xi_5}^{-1}$$

$$\begin{aligned} L &= PA + A^T P + PRP + Q \\ L_{w1} &= 2tr\left[\dot{\widetilde{W}}_{1,t}^T K_1^{-1} \widetilde{W}_{1,t}\right] + 2\Delta_t^T P \widetilde{W}_{1,t}\sigma + 2\Delta_t^T P \widetilde{W}_{1,t} D_\sigma \widetilde{V}_{1,t} \widehat{x}_t \\ L_{w2} &= 2tr\left[\dot{\widetilde{W}}_{2,t}^T K_2^{-1} \widetilde{W}_{2,t}\right] + 2\Delta_t^T P \widetilde{W}_{2,t}\phi u - 2tr\left[\sum_{i=1}^{q} \left(D_{i\phi} u_{i,t}\right) \widetilde{V}_{2,t} \widehat{x}_t \Delta_t^T P \tilde{W}_{2,t}\right] \\ L_{v1} &= 2tr\left[\dot{\widetilde{V}}_{1,t} K_3^{-1} \widetilde{V}_{1,t}^T\right] + 2\Delta_t^T P W_{1,t} D_\sigma \widetilde{V}_{1,t} \widehat{x}_t + l_1 \left\|\widetilde{V}_{1,t} \widehat{x}_t\right\|_{\Lambda_1}^2 \\ L_{v2} &= 2tr\left[\dot{\widetilde{V}}_{2,t} K_4^{-1} \widetilde{V}_{2,t}^T\right] + 2tr\left[\widehat{x}_t \Delta_t^T P W_{2,t} \sum_{i=1}^{q} \left(D_{i\phi} u_{i,t}\right) \widetilde{V}_{2,t}\right] + q l_2 \bar{u} \left\|\widetilde{V}_{2,t} \widehat{x}_t\right\|_{\Lambda_2}^2 \end{aligned}$$

R and Q are defined as in (4.88). Since we do not know Δ_t and only e_t is available, using (4.37), the term $2\widetilde{W}_{1,t}^T K_1 P \Delta_t \sigma^T$ in L_{w1} can be rewritten as

$$\begin{array}{c} 2\widetilde{W}_{1,t}^T K_1 P \Delta_t \sigma^T = 2\widetilde{W}_{1,t}^T K_1 P C_\delta^+ e_t \sigma^T + \\ 2\delta \widetilde{W}_{1,t}^T K_1 P N \Delta_t \sigma^T + 2\widetilde{W}_{1,t}^T K_1 P C_\delta^+ \xi_{2,t} \sigma^T \end{array}$$

Using again the matrix inequality (4.37), we conclude that

$$2\delta \widetilde{W}_{1,t}^T K_1 P N \Delta_t \sigma^T \le \delta \Delta_t^T P N K_1^T \Lambda_\sigma^{-1} K_1 N^T P \Delta_t$$
$$+\delta \sigma^T \widetilde{W}_{1,t}^T \Lambda_\sigma^{-1} \widetilde{W}_{1,t} \sigma$$
$$2\widetilde{W}_{1,t}^T K_1 P C_\delta^+ \xi_{2,t} \sigma^T \le \xi_{2,t}^T C_\delta^{+T} P K_1^T \Lambda_\sigma^{-1} K_1 P C_\delta^+ \xi_{2,t}$$
$$+\sigma^T \widetilde{W}_{1,t}^T \Lambda_\sigma^{-1} \widetilde{W}_{1,t} \sigma$$

If we select Λ_σ^{-1} to verify

$$C_\delta^{+T} P K_1^T \Lambda_\sigma^{-1} K_1 P C_\delta^+ = \Lambda_{\xi_2} \tag{4.107}$$

we may add the terms $2K_1 P C_\delta^+ e_t \sigma^T \widetilde{W}_{1,t}$, $\sigma^T \widetilde{W}_{1,t}^T \Lambda_\sigma^{-1} \widetilde{W}_{1,t} \sigma$ and $\delta \sigma^T \widetilde{W}_{1,t}^T \Lambda_\sigma^{-1} \widetilde{W}_{1,t} \sigma$ to L_{w1} and the term $\delta \Delta_t^T P N_\delta K_1^T \Lambda_\sigma^{-1} K_1 N^T P \Delta_t$ to the Riccati equation (combining with R), and the term $\xi_{2,t}^T C_\delta^{+T} P K_1^T \Lambda_\sigma^{-1} K_1 P C_\delta^+ \xi_{2,t}$ we can combine with Ω. So, now it is possible to obtain $L_{w1} = 0$ using only the measurable error vector e_t. The similar results can be obtained for the term $2\Delta_t^T P \widetilde{W}_{1,t} D_\sigma \widetilde{V}_{1,t} \widehat{x}_t$ as well as for the rest terms in L_{w2}, L_{v1} and L_{v2}. As for the last terms in (4.106), we have

$$\Delta_t^T \left(P L_1 \Lambda_{\xi_2}^{-1} L_1^T P - 2 P L_1 C \right) \Delta_t =$$
$$\Delta_t^T \left(P L_1 \Lambda_{\xi_2}^{-1} L_1^T P - P L_1 C - C^T L_1^T P \right) \Delta_t =$$
$$\Delta_t^T \left[\left(P L_1 \Lambda_{\xi_2}^{-1/2} - C^T \Lambda_{\xi_2}^{1/2} \right) \left(P L_1 \Lambda_{\xi_2}^{-1/2} - C^T \Lambda_{\xi_2}^{1/2} \right)^T - C^T \Lambda_{\xi_2} C \right] \Delta_t$$

$$\Delta_t^T \left(1/h^2 P L_2 \Lambda_\xi^{-1} L_2^T P - 2/h P L_2 C \right) \Delta_t =$$
$$\Delta_t^T \left(1/h^2 P L_2 \Lambda_\xi^{-1} L_2^T P - 1/h P L_2 C - 1/h C^T L_2^T P \right) \Delta_t =$$
$$\Delta_t^T \left(1/h P L_2 \Lambda_\xi^{-1/2} - C^T \Lambda_\xi^{1/2} \right) \left(1/h P L_2 \Lambda_\xi^{-1/2} - C^T \Lambda_\xi^{1/2} \right)^T \Delta_t$$
$$-\Delta_t^T C^T \Lambda_\xi C \Delta_t$$

Selecting the gain matrices as in (4.91), we may conclude that the last term is not positive. Since

$$\overline{\eta}_1 \|x_t\|_{\Lambda_f}^2 \le \overline{\eta}_1 \|\Lambda_f\| \|x_t\|^2$$

for Λ_f satisfying

$$\|\Lambda_f\| \le \frac{1}{\overline{\eta}_1} \left(\lambda_1 - \lambda_2 \|\Lambda_{\xi_2}\| \right) \tag{4.108}$$

we conclude that

$$-\left(\lambda_1 - \lambda_2 \left\|\Lambda_{\xi_2}\right\|\right) \|x_t\|^2 + \overline{\eta}_1 \|x_t\|^2_{\Lambda_f} \leq 0$$

Taking

$$\Lambda_{\xi_5} = P_1 \tag{4.109}$$

we have that

$$\left(\Delta_{t-h}^T \Lambda_{\xi_5} \Delta_{t-h} - \Delta_{t-h}^T P_1 \Delta_{t-h}\right) = 0$$

In view of **A4.8,** we get

$$\begin{aligned} \dot{V}_t \leq\ & L_{w1} + L_{w1} + L_{v1} + L_{v2} \\ & -\Delta_t^T Q_1 \Delta_t + (\overline{\eta} + \Upsilon_1 + 10\Upsilon_2) \end{aligned}$$

Using the updating law as in (4.89), it follows

$$L_{w1} = 0,\ L_{w2} = 0,\ L_{v1} = 0,\ L_{v2} = 0$$

Finally,

$$\dot{V}_t \leq -\left[\Delta_t^T Q_1 \Delta_t - \overline{d}\right]$$

where $\overline{d}$ is defined as in (4.94). The point a) follows directly from the last inequality.

Integrating (??) from 0 up to T yields

$$V_T - V_0 \leq -\int_0^T \Delta^T Q_1 \Delta s_t dt + \overline{d}$$

and hence,

$$\int_0^T \Delta^T Q_1 \Delta s_t dt \leq V_0 - V_T + \overline{d}T \leq V_0 + \overline{d}T$$

Since $W_{1,0} = W_1^*$ and $W_{2,0} = W_2^*$, V_0^0 and V_0 are bounded, (4.93) is obtained and b) is proved. ■

Remark 4.9 *The proved boundness property is global and the initial observer error availability is not required.*

Remark 4.10 *If we deal with a system without any unmodeled dynamics, i.e., neural network matches the given plant exactly* $\left(\overline{d}=0\right)$ *and without any external disturbances* $(\Upsilon_1 = \Upsilon_2 = 0)$, *the suggested neuro-observer (4.13), with the matrix gain given by (4.91) guarantees "the stability in average", that is,*

$$\limsup_{T\to\infty} \frac{1}{T} \int_0^T \Delta_t^T Q_1 \Delta_t dt = 0$$

that implies

$$\limsup_{t\to\infty} \Delta_t = 0$$

Remark 4.11 *One can see that the error upper bound (4.93) is suitable for any positive value of time delay* h *and the gain matrices* L_1 *and* $L_2 h^{-1}$ *(4.91) are independent on* h.

Remark 4.12 *Similar to the high-gain observers [21], the proved theorem stays only the fact that the estimation error is bounded asymptotically and does not say anything about a bound for a finite time that obligatory demands fulfilling a local uniform observability condition [7]. In our case, some observability properties are contained in* ***A5*** *(for example, if* $C = 0$ *this condition can not be fulfilled for any matrix* A*).*

The corresponding structure of this neuro-observer is shown in Fig. 4.1.

To check the estimation quality for the finite time intervals, we will present below two numerical examples with the corresponding simulation results.

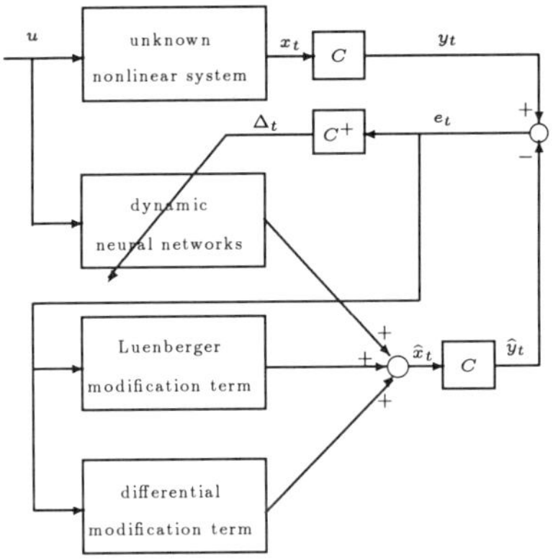

FIGURE 4.1. The general structure of the neuro-observer

4.4 Application

Example 4.1 *Consider the single-link robot rotating in a vertical plane. Its dynamics can be described as follows:*

$$\begin{bmatrix} \dot{x}_1 \\ \dot{x}_2 \end{bmatrix} = \begin{bmatrix} x_2 \\ -\sin(x_1) + u \end{bmatrix} + \Delta f + \begin{bmatrix} w_1 \\ w_2 \end{bmatrix} \qquad (4.110)$$
$$y_t = \cos(x_1) + x_2 + \Delta h + w_3$$

where the unmodeled dynamics are given by the terms

$$\Delta f(\cdot)^T := -0.05\left[x_1 \cos(x_1); x_2 \sin(x_2)\right]$$

and

$$\Delta h(\cdot) := -0.02 x_2 \sin(x_1)$$

satisfying "the strip bound condition" ***A4.1****. The signals* w_1, w_2 *are the external state perturbations modelling square wave and saw-tooth functions,* w_3 *is the output perturbation of a "white noise" nature. The control* u *is selected to be equal to zero, so the "the free rotation" is considered.*

This example is similar to one in [29], but we consider a more general case: the output equation is also nonlinear and there are unbounded unmodeled dynamics and external disturbances. We can easily check that assumptions ***A4.1-A4.3*** *are fulfilled.*

1) ***The Robust Asymptotic Observer.*** *According to (4.13) construct the following the robust observer:*

$$\begin{bmatrix} \dot{\widehat{x}}_1 \\ \dot{\widehat{x}}_2 \end{bmatrix} = \begin{bmatrix} \widehat{x}_2 \\ -\sin(\widehat{x}_1) + u \end{bmatrix} + K_t\left(y_t - \cos(\widehat{x}_1) + \widehat{x}_2\right) \tag{4.111}$$

where the gain matrix K_t is computed according to (4.51). Select

$$A_0 = \begin{bmatrix} -2 & 0 \\ 0 & -2 \end{bmatrix}, \; Q = \begin{bmatrix} 0.2 & 0 \\ 0 & 0.2 \end{bmatrix}$$

$$\Lambda = \Pi = I$$

$$Q_0 = \begin{bmatrix} 1.2 & 0 \\ 0 & 1.2 \end{bmatrix}, \; R_0 = \begin{bmatrix} 2.5 & 0 \\ 0 & 2.5 \end{bmatrix}$$

that leads to the solution P_t of the corresponding differential Riccati Equation (4.18) with the initial condition

$$P_0 = \begin{bmatrix} 0.3 & -0.01 \\ -0.01 & 0.3 \end{bmatrix}$$

which corresponds to the solution of the algebraic matrix Riccati equation when the left-hand side of this equation is equal to zero.

Using the fifth-order Runge-Kutta integration scheme under the initial condition equal to

$$x_1(0) = 2, x_2(0) = 1, \widehat{x}_1(0) = 5, \widehat{x}_2(0) = 5$$

the trajectories of (4.111) and (4.110) shown in the Figure 4.2 and the Figure 4.3 are obtained. The time evolution for the corresponding elements of P_t is shown in the Figure 4.4The time evolution of the two performance indices

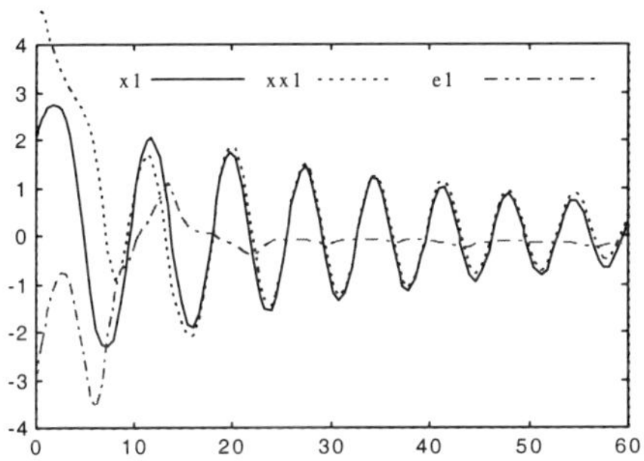

FIGURE 4.2. Robust nnlinear observing results for x_1.

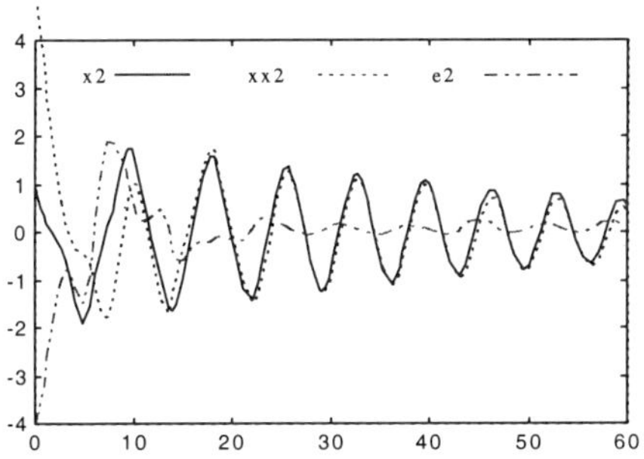

FIGURE 4.3. Robust nonlinear observing result for x_2.

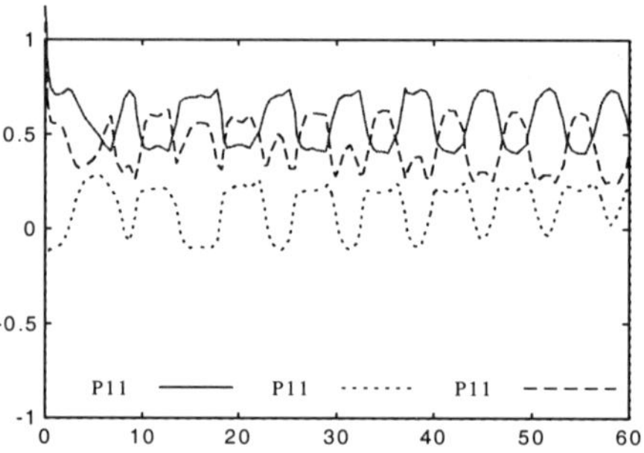

FIGURE 4.4. Time evolution of P_t.

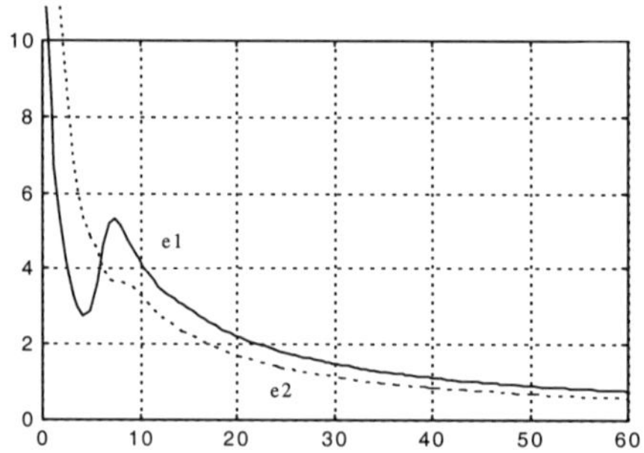

FIGURE 4.5. Performance indexes.

$$J_T^{\Delta_1} := \frac{1}{T}\int_0^T \Delta_{1,t}^2 dt, \qquad J_T^{\Delta_2} := \frac{1}{T}\int_0^T \Delta_{2,t}^2 dt$$

connected with the main performance index (4.16) is illustrated in the Figure 4.5.

2) ***The Robust Neuro Observer with the Additional Time-delay Term.*** *We consider the same single-link robot as before, but assume that now the nonlinear plant is completely unknown. Select the neuro observer as in (6.36), that is,*

$$\begin{gathered}\dot{\widehat{x}}_t = A\widehat{x}_t + W_{1,t}\sigma(V_{1,t}\widehat{x}_t) + \\ L_1\left[y_t - \widehat{y}_t\right] + L_2/h\left[(y_t - y_{t-h}) - (\widehat{y}_t - \widehat{y}_{t-h})\right], \\ \widehat{y}_t = [1,0]\,\widehat{x}_t\end{gathered}$$

The sigmoid functions are

$$\sigma_i(x_i) = \frac{2}{1+e^{-2x_i}} - 0.5$$

and

$$A = \begin{bmatrix} -15 & 0 \\ 0 & -10 \end{bmatrix}$$

$$\widehat{x}_0 = [-5,-5]^T, \gamma(u_t) = 0$$

$$W_{1,t} \in R^{2\times 3}, V_{1,t} \in R^{3\times 2}, W_{1,0} = V_{10}^T = \begin{bmatrix} 1 & 1 & 2 \\ 1 & 2 & 1 \end{bmatrix}$$

For

$$Q_1 = \Lambda_\phi = \Lambda\sigma = P_1 = I$$

and

$$\overline{W}_1 = \overline{W}_1 = \begin{bmatrix} 3 & 1 \\ 1 & 3 \end{bmatrix}$$

the solution of the algebraic Riccati equation (4.87) is

$$P = \begin{bmatrix} 0.28 & 0.09 \\ 0.09 & 0.11 \end{bmatrix}$$

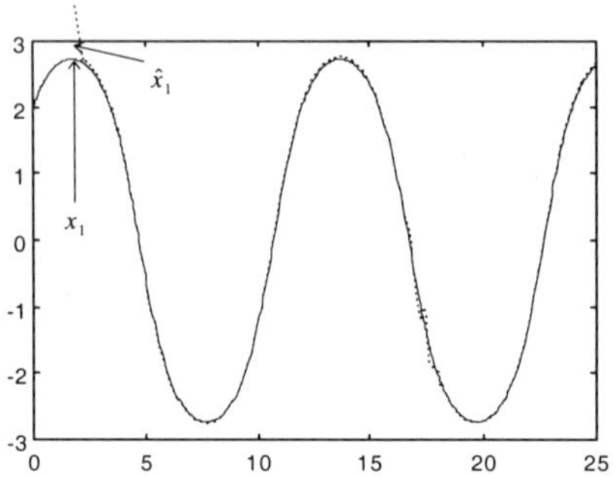

FIGURE 4.6. Estimates for x_1.

Here the following parameters have been used

$$\Lambda_{\xi_2} = 0.3, \Lambda_{\xi} = 1$$

$$h = 0.1, L_1 = \begin{bmatrix} 1.45 \\ -1.2 \end{bmatrix}, L_2 = \begin{bmatrix} 0.48 \\ -0.4 \end{bmatrix}, \delta = 0.01$$

The weights are updated according to (4.89) with $\overline{\eta} = 2$ *and*

$$K_1 = K_2 = K_3 = K_4 = 3I$$

The use of the fifth-order Runge-Kutta integration scheme under the initial condition equal to

$$x_1(0) = 2, x_2(0) = 1, \widehat{x}_1(0) = 5, \widehat{x}_2(0) = 5$$

leads to the observation results which are shown in Figure 4.6 and Figure 4.7.

Example 4.2 *Let consider the Van der Pol oscillator [14] subject to external perturbations which dynamics is given by*

$$\begin{cases} \dot{x}_1 = x_2 \\ \dot{x}_2 = 1.5\left(1 - x_1^2\right) x_2 - x_1 + \xi_{1,t} \end{cases}$$
$$y = x_1 + \xi_{2,t}.$$

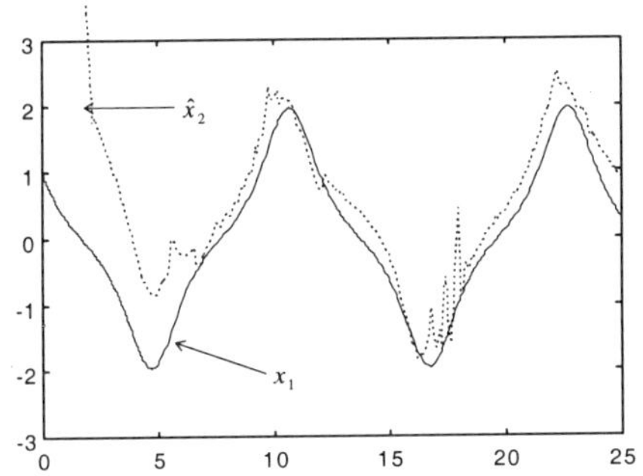

FIGURE 4.7. Estimates for x_2.

The same neuro observer is selected as

$$\dot{\widehat{x}}_t = A\widehat{x}_t + W_{1,t}\sigma(V_{1,t}\widehat{x}_t) + L_1\left[y_t - \widehat{y}_t\right] + L_2/h\left[(y_t - y_{t-h}) - (\widehat{y}_t - \widehat{y}_{t-h})\right],$$
$$\widehat{y}_t = [1,0]\,\widehat{x}_t$$

and updating law are chosen as in the Theorem 4.2. Its parameters are as follows,

$$\sigma_i(x_i) = \frac{2}{1+e^{-2x_i}} - 0.5, \quad \widehat{x}_0 = [-5,-5]^T$$
$$\overline{W}_1 = \overline{W}_1 = \begin{bmatrix} 3 & 1 \\ 1 & 3 \end{bmatrix}, \quad A = \begin{bmatrix} -15 & 0 \\ 0 & -10 \end{bmatrix}$$
$$Q_1 = \Lambda_\phi = \Lambda\sigma = P_1 = I, \quad \Lambda_{\xi_2} = 0.3, \Lambda_\xi = 1$$
$$h = 0.1, L_1 = \begin{bmatrix} 1.45 \\ -1.2 \end{bmatrix}, L_2 = \begin{bmatrix} 0.48 \\ -0.4 \end{bmatrix}, \delta = 0.01$$
$$W_{1,t} \in R^{2\times 3}, V_{1,t} \in R^{3\times 2}, W_{1,0} = V_{10}^T = \begin{bmatrix} 1 & 1 & 2 \\ 1 & 2 & 1 \end{bmatrix}$$

the solution of the algebraic Riccati equation (6.43) is

$$P = \begin{bmatrix} 0.28 & 0.09 \\ 0.09 & 0.11 \end{bmatrix}$$

The weights are updated according to (4.89) with $\overline{\eta} = 2$ *and*

$$K_1 = K_2 = K_3 = K_4 = 3I$$

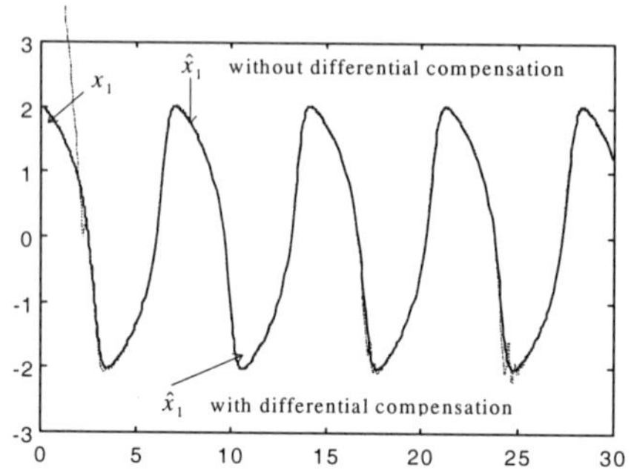

FIGURE 4.8. x_1 behaviour of neuro-observer (smooth noise).

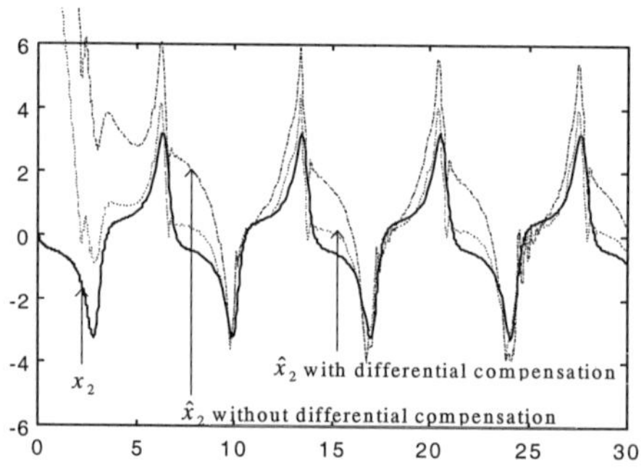

FIGURE 4.9. x_2 behaviour of neuro-observer (smooth noise).

Under the smooth noises selected as follows:

$$\xi_{1,t} = 0.1 \sin t, \; \xi_{2,t} = 0.1 \cos 5t$$

the behavior of this neuro-observer is shown in the Figure 4.8 and the Figure 4.9. For small white noise (variance is 0.005*), the corresponding trajectories are shown in the Figure 4.10 and the Figure 4.11. The results obtained above for the suggested neuro observer are compared with ones corresponding to the high-gain observer [21]*

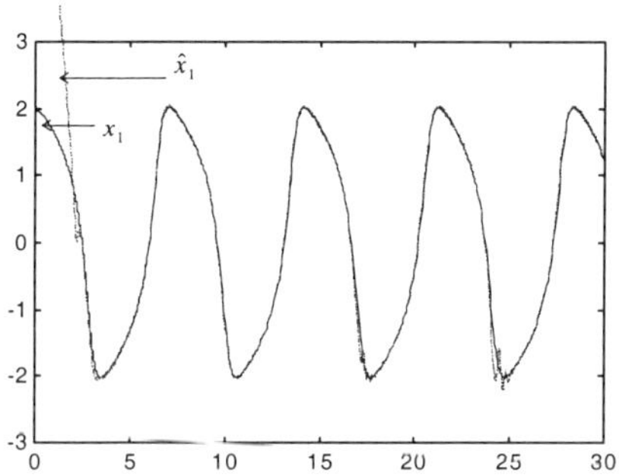

FIGURE 4.10. x_1 behaviour of neuro-observer (white noise).

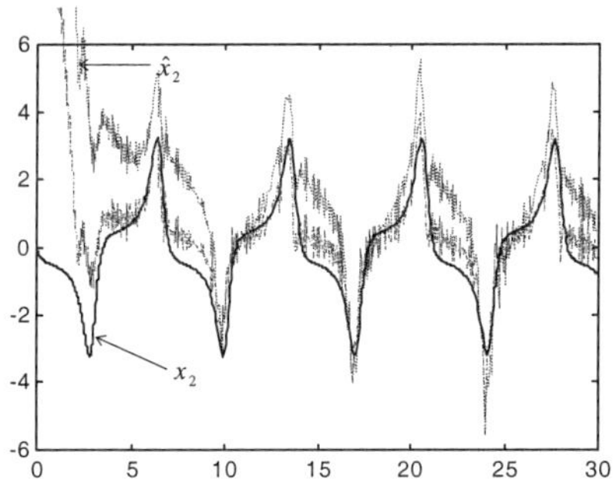

FIGURE 4.11. x_2 behaviour of neuro-observer (white noise).

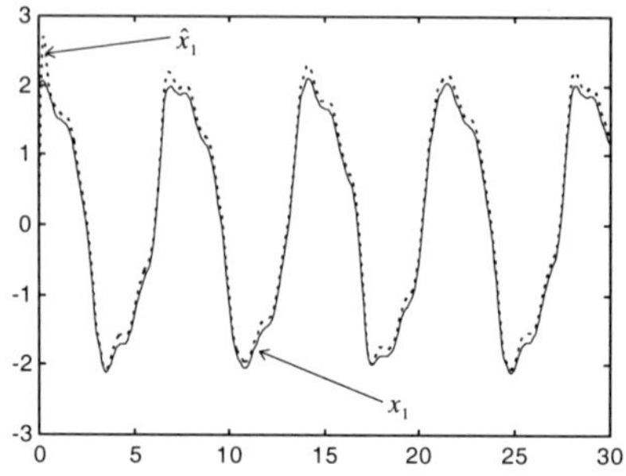

FIGURE 4.12. High-gain observer for x_1 (smooth noise)

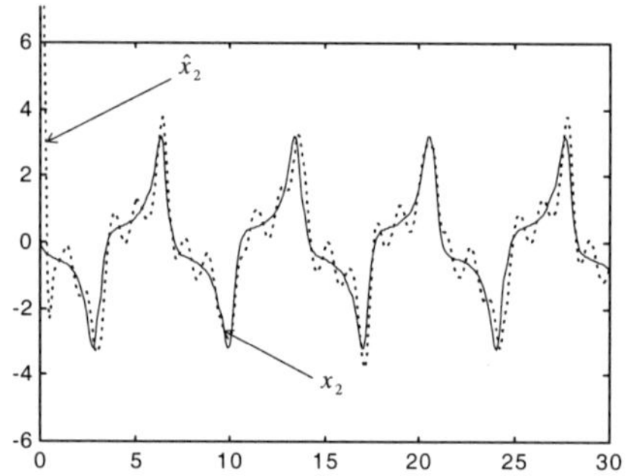

FIGURE 4.13. High-gain observer for x_2 (smooth noise)

with the following structure

$$\begin{cases} \dot{\widehat{x}}_1 = x_2 + \frac{1}{\varepsilon}(y - \widehat{y}) \\ \dot{\widehat{x}}_2 = \frac{1}{\varepsilon^2}(y - \widehat{y}) \\ \widehat{y} = \widehat{x}_1 \end{cases} \tag{4.112}$$

The selection of $\varepsilon = 20$ *(this is the best parameter obtained by the computer simulation) leads to the observation results shown in Figure 4.12 and Figure 4.13.As it can be seen from the figures presented above, the robust observer designed in this chapter is very compatible with the high-gain observer commonly used for the state estimation of unknown nonlinear systems.*

Example 4.3 *Consider the following example with external disturbances and assuming that* x_1, x_2 *are not measurable:*

$$\begin{bmatrix} \dot{x}_1 \\ \dot{x}_2 \end{bmatrix} = \begin{bmatrix} \alpha_1 x_1 \\ \alpha_2 x_2 \end{bmatrix} + \begin{bmatrix} \beta_1 sign(x_2) \\ \beta_1 sign(x_1) \end{bmatrix} + \begin{bmatrix} u_1 \\ u_2 \end{bmatrix} + \begin{bmatrix} cos(x_1) \\ 3sin(x_2) \end{bmatrix}$$
$$y_t = x_1 + x_2 + \xi_3 \tag{4.113}$$

where ξ_3 *is the output perturbation given as the Matlab Function*[1] *"chirp signal" with the frequency changing between* $0.1Hz$ *and* $1Hz$ *and the amplitude equal to* 0.3. *The given nonlinear system (4.113), even simple, is interesting enough: it has multiple isolated equilibria. We can easily check that the assumptions* ***A4.1, A4.5*** *and* ***A4.6*** *are fulfilled. According to the formula (4.54), construct the neuro-observer as follows:*

$$\begin{bmatrix} \dot{\widehat{x}}_1 \\ \dot{\widehat{x}}_2 \end{bmatrix} = A \begin{bmatrix} x_1 \\ x_2 \end{bmatrix} + W_1 \begin{bmatrix} \sigma(\widehat{x}_1) \\ \sigma(\widehat{x}_2) \end{bmatrix}$$
$$+W_2 \begin{bmatrix} \phi(\widehat{x}_1) & 0 \\ 0 & \phi(\widehat{x}_2) \end{bmatrix} \begin{bmatrix} u_1 \\ u_2 \end{bmatrix} + K_t \left(y_t - \widehat{y}_t \right)$$
$$\widehat{y}_t = \widehat{x}_1 + \widehat{x}_2$$

We select

$$x_1(0) = 10, \quad \widehat{x}_1(0) = -1, \quad x_2(0) = -10, \quad \widehat{x}_2(0) = -2$$

$$A = \begin{bmatrix} -3 & 0 \\ 0 & -3 \end{bmatrix}$$

$$\Lambda_{\Delta c} = 0, \Lambda_{\xi_2} = 0.5 * eye(2)$$

[1] Matlab is a tradmark of Mathowrks Inc.

The gain matrix K_t is calculated according to (4.76):

$$K_t = P \begin{bmatrix} 2 & 0 \\ 0 & 2 \end{bmatrix} \begin{bmatrix} 1 \\ 1 \end{bmatrix}$$

where P is the solution of the differential matrix Riccati equation

$$0 = PA + A^T P + PR_0 P + Q$$

with

$$Q_0 = \begin{bmatrix} 0.2 & 0 \\ 0 & 0.2 \end{bmatrix}, \ \delta = 0.001I$$

$$D_\sigma = D_\phi = 0.1I, \ R_0 = \begin{bmatrix} 20 & 5 \\ 5 & 20 \end{bmatrix}$$

The obtained solution is

$$P = \begin{bmatrix} 1.2 & -0.3 \\ -0.2 & 1.2 \end{bmatrix}$$

We also have

$$C_0^+ = [0.5, 0.5]^T$$

$$N = \begin{bmatrix} 0.5005 & 0.5 \\ 0.5 & 0.5005 \end{bmatrix}$$

The nonlinearities σ_i and ϕ_i are chosen as sigmoidal:

$$\sigma_i(x_i) = 2/\left(1 + e^{-2x_i}\right) - 0.5$$
$$\phi_i(x_i) = 0.2/\left(1 + e^{-0.2x_i}\right) - 0.05$$

$$M_{1,t} = \widetilde{y}_t P_t \begin{bmatrix} \sigma(\widehat{x}_1) & \sigma(\widehat{x}_1) \\ \sigma(\widehat{x}_2) & \sigma(\widehat{x}_2) \end{bmatrix}$$
$$M_{2,t} = \widetilde{y}_t P_t \begin{bmatrix} \phi(\widehat{x}_1)u_1 & \phi(\widehat{x}_1)u_1 \\ \phi(\widehat{x}_2)u_2 & \phi(\widehat{x}_2)u_2 \end{bmatrix}$$

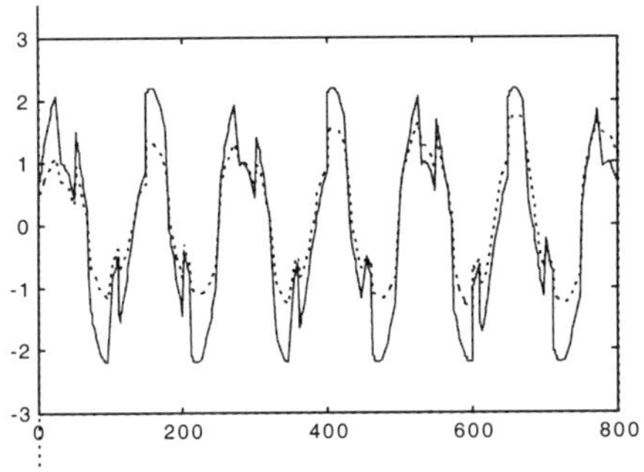

FIGURE 4.14. Neuro-observer results for x_1.

where $\widetilde{y}_t$ *is*

$$\widetilde{y}_t := C_0\widehat{x}_t - y_t = C_0\Delta_t - \left(\triangle C x_t + \xi_{2,t}\right)$$

The initial weight matrix of the neural network is equal to

$$W_{1,0} = W_1^* = \begin{bmatrix} 0.1 & 2 \\ 5 & 0.2 \end{bmatrix}$$

$$W_{2,0} = W_2^* = \begin{bmatrix} 0.1 & 0 \\ 0 & 0.1 \end{bmatrix}$$

To adapt on-line the neuro-observer weights, we use the learning algorithm (4.79). The input signals u_1 *and* u_2 *are chosen as sine wave and saw-tooth function. The simulation results are shown as Figure 4.14, Figure 4.15, Figure 4.16 and Figure 4.17. The solid lines correspond to nonlinear system state responses , and the dashed line - to neuro-observer. The abscissa values correspond to the number of iterations. It can be seen that the neural network state time evolution follows the given nonlinear system in a good manner.*

4.5 Concluding Remarks

In this chapter we have shown that the use of the observers with Luneburger structure and with a special choice of the gain matrix provides good enough observation

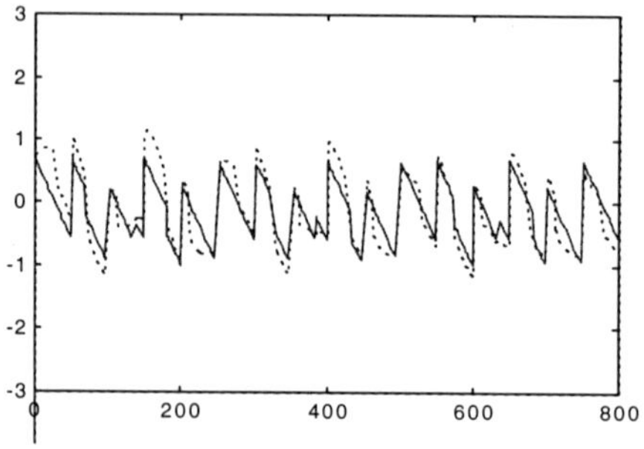

FIGURE 4.15. Neuro-observer results for x_2.

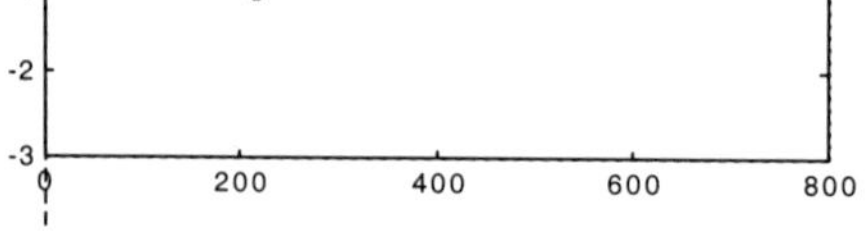

FIGURE 4.16. Observer errors.

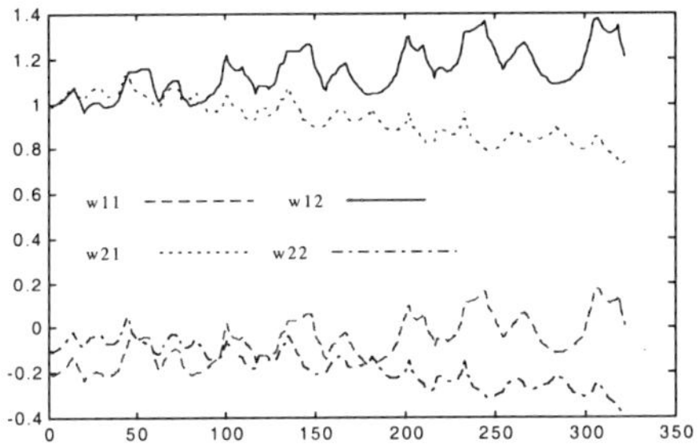

FIGURE 4.17. Weight W_1.

process within a wide class of nonlinear systems containing both unmodeled dynamics and external perturbations of state and output signals. This class includes systems with Lipschitz nonlinear part and with unmodeled dynamics satisfying "strip bound conditions". External perturbations are assumed to have a bounded power.

The gain matrix providing the property of robustness for this observer is constructed with the use of the solution of the corresponding differential Riccati equation containing time-varying parameters which are dependent on the on-line observations. An important feature of the suggested observer is the incorporation of the pseudoinverse operation applied to a specific matrix constructed in time of the estimating process.

The new differential learning law, containing the dead-zone gain coefficient, is suggested to implement this neuro-observer. This learning process provides the boundness property for the dynamic neural network weights as well for estimation error trajectories.

4.6 References

[1] A.Albert, "Regression and the Moore-Penrose Pseudoinverse", Academic Press, 1972.

[2] T.Basar and P.Bernhard, "H^{∞} – *Optimal Control and Related Minimax Design Problems (A Dynamic Game Approach)*", Birkhauser, Boston, 1991.

[3] W.T.Baumann and W.J.Rugh, "Feedback control of nonlinear systems by extended linearization", *IEEE Trans. Automat. Contr.*, vol.31, 40-46, 1986.

[4] Harry Berghuis and Nenk Nijmeijier, "Robust Control of Robots via Linear Estimated State Feedback", *IEEE Trans. Automat. Contr.*, vol.39, 2159-2162, 1994.

[5] C.A.Desoer ann M.Vidyasagar, *Feedback Systems: Input-Output Properties*, New York: Academic, 1975

[6] E.A.Coddington and N.Levinson. *Theory of Ordinary Differential Equations.* Malabar, Fla: Krieger Publishing Company, New York, 1984.

[8] Gantmacher F.R. *Lectures in Analytical Mechanics.* MIR, Moscow, 1970.

[7] R.A.Garcia and C.E.D'Attellis, "Trajectory tracking in nonlinear systems via nonlinear reduced-order observers", *Int. J. Control,* vol.62, 685-715, 1995.

[8] J.P.Gauthier, H.Hammouri and S.Othman, "A simple observer for nonlinear systems: applications to bioreactors", *IEEE Trans. Automat. Contr.*, vol.37, 875-880, 1992.

[9] J.P.Gauthier and G.Bornard, "Observability for and $u(t)$ of a Class of Nonlinear Systems", *IEEE Trans. Automat. Contr.*, vol.26, 922-926, 1981.

[10] G.Giccarella, M.D.Mora and A.Germani, "A Luenberger-like observer for nonlinear system", *Int. J. Control,* vol.57, 537-556, 1993.

[11] Isidori A., *Nonlinear Control Systems*, 3rd ed. New York: Springer-Verlag, 1991.

[11] W.L.Keerthipala, H.C.Miao and B.R.Duggal,"An efficient observer model for field oriented induction motor control", *Proc. IEEE SMC'95*, 165-170, 1995.

[12] Y.H.Kim, F.L.Lewis and C.T.Abdallah,"Nonlinear observer design using dynamic recurrent neural networks", *Proc. 35th Conf. Decision Contr.*, 1996.

[13] Lim S. Y., Dawson D. M and Anderson K., "Re-Examining the Nicosia-Tomei Robot Observer-Controller from a Backstepping Perspective", IEEE Trans. Autom. Contr. Vol 4. No. 3, 1996, pp. 304-310.

[14] Martinez-Guerra R. and De Leon-Morales J., "Nonlinear Estimators: A Differential Algebraic Approach", Appl. Math. Lett. 9, 1996, pp. 21-25.

[13] F.L.Lewis, A.Yesildirek and K.Liu, "Neural net robot controller with guaranteed tracking performance", *IEEE Trans. Neural Network*, Vol.6, 703-715, 1995.

[14] D.G.Luenberger, Observing the State of Linear System, *IEEE Trans. Military Electron*, Vol.8, 74-90, 1964

[15] H.Michalska and D.Q.Mayne, "Moving horizon observers and observer-based control", *IEEE Trans. Automat. Contr.*, vol.40, 995-1006, 1995.

[16] S.Nicosia and A.Tornambe, High-Gain Observers in the State and Parameter Estimation of Robots Having Elastic Joins, *System & Control Letter*, Vol.13, 331-337, 1989

[17] A.J.Krener and A.Isidori, "Linearization by output injection and nonlinear observers", *Systems and Control Letters,* vol.3, 47-52, 1983.

[18] R.Marino and P.Tomei, "Adaptive observer with arbitrary exponential rate of convergence for nonlinear system", *IEEE Trans. Automat. Contr.*, vol.40, 1300-1304, 1995.

[19] H.W.Knobloch, A.Isidori and D.FLockerzi, *"Topics in Control Theory"*, Birkhauser Verlag, Basel- Boston- Berlin, 1993.

[20] J. de Leon, E.N.Sanchez and A.Chataigner, "Mechanical system tracking using neural networks and state estimation simultaneously", *Proc.33rd IEEE CDC*, 405-410 1994.

[21] A.S.Poznyak and E.N.Sanchez, "Nonlinear system approximation by neural networks: error stability analysis", *Intelligent Automation and Soft Computing*, vol.1, 247-258, 1995.

[22] Alexander S.Poznyak and Wen Yu, Robust Asymptotic Newuro Observer with Time Delay, *International Journal of Robust and Nonlinear Control*, accepted for publication.

[23] Antonio Osorio, Alexander S. Poznyak and Michael Taksar, "Robust Deterministic Filtering for Linear Uncertain Time-Varying Systems", *Proc. of American Control Conference*, Albuquerque, New Mexico, 1997

[24] Zhihua Qu and John Dorsey, "Robust Tracking Control of Robots by a Linear Feedback Law", *IEEE Trans. Automat. Contr.*, vol.36, 1081-1084, 1991.

[25] J.Tsinias, "Further results on observer design problem", *Systems and Control Letters,* vol.14, 411-418, 1990.

[26] A.Tornambe, High-Gains Observer for Nonlinear Systems, *Int. J. Systems Science*, Vol.23, 1475-1489, 1992.

[27] A.Tornambe, "Use of asymptotic observers having high-gain in the state and parameter estimation", *Proc. 28th Conf. Decision Contr.*, 1791-1794, 1989.

[28] B.L.Walcott and S.H.Zak, "State observation of nonlinear uncertain dynamical system", *IEEE Trans. Automat. Contr.*, vol.32, 166-170, 1987.

[29] B.L.Walcott, M.J.Corless and S.H.Zak, "Comparative study of nonlinear state observation technique", *Int. J. Control,* vol.45, 2109-2132, 1987.

[30] H.K.Wimmer, Monotonicity of Maximal Solutions of Algebraic Riccati Equations, *System and Control Letters*, Vol.5, pp317-319, 1985

[31] J.C.Willems, "Least Squares Optimal Control and Algebraic Riccati Equations", *IEEE Trans. Automat. Contr.*, vol.16, 621-634, 1971.

[32] T.C.Wit and J.E.Slotine, "Sliding observers for robot manipulators", *Automatica,* vol.27, 859-864, 1991.

[33] M.Zeitz, "The extended Luenberger observer for nonlinear systems", *Systems and Control Letters,* vol.9, 149-156, 1987.

5

Passivation via Neuro Control

In this chapter an adaptive technique is suggested to provide the passivity property for a class of partially known SISO nonlinear systems. A simple differential neural network (DNN), containing only two neurons, is used to identify the unknown nonlinear system. By means of a Lyapunov-like analysis we derive a new learning law for this DNN guarantying both successful identification and passivation effects. Based on this adaptive DNN model we design an adaptive feedback controller serving for wide class of nonlinear systems with a priory incomplete model description. Two typical examples illustrate the effectiveness of the suggested approach. The presented materials reiterate the results of [15]

5.1 Introduction

Passivity is one of the important properties of dynamic systems which provides a special relation between the input and the output of a system and is commonly used in the stability analysis and stabilization of a wide class of nonlinear systems [4, 12]. Shortly speaking, if a nonlinear system is *passive* it can be stabilized by any negative linear feedback even in the lack of a detail description of its mathematical model (see Figure 6.2). This property seems to be very attractive in different physical applications. In view of this, the following approach for designing a feedback controller for nonlinear systems is widely used: first, a special internal nonlinear feedback is introduced to passify the given nonlinear system; second, a simple external negative linear feedback is introduced to provide a stability property for the obtained closed-loop system (see Figure 6.3). The detailed analysis of this method and the corresponding synthesis of passivating nonlinear feedbacks represent the foundation of *Passivity Theory* [1],[12].

In general, Passivity Theory deals with controlled systems whose nonlinear properties are poorly defined (usually by means of sector bounds). Nevertheless, it offers

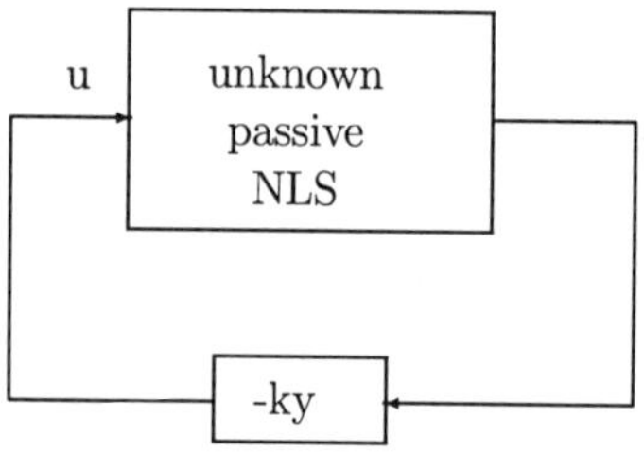

FIGURE 5.1. The general structure of passive control.

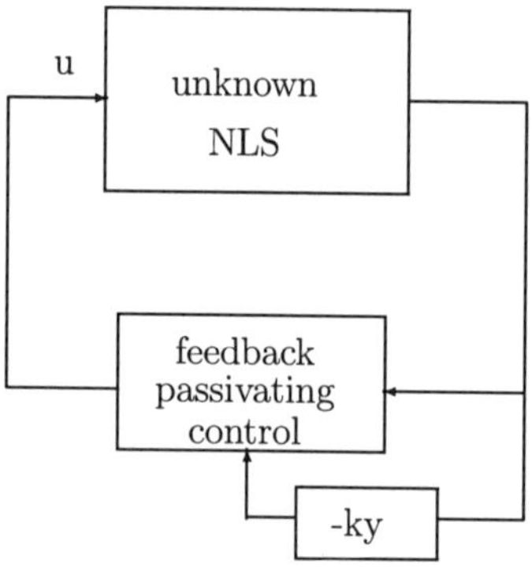

FIGURE 5.2. The structure of passivating feedback control.

an elegant solution to the problem of absolute stability of such systems. The passivity framework can lead to general conclusions on the stability of broad classes of nonlinear control systems, using only some general characteristics of the input-output dynamics of the controlled system and the input-output mapping of the controller. For example, if the system is passive and it is zero-state detectable, any output feedback stabilizes the equilibrium of the nonlinear system [12].

When the system dynamics are totally or partially unknown, the passivity feedback equivalence turns out to be an important problem. This property can be provided by a special design of robust passivating controllers (adaptive [7, 8] and non-adaptive [19, 11] passivating control). But all of them require more detailed knowledge on the system dynamics. So, to be realized successfully, an adaptive passivating control needs the structure of the system under consideration as well as the unknown parameters to be linear. If we deal with the non-adaptive passivating control, the nominal part (without external perturbations) of the system is assumed to be completely known.

If the system is considered as a "black-box" (only some general properties are assumed to be verified to guarantee the existence of the solution of the corresponding ODE-models), the learning-based control using *Neural Networks* has emerged as a viable tool [7]. This model-free approach is presented as a nice feature of Neural Networks, but the lack of model for the controlled plant makes hard to obtain theoretical results on the stability and performance of a nonlinear system closed by a designed neuro system. In the engineering practice, it is very important to have any theoretical guarantees that the neuro controller can stabilize a given system before its application to a real industrial or mechanical plant. That's why neuro controller design can be considered as a challenge to a modern control community.

Most publications in nonlinear system identification and control use *static (feed-forward) neural networks*, for example, Multilayer Perceptrons (MLP), which are implemented for the approximation of nonlinear function in the right-hand side of dynamic model equations [11]. The main drawback of these neural networks is that the weight updates do not use any information on a local data structure and the applied function approximation is sensitive to the training data [7]. *Dynamic Neural*

Networks (DNN) can successfully overcome this disadvantage as well as providing adequate behavior in the presence of unmodeled dynamics, because their structure incorporate feedback. They have powerful representation capabilities. One of best known DNN was introduced by Hopfield [5].

For this reason the framework of neural networks is very convenient for passivation of unknown nonlinear systems. Based on the static neural networks, in [2] an adaptive passifying control for unknown nonlinear systems is suggested. As we state before, there are many drawbacks on using static neural networks for the control of dynamic systems.

In this chapter we use DNN to passify the unknown nonlinear system. A special storage function is defined in such a way that the aims of identification and passivation can be reached simultaneously. It is shown in [18], [13] and [29] that the Lyapunov-like method turns out to be a good instrument to generate a learning law and to establish error stability conditions. By means of a Lyapunov-like analysis we derive a weight adaptation procedure to verify passivity conditions for the given closed-loop system. Two examples are considered to illustrate the effectiveness of the adaptive passivating control.

5.2 Partially Known Systems and Applied DNN

As in [1] and [2], let us consider a single input-single output (SISO) nonlinear system (NLS) given by

$$\begin{aligned} \dot{z} &= f_0(z) + p(z, y)y \\ \dot{y} &= a(z, y) + b(z, y)u \end{aligned} \tag{5.1}$$

where

$\zeta := \left[z^T, y\right]^T \in \Re^n$ is the state at time $t \geq 0$,

$u \in \Re$ is the input and $y \in \Re$ is the output of the system.

The functions $f_0(\cdot)$ and $p(\cdot)$ are assumed to be C^1-vector fields and the functions $a(\cdot, \cdot)$ and $b(\cdot, \cdot)$ are C^1-real functions ($b(z, y) \neq 0$ for any z and y). Let it be

$$f_0(0) = 0$$

We also assume that the set U_{ad} of admissible inputs u consists of all $\Re$-valued piecewise continuous functions defined on $\Re$, and verifying the following property: for any the initial conditions $\zeta^0 = \zeta(0) \in \Re^n$ the corresponding output

$$y(t) = h(\Phi(t, \zeta^0, u))$$

[1] of this system (5.1) satisfies

$$\int_0^t \mid y(s)u(s) \mid ds < \infty, \text{ for all } t \geq 0$$

i.e., the "energy" stored in system (5.1) is bounded.

Definition 8 ***Zero dynamics*** *of the given nonlinear system (5.1) describes those internal dynamics which are consistent with the external constraint* $y = 0$, *i.e., the zero dynamics verifies the following ODE*

$$\dot{z} = f_0(z) \tag{5.2}$$

Definition 9 *[1, 4] A system (5.1) is said to be* C^r-***passive*** *if there exists a* C^r-*nonnegative function* $V : \Re^n \rightarrow \Re$, *called* ***storage function***, *with* $V(0) = 0$, *such that, for all* $u \in U_{ad}$, *all initial conditions* ζ^0 *and all* $t \geq 0$ *the following inequality holds:*

$$\dot{V}(\zeta) \leq yu \tag{5.3}$$

If

$$\dot{V}(\zeta) = yu \tag{5.4}$$

then the system (5.1) is said to be C^r-***lossless***. *If, further, there exists a positive definite function* $S : \Re^n \rightarrow \Re$ *such that*

$$\dot{V}(\zeta) = yu - S \tag{5.5}$$

then the system is said to be C^r-***strictly passive***.

---

[1] $\Phi(t, \zeta^0, u)$ denotes the flow of $\begin{bmatrix} f_0(z) + p(z,y) & , & a(z,y) + b(z,y)u \end{bmatrix}^T$ corresponding to the initial condition $\zeta^0 = \left[(z^0)^T, y^0\right]^T \in R^n$ and to $u \in U_{ad}$.

For the nonlinear system (5.1) considered in this paper, the following assumptions are assumed to be fulfilled:

H1: *The zero dynamics* $f_0(z)$ *and the function* $b(z,y)$ *are completely known.*

H2: $f_0(\cdot)$ *satisfies global Lipschitz condition, i.e. , for any* $z_1, z_2 \in \Re^{n-1}$

$$\|f_0(z_1) - f_0(z_2)\| \leq L_{f0} \|z_1 - z_2\|, \quad L_{f0} > 0$$

H3: *The zero dynamics in (5.1) is Lyapunov stable, i.e., there exists a function* $\mathcal{W}_0 : \Re^{n-1} \to \Re^+$, *with* $\mathcal{W}_0(0) = 0$, *such that for all* $z \in \Re^{n-1}$

$$\frac{\partial \mathcal{W}_0(z)}{\partial z} f_0(z) \leq 0$$

H4: *The unknown part of the system (5.1) is related to the functions* $a(z,y)$, $p(z,y)$ *with known upper bounds, i.e.,*

$$\|a(z,y)\| \leq \overline{a}(z,y), \quad \|p(z,y)\| \leq \overline{p}(z,y)$$

where $\overline{a}(z,y)$ *and* $\overline{p}(z,y)$ *are Lipschitz functions (selected by the designer) and satisfy the following "strip conditions":*

$$\begin{aligned} \overline{a}(z,y) &\leq a_0 + a_1 (\|z\| + |y|) \\ \overline{p}(z,y) &\leq p_0 + p_1 (\|z\| + |y|) \end{aligned}$$

Following to [29] and [13], to identify this partially unknown nonlinear system we propose DNN having the structure as follows:

$$\begin{aligned} \dot{\widehat{z}} &= f_0(\widehat{z}) + [W_1 \varphi_1(\widehat{z}, \widehat{y}) + \psi_1] y \\ \dot{\widehat{y}} &= W_2 \varphi_2(\widehat{z}, \widehat{y}) + \psi_2 + b(z,y)u \end{aligned} \tag{5.6}$$

Here

$(\widehat{z}, \widehat{y}) \in \Re^n$ is the state of the DNN,

$W_1 \in \Re^{(n-1)\times n}$, $W_2 \in \Re^{1\times n}$ are the weight matrix and vector, correspondingly,

the functions $\psi_1 \in \Re^{n\times 1}$ and $\psi_2 \in \Re^{n\times 1}$ are the "output threshold" of each neuron, the activation functions $\varphi_1(\cdot,\cdot) \in \Re^{n\times 1}$ and $\varphi_2(\cdot,\cdot) \in \Re^{n\times 1}$ are defined as follows:

$$\varphi_i(\alpha,\beta) = [\tanh(k_i \cdot \alpha_1), ..., \tanh(k_i \cdot \alpha_{n-1}), \tanh(k_i \cdot \beta)]^T$$
$$k_i \in R,\ i = 1, 2$$

As it is seen from (5.6), the structure of the DNN is constructed using the known part $f_0(z)$, $b(z,y)$ and the unknown part is identified by two neurons:

the neuron $[W_1\varphi_1(\widehat{z},\widehat{y}) + \psi_1]$ corresponding to the function $p(z,y)$

and $[W_2\varphi_2(\widehat{z},\widehat{y}) + \psi_2]$ corresponding to the function $a(z,y)$.

Sure, that in general case, the given unknown NLS (5.1) does not obligatory belong to the class of systems which can be exactly modelled by the equation (5.6). Hence, for any $t \geq 0$ and for any fixed initial weights of the DNN, denoted by W_1^* and W_2^*, there exists, so called, *the identification error* (ν_1, ν_2) defined as

$$\begin{aligned} \nu_1 &:= \dot{z} - f_0(z) + [W_1^*\varphi_1(z,y) + \psi_1]\, y \\ \nu_2 &:= \dot{y} - W_2^*\varphi_2(z,y) + \psi_2 + b(z,y)u \end{aligned} \tag{5.7}$$

In view of (5.1) and (5.7), we can get the algebraic presentation for the identification error:

$$\begin{aligned} \nu_1 &= B_1(z,y)y - \psi_1 y \\ \nu_2 &= B_2(z,y) - \psi_2 \end{aligned} \tag{5.8}$$

where:

$$\begin{aligned} B_1(z,y) &:= p(z,y) - W_1^*\varphi_1(z,y) \\ B_2(z,y) &:= a(z,y) - W_2^*\varphi_2(z,y) \end{aligned}$$

From **H4** and the boundness of the functions φ_1 and φ_2 we can conclude that B_1 and B_2 are also bounded and satisfy

$$\begin{aligned} \|B_1(z,y)\| &\leq \overline{p}(z,y) + \|W_1^*\| \cdot \|\varphi_1\| =: \overline{B}_1(z,y) \\ \|B_2(z,y)\| &\leq \overline{a}(z,y) + \|W_2^*\| \cdot \|\varphi_2\| =: \overline{B}_2(z,y) \end{aligned} \tag{5.9}$$

Now we are ready to discuss the following problem: how incorporate this DNN into the internal feedback of the given NLS such that the obtained closed-loop system possesses the passivity property.

5.3 Passivation of Partially Known Nonlinear System via DNN

In order to simplify the notation, the next expressions will be in use:

$$\varphi_i := \varphi_i(z,y), \quad \widehat{\varphi}_i := \varphi_i(\widehat{z},\widehat{y})$$
$$\widetilde{\varphi}_i := \varphi_i(\widehat{z},\widehat{y}) - \varphi_i(z,y) \quad (i=1,2)$$

$$\Delta := \begin{bmatrix} \Delta_z \\ \Delta_y \end{bmatrix} = \begin{bmatrix} \widehat{z}-z \\ \widehat{y}-y \end{bmatrix} \tag{5.10}$$

For any vector $\omega \in \Re^m$ and any positive integer $m = 1,2,...$ we denote

$$|\omega| := \begin{bmatrix} |\omega_1| & |\omega_2| & \cdots & |\omega_m| \end{bmatrix}^T$$

$$diag(\omega) := diag\{\omega_1,\ \omega_2,\ \cdots,\ \omega_m\} \in \Re^{m\times m}$$

and for a scalar $\kappa \in \Re$ and a positive integer $l = 1,2,...$ we will write

$$vec_l(\kappa) := \begin{bmatrix} \kappa & \kappa & \cdots & \kappa \end{bmatrix}^T \in \Re^l$$

Consider the class of nonlinear systems given by

$$\begin{array}{l} \dot{z} = f_0(z) + [W_1^* \varphi_1(z,y) + \psi_1]\, y + \nu_1 \\ \dot{y} \ -W_2^* \varphi_2(z,y) + \psi_2 + b(z,y)u + \nu_2 \end{array} \tag{5.11}$$

satisfying the assumptions H1- H4 where the unmodeled dynamics (ν_1, ν_2) is defined by (5.8).

The following theorem give the main result on the passivation of partially unknown nonlinear system via DNN.

Theorem 5.1 *Let the nonlinear system (5.11) be identified by DNN (5.6) with the following differential learning law*

$$\begin{array}{ll} \dot{W}_1^T = \eta_1 \left(-2\widehat{\varphi}_1 y \Delta_z^T P_z + \varphi_1 y \frac{\partial W_0(z)}{\partial z} \right), & W_1(0) = W_1^* \\ \dot{W}_2^T = \eta_2 \left(-2\widehat{\varphi}_2 \Delta_y P_y + \varphi_2 y \right), & W_2(0) = W_2^* \end{array} \tag{5.12}$$

where $P_z \in R^{(n-1)\times(n-1)}$ *is a positive solution to the following matrix Riccati equation*

$$P_z A + A P_z + P_z \Lambda_{\tilde{f}'}^{-1} P_z + I_z \cdot L_{\tilde{f}'}^2 \left\| \Lambda_{\tilde{f}'} \right\| = 0 \tag{5.13}$$

and P_y *is a positive constant. Let also the output thresholds of DNN nodes are adjusted according to*

$$\begin{array}{c} \psi_1 = -sign\left(diag\left(\Delta_z P_z\right)\right)\left[\|W_1^*\| \left|\widetilde{\varphi}_1\right| + vec_{n-1}(\overline{B}_1)\right] sign(y) \\ \psi_2 = -sign\left(\Delta_y P_y\right)\left[\|W_2^*\| \cdot \|\widetilde{\varphi}_2\| + \overline{B}_2)\right] \end{array} \tag{5.14}$$

If the control law is constructed as follows:

$$u = b^{-1}(z,y)\left[v - \frac{\partial \mathcal{W}_0(z)}{\partial z} W_1 \varphi_1 - \left[\overline{B}_2 + \left\|\frac{\partial \mathcal{W}_0(z)}{\partial z}\right\| \overline{B}_1\right] sign(y) - W_2 \varphi_2\right] \tag{5.15}$$

then such closed-loop system turns out to be passive (with respect to the new input v*) with the storage function*

$$\begin{array}{c} V = \Delta^T P \Delta + \frac{1}{2} y^2 + \mathcal{W}_0(z) + \frac{\eta_2^{-1}}{2} \widetilde{W}_2 \widetilde{W}_2^T + \frac{1}{2} tr\left\{\widetilde{W}_1 \eta_1^{-1} \widetilde{W}_1^T\right\} \\ P = \begin{bmatrix} P_z & 0 \\ 0 & P_y \end{bmatrix} \in R^{n\times n}, 0 < \eta_1 \in R^{n\times n}, 0 < \eta_2 \in R. \end{array} \tag{5.16}$$

i.e., for any $t \geq 0$

$$\dot{V} \leq vy$$

Proof. Let us define

$$\widetilde{W}_i := W_i - W_i^*, \;\; i = 1, 2 \tag{5.17}$$

Now we start with the calculation of the derivative of the storage function (5.16) along the trajectory of the systems (5.11) and (5.6):

$$\dot{V} = 2\Delta^T P \dot{\Delta} + \frac{\partial \mathcal{W}_0(z)}{\partial z} \dot{z} + y \dot{y} + \eta_2^{-1} \cdot \widetilde{W}_2 \dot{\widetilde{W}}_2^T + tr\left\{\widetilde{W}_1 \eta_1^{-1} \dot{\widetilde{W}}_1^T\right\} \tag{5.18}$$

We are going to calculate the upper bounds for $\dot{V}$ in such a way that this bound is going to be a function of known data. To do this, we should use the assumptions H1-H4. Now let us start with the term

$$2\Delta^T P \dot{\Delta} = 2\Delta_z^T P_z \dot{\Delta}_z + 2\Delta_y^T P_y \dot{\Delta}_y$$

From (5.11), (5.6), (5.17) and (5.10), it follows

$$
\begin{aligned}
2\Delta_z^T P_z \dot{\Delta}_z &= \hat{f}_0 - f_0 + \left(\widetilde{W}_1 \widehat{\varphi}_1 + W_1^* \widetilde{\varphi}_1\right) y - \nu_1 \\
2\Delta_y^T P_y \dot{\Delta}_y &= \widetilde{W}_2 \widehat{\varphi}_2 + W_2^* \widetilde{\varphi}_2 - \nu_2
\end{aligned}
\tag{5.19}
$$

The equation (5.19) can be rewritten as

$$
\begin{aligned}
2\Delta_z^T P_z \dot{\Delta}_z = {} & 2\Delta_z^T P_z \left[f_0(\hat{z}) - f_0(z) + (A\Delta_z - A\Delta_z) + \right. \\
& \left. \left(\widetilde{W}_1 \widehat{\varphi}_1 + W_1^* \widetilde{\varphi}_1\right) y - \nu_1 \right] \\
= {} & 2\Delta_z^T P_z \left[\widehat{f} + A\Delta_z + \left(\widetilde{W}_1 \widehat{\varphi}_1 + W_1^* \widetilde{\varphi}_1\right) y - \nu_1 \right]
\end{aligned}
$$

where

$$
\tilde{f}' := f_0(\hat{z}) - f_0(z) - A(\hat{z} - z)
$$

is the function, which in view of **H2**, verifies the following Lipschitz condition:

$$
\left\| \widehat{f} \right\| \le L_{\overline{f}} \left\| \Delta_z \right\|, \; L_{\overline{f}} > 0 \tag{5.20}
$$

The equation (5.8) implies

$$
\begin{aligned}
2\Delta_z^T P \dot{\Delta}_z = {} & 2\Delta_z^T P_z \left[\tilde{f}' + A\Delta_z \right] + \\
& + 2\Delta_z^T P_z \widetilde{W}_1 \widehat{\varphi}_1 y + \\
& + 2\Delta_z^T P_z \left[W_1^* \widetilde{\varphi}_1 - B_1 + \psi_1 \right] y
\end{aligned}
$$

and taking into account the inequality (5.20) we can estimate the first term from the right-hand side of the term $2\Delta_z^T P_z \left[\tilde{f}' + A\Delta_z \right]$ as

$$
2\Delta_z^T P_z \left[\tilde{f}' + A\Delta_z \right] \le \Delta_z^T \left[P_z A + A P_z + P_z P_z + I_z \cdot L_{\tilde{f}'}^2 \left\| \Lambda_{\tilde{f}'} \right\| \right] \Delta_z
$$

The following estimations hold:

$$
\begin{aligned}
& 2\Delta_z^T P_z \left[W_1^* \widetilde{\varphi}_1 - B_1 + \psi_1 \right] y \le \\
& \le 2 \left| \Delta_z^T P_z \right| \left(\left\| W_1^* \right\| \left| \widetilde{\varphi}_1 \right| + vec_{n-1}\left(\overline{B}_1 \right) \right) |y| + 2\Delta_z^T P_z \psi_1 y = \\
& = 2\Delta_z^T P_z \left[sign\left(diag\left(\Delta_z^T P_z \right) \right) \left(\left\| W_1^* \right\| \left| \widetilde{\varphi}_1 \right| + vec_{n-1}\left(\overline{B}_1 \right) \right) sign\left(y \right) + \psi_1 \right] y.
\end{aligned}
$$

The upper bound for $2\Delta_z^T P\dot{\Delta}_z$ is

$$\begin{gathered} 2\Delta_z^T P\dot{\Delta}_z \le \Delta_z^T \left[P_z A + AP_z + P_z P_z + I_z \cdot L_{\tilde{f}'}^2 \left\|\Lambda_{\tilde{f}'}\right\|\right] \Delta_z + \\ 2\Delta_z^T P_z \left[sign\left(diag\left(\Delta_z^T P_z\right)\right)\left(\|W_1^*\| \left|\widetilde{\varphi}_1\right| + vec_{n-1}\left(\overline{B}_1\right)\right) sign\left(y\right) + \psi_1\right] y + \\ tr\left\{\widetilde{W}_1 \left[2\widehat{\varphi}_1 y \Delta_z^T P_z\right]\right\} \end{gathered} \tag{5.21}$$

Using (5.8), analogously to previous calculations, we can estimate the second term in (5.19) as follows:

$$2\Delta_y P_y \dot{\Delta}_y \le \tilde{W}_2 \hat{\varphi}_2 2\Delta_y P_y + 2\Delta_y P_y \left[sign\left(\Delta_y P_y\right)\left(\|W_2^*\| \tilde{\varphi}_2 + \overline{B}_2\right) + \psi_2\right]. \tag{5.22}$$

The upper bound of the term $\frac{\partial \mathcal{W}_0(z)}{\partial z}\dot{z}$ is

$$\begin{gathered} \frac{\partial \mathcal{W}_0(z)}{\partial z}\dot{z} \le \frac{\partial \mathcal{W}_0(z)}{\partial z} f_0 + tr\left\{\widetilde{W}_1 \left[-\varphi_1 y \frac{\partial \mathcal{W}_0(z)}{\partial z}\right]\right\} + \\ + \left[\left\|\frac{\partial \mathcal{W}_0(z)}{\partial z}\right\| \overline{B}_1 sign\left(y\right) + \frac{\partial \mathcal{W}_0(z)}{\partial z} W_1 \varphi_1\right] y \end{gathered} \tag{5.23}$$

The term $y\dot{y}$ is estimated by the following way:

$$y\dot{y} \le \left[W_2\varphi_2 + bu + \overline{B}_2 sign\left(y\right)\right] y + \tilde{W}_2\left[-\varphi_2 y\right] \tag{5.24}$$

Combining (5.18) with the estimations (5.21), (5.22), (5.23) and (5.24), we finally obtain the upper bound for the derivative of the storage function (5.16) in the following form:

$$\begin{gathered} 2\Delta^T P\dot{\Delta} \le \Delta_z^T \left[P_z A + AP_z + P_z \Lambda_{\overline{f}}^{-1} P_z + I_z \cdot L_{\tilde{f}'}^2 \left\|\Lambda_{\overline{f}}\right\|\right] \Delta_z + \\ 2\Delta_z^T P_z \left[sign\left(diag\left(\Delta_z^T P_z\right)\right)\left(\|W_1^*\| \left|\widetilde{\varphi}_1\right| + vec_{n-1}\left(\overline{B}_1\right)\right) sign\left(y\right) + \psi_1\right] y + \\ 2\Delta_y P_y \left[sign\left(\Delta_y P_y\right)\left(\|W_2^*\| \cdot \|\tilde{\varphi}_2\| + \overline{B}_2\right) + \psi_2\right] + \\ tr\left\{\widetilde{W}_1 \left[\eta_1^{-1} \dot{\widetilde{W}}_1^T + 2\widehat{\varphi}_1 y \Delta_z^T P_z - \varphi_1 y \frac{\partial \mathcal{W}_0(z)}{\partial z}\right]\right\} + \\ \tilde{W}_2 \left[\eta_2^{-1} \dot{\tilde{W}}_2^T + \hat{\varphi}_2 2\Delta_y P_y - \varphi_2 y\right] + \\ \frac{\partial \mathcal{W}_0(z)}{\partial z} f_0 + \left[\left\|\frac{\partial \mathcal{W}_0(z)}{\partial z}\right\| \overline{B}_1 sign\left(y\right) + \frac{\partial \mathcal{W}_0(z)}{\partial z} W_1 \varphi_1\right] y + \\ \left[W_2\varphi_2 + bu + \overline{B}_2 sign\left(y\right)\right] y \end{gathered} \tag{5.25}$$

From this expression, it follows that

the first right-hand term contains the algebraic Riccati equation inside of the square brackets and, hence, is equal to zero identically,

the second and *third terms* are cancellated by the output thresholds ψ_1 and ψ_2 selected as in (5.14),

since $\dot{\widetilde{W}}=\dot{W}_i$ $(i=1,2)$, the learning law for the weights adjusting cancels *the fourth* and *fifth terms* of (5.25).

By the assumption **H3** and imposing the control law as in (5.15), we finally get

$$\dot{V} \leq vy.$$

The theorem is proved.■

To clarify the main contribution of this paper, formulated in the proved theorem, the following remarks seems to be useful.

5.3.1 *Structure of Storage Function*

The storage function (5.16)consists of three parts:

- the first one $\Delta^T P \Delta$ makes the identification error smaller;
- the second part

$$y^2 + W_0(z)$$

  are the terms related to the passivity property;
- the third one

$$\frac{1}{2} tr \left\{ \widetilde{W}_1 \eta_1^{-1} \widetilde{W}_1^T \right\} + \frac{1}{2} \eta_2^{-1} \widetilde{W}_2 \widetilde{W}_2^T$$

  is used to generate the learning law for DNN.

5.3.2 *Thresholds Properties*

The output thresholds ψ_1 and ψ_2 are introduced to cancel the influence of the uncertain terms. These thresholds are functions of the bounding functions $\overline{B}_1$ and $\overline{B}_2$, hence, it is preferable to select these functions in such a way that they are sharp (reachable). Such selection can significantly improve the performance of the interconnected system.

5.3.3 *Stabilizing Robust Linear Feedback Control*

The control law u is constructed based on the information received from the updated neurons of DNN $(W_1\varphi_1(z,y)\ ,\ W_2\varphi_2(z,y))$ The neural part of the control law identifies the uncertain terms of the system. So, the control law passify the system and also cancel the influence of the uncertain terms simultaneously. Selecting a feedback as a negative linear one, i.e.,

$$v = -ky,\ k > 0$$

we obtain

$$\dot{V} \leq vy = -ky^2 \leq 0$$

So, the closed-loop system turns out to be stable for any negative linear feedback.

5.3.4 *Situation with Complete Information*

If the nonlinear functions $f_0(z)$, $p(z,y)$, $a(z,y)$ and $b(z,y)$ are known, then the control law (5.15) can be constructed replacing the neuron part $W_1\varphi_1$ by the known function $p(z,y)$, and $W_2\varphi_2$ replaced by $a(z,y)$. The functions $\overline{B}_1$ and $\overline{B}_2$ may be replaced by zeros since this functions are related to the uncertainty (unmodeled dynamics). The control law, which makes the system with known model passive from input v to output y, now is

$$u = b^{-1}(z,y)\left[v - \frac{\partial \mathcal{W}_0(z)}{\partial z}p(z,y) - a(z,y)\right]$$

The corresponding storage function is

$$V_c = \frac{1}{2}y^2 + \mathcal{W}_0(z)$$

These facts coincide with the results in [1].

5.3.5 *Two Coupled Subsystems Interpretation*

We can consider the given NLS (5.1) as a collection of two coupled subsystems. One of them (say, subsystem F) is

$$\dot{y} = a(z,y) + b(z,y)u$$

with the input is u and the output y. The other one (subsystem G) is

$$\dot{z} = f_0(z) + p(z,y)y$$

with the corresponding input y and output z related to the internal dynamics of the whole system. The input of the entire system (5.1, F coupled with G) is u and its output is y. So, in this sense we can say that the function $p(z,y)$ is the "coupling term" of this NLS. In view of this remark, we can say that the uncertainties, described before, are related to the coupling term $p(z,y)$.

5.3.6 Some Other Uncertainty Descriptions

Case 1: Uncertainty in the term $p(z,y)$

If the functions of NLS (5.1) $f_0(z)$, $a(z,y)$, $b(z,y)$ are known. and the function $p(z,y)$ is unknown but bounded, i.e.,

$$\|p(z,y)\| \leq \overline{p}(z,y)$$

($\overline{p}(z,y)$ is selected by the designer), the particular DNN for such uncertain system can be selected as

$$\dot{\widehat{z}} = f_0(\widehat{z}) + [W_1\varphi_1(\widehat{z},y) + \psi_1]\, y$$

with the learning law

$$\dot{W}_1^T = \eta_1 \left(-2\varphi_1(\widehat{z},y)\Delta_z^T P_z + \varphi_1(z,y)\frac{\partial W_0(z)}{\partial z} \right) y, \; W_1(0) = W_1^* \tag{5.26}$$

and the output threshold given by

$$\psi_1 = -sign\left(diag\left(\Delta_z P_z\right)\right)\left[\|W_1^*\| \cdot |\varphi_1(\widehat{z},y) - \varphi_1(z,y)| + vec_{n-1}(\overline{B}_1)\right] sign(y) \tag{5.27}$$

The passifying control law is

$$u = b^{-1}(z,y)\left[v - \frac{\partial W_0(z)}{\partial z}W_1\varphi_1(z,y) - \left[\left\|\frac{\partial W_0(z)}{\partial z}\right\|\overline{B}_1\right] sign(y) - a(z,y)\right] \tag{5.28}$$

with the storage function as

$$V_p = \Delta_z^T P_z \Delta_z + \mathcal{W}_0(z) + \frac{1}{2}y^2 + tr\left\{ \widetilde{W}_1 \eta_1^{-1} \dot{\widetilde{W}}_1^T \right\} \tag{5.29}$$

On the other hand, the coupling term $p(z, y)$ can be expressed as

$$p(z, y) = p_0(z, y) + \delta_p(z, y)$$

where $p_0(z, y)$ is a known part and $\delta_p(z, y)$ is an unknown one, satisfying the constraint

$$\|\delta_p(z, y)\| \leq \overline{\delta}_p(z, y)$$

In this case the corresponding DNN can be constructed as

$$\dot{\widehat{z}} = f_0(\widehat{z}) + [p_0(z, y) + W_1 \varphi_1(\widehat{z}, y) + \psi_1]\, y$$

with the function $\overline{B}_1$ changed to

$$\overline{B}_1 = \overline{\delta}_p(z, y) + \|W_1^*\| \cdot \|\varphi_1\|$$

The control and learning laws, as well as the threshold and the storage function, remain as in (5.28,5.26,5.27 and 5.29). So, we have two alternatives for the uncertainty description in the coupling term $p(z, y)$. But in both cases, the suggested passifying control law (5.28) turns out to be robust with respect to the uncertainty in this coupling term.

Case 2: Uncertainty in the term $a(z, y)$

The main result of this paper, formulated in the theorem given above, concerns the uncertainty in the terms $p(z, y)$ and $a(z, y)$, As a partial case, we can formulate the main result for the situation when the uncertainties are involved only in the term $a(z, y)$. If the functions $f_0(z)$, $p(z, y)$ of the NLS (5.1) and $b(z, y)$ are known and $a(z, y)$ is unknown but it is bounded as

$$\|a(z, y)\| \leq \overline{a}(z, y)$$

where $\overline{a}(z, y)$ is selected by the designer, then DNN, identifying the unknown part, can be constructed as

$$\dot{\widehat{y}} = W_2\varphi_2(z, \widehat{y}) + \psi_2 + b(z, y)u$$

with the weights adjusting according to

$$\dot{W}_2^T = \eta_2 \left(-2\varphi_2(z, \widehat{y})\Delta_y P_y + \varphi_2(z, y)y\right), \qquad W_2(0) = W_2^* \tag{5.30}$$

and with the output thresholds tuned as

$$\psi_2 = -sign\left(\Delta_y P_y\right)\left[\|W_2^*\| \cdot \|\varphi_2(z, \widehat{y}) - \varphi_2(z, y)\| + \overline{B}_2)\right] \tag{5.31a}$$

The control law

$$u = b^{-1}(z, y)1\left[v - \frac{\partial W_0(z)}{\partial z}p(z, y) - \overline{B}_2 sign(y) - W_2\varphi_2\right] \tag{5.32}$$

passify the NLS with the storage function

$$V_a = \Delta_y^T P_y \Delta_y + \frac{1}{2}y^2 + \mathcal{W}_0(z) + \frac{\eta_2^{-1}}{2}\widetilde{W}_2\widetilde{W}_2^T \tag{5.33}$$

As in the former case, we can model the uncertainty of the function $a(z, y)$ as

$$a(z, y) = a_0(z, y) + \delta_a(z, y)$$

where $a_0(z, y)$ is the known part of $a(z, y)$ and $\delta_a(z, y)$ is the unknown one, satisfying

$$\|\delta_a(z, y)\| \leq \overline{\delta}_a(z, y)$$

Then the DNN is constructed using only $a_0(z, y)$ and can have the following structure:

$$\dot{\widehat{y}} = a_0(z, y) + W_2\varphi_2(z, \widehat{y}) + \psi_2 + b(z, y)u \tag{5.34}$$

The function $\overline{B}_2$ changes to

$$\overline{B}_2 = \overline{\delta}_a(z, y) + \|W_2^*\| \cdot \|\varphi_2\|$$

the control law, the learning law, the threshold and the storage functions remain as in (5.32), (5.30), (5.31a) and (5.33). For example, for a single link manipulator we can assume that the friction term is only single uncertainty of the corresponding model. So, the friction term $\delta_a(z, y)$ can be identified by DNN given by (5.34). More details, concerning this example, will be discussed in the next section.

5.4 Numerical Experiments

5.4.1 Single link manipulator

Let us consider the following nonlinear system (single link manipulator)

$$\begin{bmatrix} \dot{z}_1 \\ \dot{y} \end{bmatrix} = \begin{bmatrix} y \\ -\frac{g}{a}\cos(z_1) - \lambda(y) \end{bmatrix} + \begin{bmatrix} 0 \\ \frac{1}{ma^2} \end{bmatrix} u \tag{5.35}$$

where

m is the mass and a is the length of the link,

$\lambda(y)$ is the friction of the joint,

$z_1 \in \Re$ is the joint variable,

$y \in \Re$ is the velocity of z_1,

u torque control.

It can be easily seen that the system (5.35) can be rewritten in the form (5.1), i.e.,

$$\begin{aligned} \dot{z} &= f_0(z) + p(z,y) \\ \dot{y} &= a(z,y) + b(z,y) \end{aligned}$$

with

$$\begin{aligned} f_0(z_1) &= 0 \\ p(z,y) &= 1 \\ a(z,y) &= -ga^{-1}\cos(z_1) - \lambda(y) \\ b(z,y) &= \left(ma^2\right)^{-1} z_1 \end{aligned}$$

The zero dynamics of the system (5.35) is stable and the Lyapunov function for this dynamics is

$$W_0(z_1) = \frac{1}{2}z_1^2$$

Now, let us simulate the single link manipulator assuming that the terms $a(z,y)$ and $b(z,y)$ are unknown. Chose

$$v = -y$$

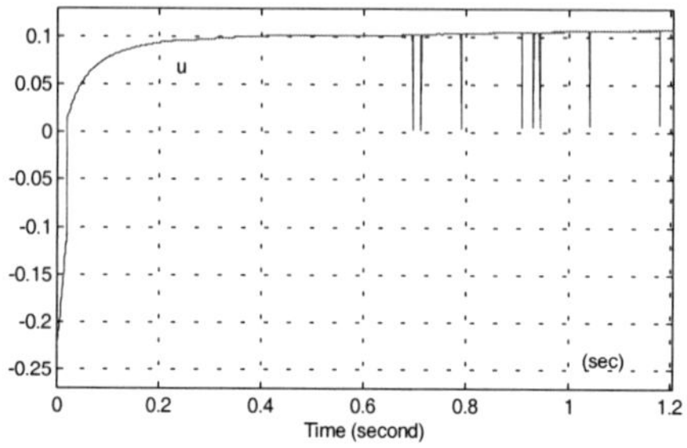

FIGURE 5.3. Control input u.

The initial conditions are selected as

$$[z_1(0),\ y(0)]^T = [1,\ 1]^T$$
$$[\hat{z}_1(0),\ \hat{y}(0)]^T = [5,\ 0]^T$$

To realize the numerical simulations the following parameters were selected:

$$m = \frac{0.5}{9.8},\quad a = 0.3,\quad \lambda(y) = 0.1(y + \tanh(50y))$$
$$A = -2,\quad W_1^* = [9.8521,\ 11.8528],\quad W_2^* = [3.6141,\ 3.5260]$$
$$P_z = 0.5359,\quad \Lambda_{\tilde{f}'} = 1,\quad L_{\tilde{f}'} = 2$$
$$\overline{B}_1 = 15.4127\,\|\varphi_1\| + 3, \qquad \overline{B}_2 = 5.0492\,\|\varphi_2\| + |y| + 4$$
$$\eta_1 = diag\{2, 2\},\quad \eta_2 = 2,\quad k_1 = 1,\quad k_2 = 1$$

The results are shown in Figure 6.4 - Figure 6.6.

5.4.2 Benchmark problem of passivation

Let us consider the following benchmark nonlinear system [8]

$$\begin{bmatrix} \dot{z}_1 \\ \dot{z}_2 \\ \dot{y} \end{bmatrix} = \begin{bmatrix} -z_1 \\ 0 \\ 0 \end{bmatrix} + \begin{bmatrix} -5z_1^2 \\ -1.5 \\ 0 \end{bmatrix} y + \begin{bmatrix} 0 \\ 0 \\ 1 \end{bmatrix} u \tag{5.36}$$

A passifying controller for this system was derived in [8], [19] and [2]. We can rewrite this system (5.36) in the form (5.1) with

$$a(z, y) = 0,\quad b(z, y) = 1,\quad f_0 = [-z_1\ 0]^T,\quad p(y, z) = [-5z_1^2\ -1.5]^T$$

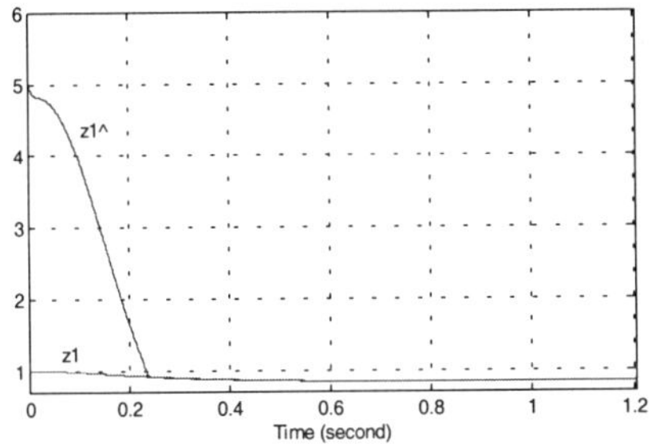

FIGURE 5.4. States z_1 and $\hat{z}_1$.

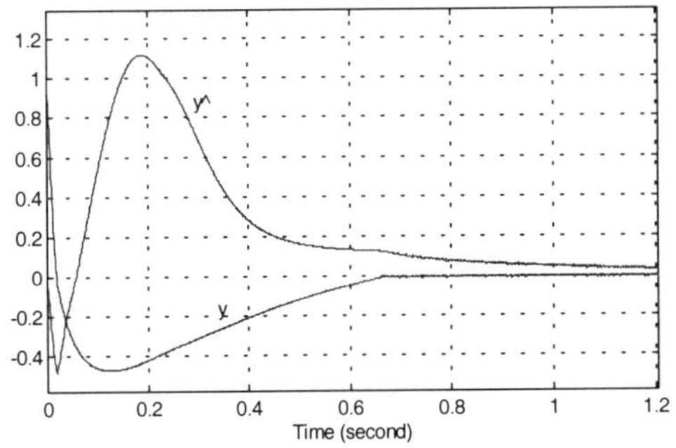

FIGURE 5.5. Output y and $\hat{y}$.

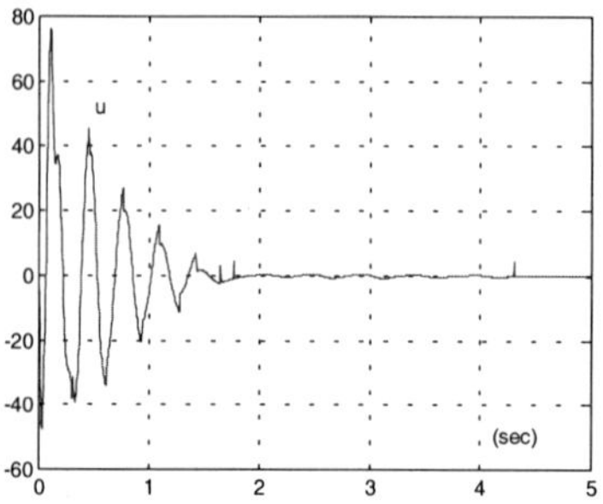

FIGURE 5.6. Control input u.

The zero dynamics of the system (5.36) is stable with the corresponding Lyapunov function equal to

$$W_0(z) = \frac{1}{2}(z_1^2 + z_2^2)$$

One assumes that $f_0(z)$ and $b(y,z)$ are known and that $p(y,z)$, jointly with $a(y,z)$, are unknown. The initial conditions are

$$[z_1(0),\ z_2(0),\ y(0)]^T = [-1,\ 1,\ -2]^T$$
$$[\hat{z}_1(0),\ \hat{z}_2(0),\ \hat{y}(0)]^T = [1,\ -1,\ 1]^T$$

The selected feedback is

$$v = -y$$

The corresponding simulation results are shown in Figure 6.7 - Figure 6.10.

The parameters were selected as follows:

$$A = \begin{bmatrix} -3 & 0 \\ 0 & -3 \end{bmatrix}, \quad W_1^* = \begin{bmatrix} 0 & 0 & 0 \\ 0 & 0 & 0 \end{bmatrix}, \quad W_2^* = [0,0,0]$$
$$P_z = \begin{bmatrix} 2.2918 & 0 \\ 0 & 2.2918 \end{bmatrix} \Lambda_{\tilde{f}'} = \begin{bmatrix} 3 & 0 \\ 0 & 3 \end{bmatrix}, \quad L_{\tilde{f}'} = 2$$
$$\overline{B}_1 = 1.5\sqrt{25 * z_1^2 + 2.25 + .01y^2}, \quad \overline{B}_2 = \sqrt{0.01 \cdot z_2^2 + 0.001\,(z_1^2 + y^2)}$$
$$\eta_1 = diag\,\{20, 20, 20\}\,, \ \ \eta_2 = 20, \ k_1 = 1, \ k_2 = 1$$

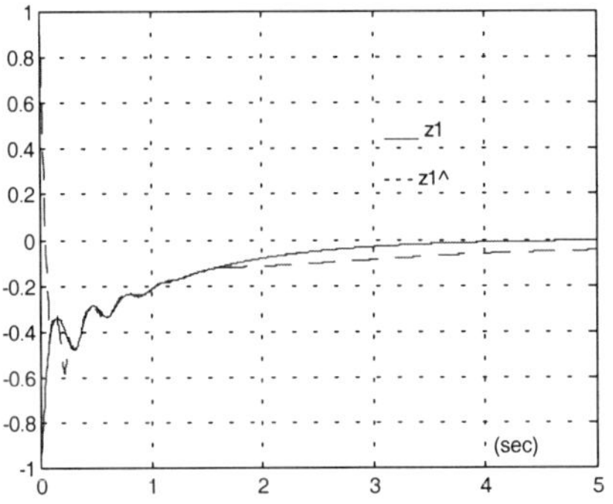

FIGURE 5.7. States $z_1$and $\hat{z}_1$.

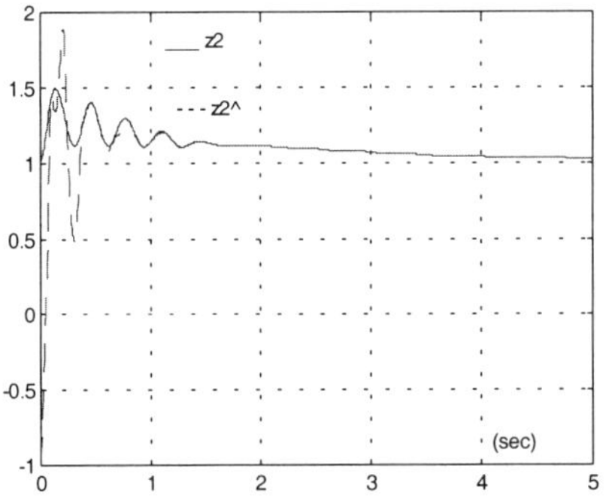

FIGURE 5.8. States z_2 and $\hat{z}_2$.

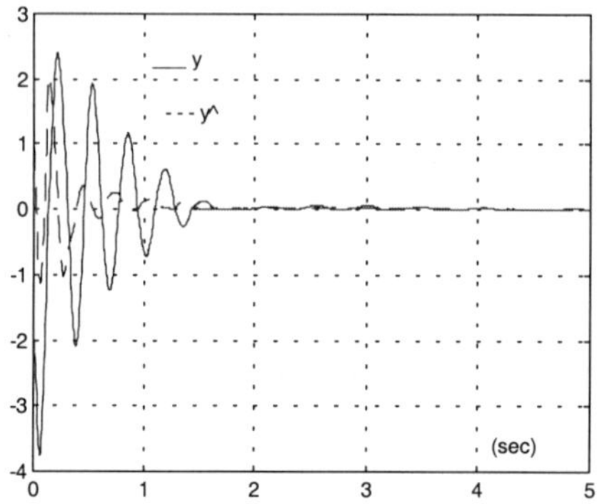

FIGURE 5.9. States y and $\hat{y}$.

As it follows from the examples presented above, the suggested approach provides the stability property for the partially known nonlinear systems closed by the simplest linear negative feedback.

5.5 Conclusions

In this chapter a new adaptive technique is suggested to provide the passivity property for a class of partially known SISO nonlinear systems. A simple dynamic neural network (DNN), containing only two neurons, is used to identify the unknown nonlinear system. The methodology, proposed in this study, can be considered as an alternative approach to the existing ones dealing with a passivity feedback equivalence within a class of uncertain nonlinear systems. The structure of the Dynamical Neural Network is constructed using only the known part of the uncertain nonlinear system. The corresponding output thresholds are adjusted in such a way to compensate the uncertainty influence. A learning law is derived by means of a Lyapunov-like analysis. The passivating feedback control law, as well as the learning law for the dynamical neural network, contains some design parameters which with an adequate selection can improve the performance of the corresponding closed-loop system.

5.6 References

[1] C.I. Byrnes, A. Isidori and J.C. Willems, "Passivity, Feedback Equivalence, and the Global Stabilization of Minimum Phase Nonlinear Systems," *IEEE Trans. Automat. Contr.*, vol.36, 1228-1240, 1991.

[2] R. Castro-Linares, Wen Yu and Alexander S. Poznyak, "Passivity Feedback Equivalence of Nonlinear Systems via Neural Network Aproximation", European Control Conference (ECC′99) Dusseldorf, 1999.

[3] S.Haykin , *Neural Networks- A comprehensive Foundation*, Macmillan College Publ. Co., New York, 1994.

[4] D. Hill and P. Moylan, "Stability results for nonlinear feedback systems," *Automatica*, Vol. 13, pp. 373-382, 1976.

[5] J.J.Hopfield, "Neurons with grade response have collective computational properties like those of a two-state neurons", *Proc. of the National Academy of Science, USA*, vol. 81, 3088-3092, 1984.

[6] K.J. Hunt, D. Sbarbaro, R. Zbikowski and P.J. Gawthrop, "Neural networks for control systems: A survey," *Automatica*, vol. 28, pp. 1083-1112, 1992.

[7] I. Kanellakopoulos, "Passive adpative control of non-linear systems," *Int. J. Adaptive Control Signal Processing*, vol. 7, pp. 339-352, 1993.

[8] P. Kokotovic, M. Krstic and I. Kanellakopoulos, "Backstepping to passivity: recursive design of adaptive systems," *Proc. of 31st IEEE Conf. on Decison and Control*, Tucson, Arizona, pp. 3276-3280, 1997.

[9] E.B.Kosmatopoulos, M.M.Polycarpou, M.A.Christodoulou and P.A.Ioannpu, "High-Order Neural Network Structures for Identification of Dynamical Systems", *IEEE Trans. on Neural Networks*, Vol.6, No.2, 442-431, 1995.

[10] F.L. Lewis and T. Parisini, "Neural network feedback control with guaranteed stability," *Int. J. Control*, vol.70, pp. 337-340, 1998.

[11] W.Lin and T.Shen, "Robust Passivity and Control of Nonlinear Systems with Structural Uncertainty," *Proc. of 36th IEEE Conf. on Decison and Control*, San Diego, California, pp. 2837-2842, 1997.

[12] W.Lin, "Feedback Stabilization of General Nonlinear Control System: A Passive System Approach," *Syst. Control Lett.*, Vol.25, 41-52, 1995.

[13] M.M. Polycarpou and M.J. Mears, "Stable adaptive tracking of uncertain systems using nonlinearly parametrized on-line approximators," *Int. J. Control*, vol.70, pp. 363-384, 1998.

[14] A.S. Poznyak, W.Yu , E.N. Sanchez and Jose P. Perez, Stability Analysis of Dynamic Neural Control, *Expert System with Applications*, Vol.14, No.1, 227-236, 1998.

[15] Juan Reyes-Reyes, Wen Yu and Alexander S.Poznyak, Passivation and Control of Partially Known SISO Nonlinear Systems via Dynamic Neural Networks, Mathematical Problems in Engineering, *Mathematical Problems in Engineering*,, Vol.6, No.1, 61-83, 2000.

[16] G.A.Rovithakis and M.A.Christodoulou, "Direct Adaptive Regulation of Unknown Nonlinear Dynamical System via Dynamical Neural Networks", *IEEE Trans. on Syst., Man and Cybern.*, Vol. 25, 1578-1594, 1995.

[17] N. Sadegh, "A multilayer nodal link perceptron network with least squares training algorithm," *Int. J. Control*, vol.70, pp. 385-404, 1998.

[18] M.A. Seron, D.J. Hill and A.L. Fradkov, "Nonlinear adaptive control of feedback passive systems," *Automatica*, vol.31, pp. 1053-1060, 1995.

[19] W. Su and L. Xie, "Robust control of nonlinear feedback passive systems," *Syst. Control Lett.*, vol.28, pp. 85-93, 1996.

[20] J. C. Willems, "Dissipative dynamical systems part I: General theory," *Arch. Ration. Mech. Anal.*, vol.45, pp. 325-351, 1972.

[21] J. C. Willems, "Dissipative dynamical systems part II: Linear systems with quadratic supply rates, *Arch. Ration. Mech. Anal.*, vol.45, pp. 352-393, 1972.

6

Neuro Trajectory Tracking

Once we know how to model a nonlinear system by a dynamic neural identifier or neuro-observer, we develop an indirect control law to track a reference nonlinear model. First we consider a neuro identifier and using the on-line adapted parameter of the dynamic neural network, we implement an nonlinear control law, which minimizes the tracking error and the input energy. Then, we assume that not all system states are measurable and implement the above discussed neuro-observer. The indirect control law has the same structure as before, but with the state space replaced by its estimate. In both cases a bound for the trajectory error is guaranteed. We propose a trajectory tracking controller based on a new adaptive neuro-observer. First, a solution of the state observation problem using dynamic neural network, for continuous, uncertain nonlinear systems, subjected to external and internal disturbances of bounded power, is discussed. The proposed robust neuro-observer has an extended Luneburger structure as before. Its weights are learned on-line by a new adaptive gradient-like technique. After, we analyze trajectory tracking for uncertain nonlinear system. The control scheme is based on the proposed neuro-observer. The trajectory to be tracked is generated by a nonlinear reference model. We derive a control law to guarantee a bound for the trajectory error. The final structure is composed by two parts: the neuro-observer and the tracking controller. Some simulation results conclude this chapter.

6.1 Tracking Using Dynamic Neural Networks

The nonlinear system to be controlled is given as

$$\begin{array}{c} \dot{x}_t = f(x_t, u_t, t), \\ x_t \in \Re^n,\ u_t \in \Re^m, \quad n \geq m \end{array} \tag{6.1}$$

We consider the neural network without hidden layer as before (see Chapter 2):

$$\dot{\widehat{x}}_t = A\widehat{x}_t + W_{1,t}\sigma\left(\widehat{x}_t\right) + W_{2,t}\phi\left(\widehat{x}_t\right)u_t \tag{6.2}$$

where $A \in \Re^{n\times n}$ is a stable matrix, $W_{1,t} \in \Re^{n\times m}$ is the weight matrix for nonlinear state feedback, $W_{2,t} \in \Re^{n\times k}$ is the input weight matrix, $\widehat{x}_t \in \Re^n$ is the state of the neural network. The matrix function $\phi(\cdot)$ is assumed to be $\Re^{k\times k}$ diagonal:

$$\phi\left(\widehat{x}_t\right) = \delta_{ij}\phi_i(\widehat{x})$$

The vector functions $\sigma(\cdot)$ are assumed to be $m-$dimensional with the elements increasing monotonically. The typical presentation of the elements $\sigma_i(\cdot)$ are as sigmoid functions. The vector $u_t \in \Re^q$ is a control action to be synthesized.

As it follows from Chapter 2, the nonlinear system (6.1) may be modeled as

$$\begin{aligned}\dot{x}_t = & Ax_t + W_{1,t}\sigma(\widehat{x}_t) + W_2\phi\left(\widehat{x}_t\right)u_t + \Delta f(x_t,u_t,t) \\ & + W_{1,t}\left(\sigma(x_t) - \sigma(\widehat{x}_t)\right) + W_{2,t}\left(\phi(x_t) - \phi(\widehat{x}_t)\right)u_t\end{aligned} \tag{6.3}$$

where $\Delta f(x_t,u_t,t)$ is an unmodeled dynamic part. If updated law of $W_{1,t}$ and $W_{2,t}$ is the same as in Chapter 2, and the control is bounded as $\|u_t\| \leq \overline{u}$, then$W_{1,t}$ and $W_{2,t}$ are bounded. Using the assumptions of **A2.1** and **A2.4,** we can conclude that the term

$$d_t := \Delta f(x_t,u_t,t) + W_{1,t}\left(\sigma(x_t) - \sigma(\widehat{x}_t)\right) + W_{2,t}\left(\phi(x_t) - \phi(\widehat{x}_t)\right)u_t$$

is bounded too. So, (6.3) can be rewritten as

$$\dot{x}_t = Ax_t + W_{1,t}\sigma(\widehat{x}_t) + W_{2,t}\phi(\widehat{x}_t)u_t + d_t \tag{6.4}$$

where d_t is bounded vector function.

Based on the neural network identifier (6.2), we will design a controller to force the nonlinear system (6.1) to track a optimal trajectory

$$x_t^* \in \Re^r$$

which is assumed to be smooth enough. This trajectory is regarded as a solution of a nonlinear reference model given by

$$\dot{x}_t^* = \varphi\left(x_t^*, t\right) \tag{6.5}$$

with a known fixed initial condition. In other words, we would like to *synchronize* our dynamics with a given reference dynamic given by (6.5). If the trajectory has points of discontinuity in some fixed moments, we can use any approximating trajectory which is smooth.

In the case of regulation problem we have

$$\varphi(x_t^*, t) = 0, \quad x^*(0) = c$$

where c is a known constant vector.

Let us define the state trajectory error as:

$$\Delta_t = x_t - x_t^* \tag{6.6}$$

From (6.4) and (6.5) we have

$$\dot{\Delta}_t = Ax_t + W_{1,t}\sigma(\widehat{x}_t) + W_{2,t}\phi(\widehat{x}_t)\, u_t + d_t - \varphi(x_t^*, t) \,. \tag{6.7}$$

Now let us assume the control action u_t is made up of two parts:

$$u_t = u_{1,t} + [W_{2,t}\phi(\widehat{x}_t)]^+ \, u_{2,t} \tag{6.8}$$

where $u_{1,t} \in \Re^n$ is direct linearization part and $u_{2,t} \in \Re^n$ is a compensation of unmodeled dynamic d_t. Here $[\cdot]^+$ is the pseudoinverse operator in Moor-Penrose sense [1] satisfying

$$A^+AA^+ = A^+, \quad AA^+A = A$$

and, in view of Householder's separative presentation,

$$A = U\begin{bmatrix} \Lambda & 0 \\ 0 & 0 \end{bmatrix}V, \quad A^+ = V^{-1}\begin{bmatrix} \Lambda^{-1} & 0 \\ 0 & 0 \end{bmatrix}U^{-1}$$

where the matrices U, V are unitary and ortolagium, i.e.,

$$UU^T = I, \quad VV^T = I, \quad UV^T = 0$$

As

$$\varphi(x_t^*, t), \quad x_t^*, \quad W_{1,t}\sigma(\widehat{x}_t), \quad W_{2,t}\phi(\widehat{x}_t)$$

are available, we can select $u_{1,t}$ satisfying

$$W_{2,t}\phi\left(\widehat{x}_t\right)u_{1,t} = \left[\varphi\left(x_t^*,t\right) - Ax_t^* - W_{1,t}\sigma(\widehat{x}_t)\right] \tag{6.9}$$

One of the way to do this is as follows:

$$u_{1,t} = \left[W_{2,t}\phi\left(\widehat{x}_t\right)\right]^+ \left[\varphi\left(x_t^*,t\right) - Ax_t^* - W_{1,t}\sigma(\widehat{x}_t)\right]$$

So, (6.6) becomes

$$\dot{\Delta}_t = A\Delta_t + u_{2,t} + d_t \tag{6.10}$$

Four approaches will be applied to construct $u_{2,t}$ to cancel the negative influence of the term d_t.

1. : **Direct compensation control for nonlinear systems with measurable state derivatives.**

   From (6.4) and (6.2) we have

   $$d_t = \left(\dot{x}_t - \dot{\widehat{x}}_t\right) - A\left(x_t - \widehat{x}_t\right)$$

   If $\dot{x}_t$ is available, we can select $u_{2,t}$ as

   $$u_{2,t} = -d_t = A\left(x_t - \widehat{x}_t\right) - \left(\dot{x}_t - \dot{\widehat{x}}_t\right) \tag{6.11}$$

   So, the ODE, which describes the state trajectory error, now is

   $$\dot{\Delta}_t = A\Delta_t \tag{6.12}$$

   Since A is (6.2) is stable, Δ_t is globally asymptotically stable.

   $$\lim_{t\to\infty} \Delta_t = 0$$

2. : **Sliding mode type control.**

If $\dot{x}_t$ is not available, the sliding mode technique may be applied. Let us define Lyapunov-like function as

$$V_t = \Delta_t P \Delta_t, \quad P = P^T > 0 \tag{6.13}$$

where P is a solution of the Lyapunov equation

$$A^T P + PA = -I \tag{6.14}$$

Using (6.10), we can calculate the time derivative of V which turns out to be equal

$$\dot{V}_t = \Delta_t^T \left(A^T P + PA\right) \Delta_t + 2\Delta_t^T P u_{2,t} + 2\Delta_t^T P d_t \tag{6.15}$$

According to sliding mode technique described in Chapter 3, we select $u_{2,t}$ as

$$u_{2,t} = -kP^{-1}\text{sign}(\Delta_t), \quad k > 0 \tag{6.16}$$

where k is a positive constant, and

$$\text{sign}(\Delta_t) := [\text{sign}(\Delta_{1,t}), \cdots, \text{sign}(\Delta_{n,t})]^T \in \Re^n$$

Compared with Chapter 3, substituting (6.14) and (6.16) into (6.15) leads to

$$\begin{aligned}\dot{V}_t &= - \|\Delta_t\|^2 - 2k \|\Delta_t\| + 2\Delta_t^T P d_t \\ &\leq - \|\Delta_t\|^2 - 2k \|\Delta_t\| + 2\lambda_{\max} (P) \|\Delta_t\| \|d_t\| \\ &= - \|\Delta_t\|^2 - 2 \|\Delta_t\| (k - \lambda_{\max} (P) \|d_t\|)\end{aligned}$$

If we select

$$k > \lambda_{\max} (P) \overline{d}$$

where $\overline{d}$ is upper bound of $\|d_t\|$,i.e.,

$$(\overline{d} = \sup_t \|d_t\|)$$

then we get

$$\dot{V}_t < 0$$

So (see Appendix B)

$$\lim_{t \to \infty} \Delta_t = 0$$

3. : **Direct compensation control with on-line derivative estimation.**

If x_t is not available, an approximate method can be used to obtain good enough estimates:

$$\dot{x}_t = \frac{x_t - x_{t-\tau}}{\tau} + \delta_t \tag{6.17}$$

where δ_t is the approximation error. According to (6.11), we can select

$$u_{2,t} = A\left(x_t - \widehat{x}_t\right) - \left(\frac{x_t - x_{t-\tau}}{\tau} - \dot{\widehat{x}}_t\right) \tag{6.18}$$

So, (6.12) becomes

$$\dot{\Delta}_t = A\Delta_t + \delta_t$$

Define Lyapunov-like function as in (6.13) whose time derivative is

$$\dot{V}_t = \Delta_t^T \left(A^T P + PA\right) \Delta_t + 2\Delta_t^T P \delta_t \tag{6.19}$$

Similar with (2.19) in Chapter 2, the term $2\Delta_t^T P\delta_t$ can be estimated as

$$2\Delta_t^T P\delta_t \leq \Delta_t^T P\Lambda P\Delta_t + \delta_t^T \Lambda^{-1} \delta_t$$

where Λ is any positive matrix. So, (6.19) becomes

$$\dot{V}_t \leq \Delta_t^T \left(A^T P + PA + P\Lambda P + Q\right) \Delta_t + \delta_t^T \Lambda^{-1} \delta_t - \Delta_t^T Q \Delta_t$$

where Q is also any positive define matrix. Because A is stable, there exists the matrices Λ and Q such that the matrix Riccati equation

$$A^T P + PA + P\Lambda^{-1} P + Q = 0$$

has a positive solution $P = P^T > 0$. Defining the following semi-norm:

$$\|\Delta\|_Q^2 = \overline{\lim_{T\to\infty}} \frac{1}{T} \int_0^T \Delta_t^T Q \Delta_t dt$$

where $Q = Q > 0$ is the given weighting matrix, the state trajectory tracking can be formulated as the following optimization problem:

$$J_{\min} = \min_{u_t} J, \quad J = \|x_t - x_t^*\|_Q^2 \tag{6.20}$$

The control law (6.18) and (6.9), based on neural network (6.2) and the nonlinear reference model (6.5), leads to the following property :

$$\dot{V_t} \leq \Delta_t^T \left(A^T P + PA + P\Lambda P + Q\right) \Delta_t + \delta_t^T \Lambda^{-1} \delta_t - \Delta_t^T Q \Delta_t = \\ \delta_t^T \Lambda^{-1} \delta_t - \Delta_t^T Q \Delta_t$$

from which we conclude that

$$\Delta_t^T Q \Delta_t \leq \delta_t^T \Lambda^{-1} \delta_t - \dot{V_t}$$

$$\int_{t=0}^{T} \Delta_t^T Q \Delta_t dt \leq \int_{t=0}^{T} \delta_t^T \Lambda^{-1} \delta_t dt - V_t + V_0 \leq \int_{t=0}^{T} \delta_t^T \Lambda^{-1} \delta_t dt + V_0$$

and, hence,

$$J = \|\Delta_t\|_Q^2 \leq \|\delta\|_{\Lambda^{-1}}^2$$

4. : **Local Optimal Control.**

If $\dot{x}_t$ is not available and $\dot{x}_t$ is not approximated as in Approach 3, in order to analyze the tracking error stability, we also introduce the quadratic Lyapunov function:

$$V_t(\Delta_t) = \Delta_t^T P \Delta_t, \quad P = P^T > 0 \tag{6.21}$$

In view of (6.10), its time derivative can be calculated as

$$\dot{V}_t(\Delta_t) = \Delta_t^T \left(A^T P + PA\right) \Delta_t + 2\Delta_t^T P u_{2,t} + 2\Delta_t^T P d_t \tag{6.22}$$

The term $2\Delta_t^T P d_t$ can be estimated as

$$2\Delta_t^T P d_t \leq \Delta_t^T P \Lambda^{-1} P \Delta_t + d_t^T \Lambda d_t \tag{6.23}$$

Substituting (6.23) in (6.22), adding and subtracting the terms

$$\Delta_t^T Q \Delta_t \text{ and } \Delta u_{2,t}^T R u_{2,t}$$

with

$$Q = Q^T > 0 \text{ and } R = R^T > 0$$

we formulate:

$$\dot{V}_t(\Delta_t) \leq \Delta_t^T \left(A^T P + PA + P\Lambda P + Q \right) \Delta_t \tag{6.24}$$
$$+2\Delta_t^T P u_{2,t} + u_{2,t}^T R u_{2,t} + d_t^T \Lambda^{-1} d_t - \Delta_t^T Q \Delta_t - u_{2,t}^T R u_{2,t}$$

We need to find a positive solution to make the first term in (6.24) equal to zero. That means that there exists a positive solutions P satisfy following matrix Riccati equation

$$A^T P + PA + P\Lambda P + Q = 0 \tag{6.25}$$

It has positive definite solution if the pair $(A, \Lambda^{1/2})$ is controllable, the pair $(Q^{1/2}, A)$ is observable, and a special *local frequency condition* (see Appendix A), its sufficient condition is fulfilled:

$$\frac{1}{4}\left(A_0^T R^{-1} - R^{-1} A_0 \right) R \left(A_0^T R^{-1} - R^{-1} A_0 \right)^T \leq A_0^T R^{-1} A_0 - Q \tag{6.26}$$

This can be realized by a corresponding selection of A and Q. So, (6.25) is established. Then, in view of this fact, the inequality (6.24) takes the form

$$\dot{V}_t(\Delta_t) \leq -\left(\|\Delta_t\|_Q^2 + \|u_{2,t}\|_R^2 \right) + \Psi(u_{2,t}) + d_t^T \Lambda^{-1} d_t \tag{6.27}$$

where the function Ψ is defined as

$$\Psi(u_{2,t}) := 2\Delta_t^T P u_{2,t} + u_{2,t}^T R u_{2,t}$$

We reformulate (6.27) as

$$\|\Delta_t\|_Q^2 + \|u_{2,t}\|_R^2 \leq \Psi(u_{2,t}) + d_t^T \Lambda^{-1} d_t - \dot{V}_t(\Delta_t)$$

Then, integrating each term from 0 to τ, dividing each term by τ, and taking the limit on $\tau \to \infty$ of these integrals' supreme, we obtain:

$$\begin{array}{c}\limsup\limits_{\tau\to\infty}\frac{1}{\tau}\int_0^{\tau}\Delta_t^TQ\Delta_t dt+\limsup\limits_{\tau\to\infty}\frac{1}{\tau}\int_0^{\tau}u_{2,t}^TRu_{2,t}dt \\ \leq \limsup\limits_{\tau\to\infty}\frac{1}{\tau}\int_0^{\tau}d_t^T\Lambda^{-1}d_t dt+\limsup\limits_{\tau\to\infty}\frac{1}{\tau}\int_0^{\tau}\Psi\left(u_{2,t}\right)dt+\limsup\limits_{\tau\to\infty}\left[-\frac{1}{\tau}\int_0^{\tau}\dot{V}\left(\Delta_t\right)\right]\end{array}$$

Using the following semi-norms definition

$$\left\|\Delta_t\right\|_Q^2=\limsup_{\tau\to\infty}\frac{1}{\tau}\int_0^{\tau}x_t^TQ_cx_tdt,\quad \left\|u_{2,t}\right\|_R^2=\limsup_{\tau\to\infty}\frac{1}{\tau}\int_0^{\tau}u_t^TR_cu_tdt$$

we get

$$\left\|\Delta_t\right\|_Q^2+\left\|u_{2,t}\right\|_R^2\leq\left\|d_t\right\|_{\Lambda^{-1}}^2+\limsup_{\tau\to\infty}\frac{1}{\tau}\int_0^{\tau}\Psi\left(u_{2,t}\right)dt$$

The right-side hand fixes a *tolerance level* for the trajectory tracking error. So, the control goal now is to minimize $\Psi\left(u_{2,t}\right)$ and $\left\|d_t\right\|_{\Lambda^{-1}}^2$. To minimize $\left\|d_t\right\|_{\Lambda^{-1}}^2$,we should minimize Λ^{-1}. From (6.26), if select A and Q such a way to guarantee the existence of the solution of (6.25), we can choose the minimal Λ^{-1} as

$$\Lambda^{-1}=A^{-T}QA^{-1}$$

To minimizing $\Psi\left(u_t\right)$, we assume that, at the given t (positive), $x^*(t)$ and $\widehat{x}(t)$ are already realized and do not depend on $u_{2,t}$. We name the $u_{2,t}^*(t)$ as *the locally optimal* control (see Appendix C), because it is calculated based only on "local" information. The solution $u_{2,t}^*$ of this optimization problem is given by

$$\begin{cases} u_{2,t}^*=\arg\min\Psi\left(u\right),\quad u\in U \\ \Psi\left(u\right)=2\Delta_t^TPu+u^TRu \\ \text{subjected} \\ A_0(u_{1,t}+u)\leq B_0 \end{cases}$$

It is typical quadratic programming problem. Without any additional constraints ($U=R^n$) the locally optimal control $u_{2,t}^*$ can be found analytically

$$u_{2,t}^*=-2R^{-1}P\Delta_t \tag{6.28}$$

that corresponds to the linear quadratic optimal control law.

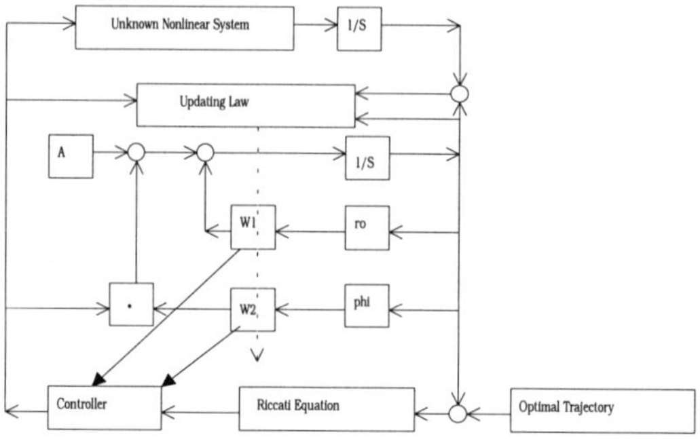

FIGURE 6.1. The structure of the new neuro-controller.

Remark 6.1 *Approach 1,2 lead to exact compensation of d_t, but Approach 1 demands the information on $\dot{x}_t$. As for the approach 2, it realizes the sliding mode control and leads to high vibrations in control that provides quite difficulties in real application.*

Remark 6.2 *Approach 3 uses the approximate method to estimate $\dot{x}_t$ and the finial error δ_t turns out to be much smaller than d_t.*

The final structure of the neural network identifier and the tracking controller is shown in Figure 6.1. The crucial point here is that the neural network weights are learned on-line.

6.2 Trajectory Tracking Based Neuro Observer

Let the class of nonlinear systems be given by

$$\begin{cases} \dot{x}_t = f(x_t, u_t, t) + \xi_{1,t} \\ y_t = Cx_t + \xi_{2,t} \end{cases} \tag{6.29}$$

where

$x_t \in \mathbb{R}^n$ is the state vector of the system,

$u_t \in \mathbb{R}^q$ is a given control action,

$y_t \in \mathbb{R}^m$ is the output vector assumed to be available at any time,

$C \in \mathbb{R}^{m \times n}$ is a *known* output matrix,

$f(\cdot) : \mathbb{R}^{n+q+1} \to \mathbb{R}^n$ is *unknown* vector valued nonlinear function describing the system dynamics and satisfying the following assumption

A6.1: For a realizable feedback control verifying

$$\|u_t(x)\|^2 \leq v_0 + v_1 \|x\|^2 \ \forall x \in \mathbb{R}^n$$

the nominal (unperturbed) closed-loop nonlinear system is quadratically stable, that is, there exists a Lyapunov (may be, unknown) function $\overline{V}_t(x) \geq 0$ such that

$$\frac{\partial \overline{V}_t}{\partial x} f(x_t, u_t(x_t)) \leq -\lambda_1 \|x_t\|^2, \quad \left\| \frac{\partial \overline{V}_t}{\partial x} \right\| \leq \lambda_2 \|x_t\|, \quad \lambda_1, \lambda_2 > 0$$

Remark 6.3 *If a closed-loop system is exponentially stable and $f(x_t, u_t(x_t))$ is uniformly (on t) Lipshitz in x_t, then the converse Lyapunov theorem [8] implies **A6.1.** But assumption **A6.1** is weaker and easy to be satisfied.*

The vectors $\xi_{1,t}$ and $\xi_{2,t}$ represent external unknown bounded disturbances.

A6.2.

$$\|\xi_{i,t}\|^2_{\Lambda_{\xi_i}} = \Upsilon_i < \infty, \ 0 < \Lambda_{\xi_i} = \Lambda_{\xi_i}^T, \ i = 1, 2 \tag{6.30}$$

Normalizing matrices Λ_{ξ_i} (introduced to insure the possibility to work with components of different physical nature) are assumed to be a priori given.

Following to standard techniques [18], if the nonlinear system (without unmodeled dynamics and external disturbances) model is known, the structure for the corresponding nonlinear observer can be suggested as follows:

$$\frac{d}{dt}\hat{x}_t = f(\hat{x}_t, u_t, t) + L_{1,t}[y_t - C\hat{x}_t] \tag{6.31}$$

The first term in the right-hand side of (6.31) repeats the known dynamics of the nonlinear system and the second one is intended to correct the estimated trajectory based on current residual values.

If $L_{1,t} = L_{1,t}(\hat{x}_t)$, this observer is named a "differential algebra" type observer (see [7], [16], and [2]). In the case of $L_{1,t} = L_1 = Const$, it is usually named a "high-gain" type observer studied in [21], [30].

Applying the observer (6.31) to a class of mechanical systems when only position measurements available (velocities are unmeasurable), as a rule, the corresponding velocity estimates turn out to be not so good because of the following effect: the original dynamic mechanical system, in general, is given as

$$\begin{cases} \ddot{z}_t = F(z_t, \dot{z}_t, u_t, t) \\ y = z_t \end{cases}$$

or, in equivalent standard Cauchy form,

$$\begin{cases} \dot{x}_{1,t} = x_{2,t} \\ \dot{x}_{2,t} = \tilde{F}(x_t, u_t, t) \\ y_t = x_{1,t} \end{cases}$$

leading to the corresponding nonlinear observer (6.31) as

$$\frac{d}{dt}\begin{pmatrix} \hat{x}_{1,t} \\ \hat{x}_{2,t} \end{pmatrix} = \begin{pmatrix} \hat{x}_{2,t} \\ \tilde{F}(\hat{x}_t, u_t, t) \end{pmatrix} + \begin{pmatrix} L_{1,1,t} \\ L_{1,2,t} \end{pmatrix} [y_t - \hat{x}_{1,t}] \tag{6.32}$$

which means that observable state components are estimated very well such that the residual term $[y_t - \hat{x}_{1,t}]$ turns out to be very small and has no effect in (6.32). One of the possible solutions of this problem is to add a new time delayed term

$$L_{2,t}\left[h^{-1}(y_t - y_{t-h}) - Ch^{-1}(\hat{x}_t - \hat{x}_{t-h})\right]$$

which can be considered as a "derivative estimation error" and used for tuning the velocity estimations. This new modified observer can be described as

$$\begin{array}{c} \frac{d}{dt}\hat{x}_t = f(\hat{x}_t, u_t, t) + L_{1,t}[y_t - C\hat{x}_t] \\ +L_{2,t}h^{-1}[(y_t - y_{t-h}) - C(\hat{x}_t - \hat{x}_{t-h})] \end{array} \tag{6.33}$$

If we have no complete information on the nonlinear function $f(\hat{x}_t, u_t, t)$, it seems natural to construct its estimate as $\hat{f}(\hat{x}_t, u_t, t \mid W_t)$ depending on parameters W_t. These parameters should be adjusted on-line to obtain the best nonlinear approximation of unknown dynamic operator. That leads to the following observer scheme:

$$\begin{array}{c} \frac{d}{dt}\hat{x}_t = \hat{f}(\hat{x}_t, u_t, t \mid W_t) + L_{1,t}[y_t - C\hat{x}_t] \\ +L_{2,t}h^{-1}[(y_t - y_{t-h}) - C(\hat{x}_t - \hat{x}_{t-h})] \end{array} \tag{6.34}$$

supplied with a special *updating* (learning) law

$$\dot{W}_t = \Phi\left(W_t, \hat{x}_t, u_t, t, y_t\right) \tag{6.35}$$

Such "robust adaptive observer" seems to be a more advanced device, which provides a good estimation under the absence of dynamic model and incomplete state measurement.

Starting from this point throughout of this chapter we will consider the controlled nonlinear dynamics closed by *the nonlinear feedback using the current state estimates*, that is, the following assumption will be in force

A6.3:

$$u_t = u\left(\widehat{x}_t, t\right)$$

such a way that, in view of A6.1,

$$\left\|u_t\right\|^2 = \left\|u\left(\widehat{x}_t, t\right)\right\|^2 \leq v_0 + v_1 \left\|\widehat{x}_t\right\|^2 \ \forall \widehat{x}_t \in \mathbb{R}^n$$

In the next subsection a special observer structure based on *Dynamic Neural Network* (see [10], [24], [21], [25] and [26]) is introduced.

6.2.1 Dynamic Neuro Observer

The *robust neuro observer*, considered in this chapter, uses the structure of dynamic (Hopfield's type) neural networks as in [32], [18] and [24], [21], [25] and [26]. It looks as a Luneburger-like "second order" observer with a new additional time-delay term (see Fig.1):

$$\begin{cases} \frac{d}{dt}\widehat{x}_t = A\widehat{x}_t + W_{1,t}\sigma(V_{1,t}\widehat{x}_t) + W_{2,t}\phi(V_{2,t}\widehat{x}_t)u_t \\ +L_1\left[y_t - \widehat{y}_t\right] + L_2/h\left[(y_t - y_{t-h}) - (\widehat{y}_t - \widehat{y}_{t-h})\right] \\ \widehat{y}_t = C\widehat{x}_t \end{cases} \tag{6.36}$$

Here

the vector $\widehat{x}_t \in \mathbb{R}^n$ is the state of the neural network,

$u_t \in \mathbb{R}^q$ the input,

$A \in \mathbb{R}^{n\times n}$ is a Hurtwitz (stable) constant matrix,

the matrices $W_{1,t} \in \mathbb{R}^{n\times m}$ and $W_{2,t} \in \mathbb{R}^{n\times k}$ are the weights of the output layers,
$V_1 \in \mathbb{R}^{m\times n}$ and $V_2 \in \mathbb{R}^{q\times n}$ are the weights of the hidden layer,
$\sigma(\cdot) : \mathbb{R}^n \to \mathbb{R}^m$ is a sigmoidal vector function,
$\phi(\cdot) : \mathbb{R}^q \to \mathbb{R}^{k\times q}$ is a matrix valued function,
$L_1 \in \mathbb{R}^{n\times m}$ and $L_2 \in \mathbb{R}^{n\times m}$ are first and second order gain matrices,
the scalar $h > 0$ characterizes the time delay used in this procedure.

Remark 6.4 *The most simple structure without hidden layers (containing only input and output layers), corresponds to the case*

$$m = n,\ V_1 = V_2 = I, \quad L_2 = 0 \tag{6.37}$$

This single-layer dynamic neural networks with Luenberger-like observer was considered in [10].

Remark 6.5 *The structure of the observer (6.36) has three parts:*

- *the neural networks identifier*

$$A\widehat{x}_t + W_{1,t}\sigma(V_{1,t}\widehat{x}_t) + W_{2,t}\phi(V_{2,t}\widehat{x}_t)u_t$$

- *the Luenberger tuning term*

$$L_1\left[y_t - \widehat{y}_t\right]$$

- *the additional time-delay term*

$$L_2 h^{-1}\left[(y_t - y_{t-h}) - (\widehat{y}_t - \widehat{y}_{t-h})\right]$$

  where $(y_t - y_{t-h})/h$ *and* $(\widehat{y}_t - \widehat{y}_{t-h})/h$ *are introduced to estimate* $\dot{y}_t$ *and* $\dot{\widehat{y}}_t$, *correspondingly.*

6.2.2 *Basic Properties of DNN-Observer*

Define the estimation error as:

$$\Delta_t := \widehat{x}_t - x_t \tag{6.38}$$

Then, the output error is

$$e_t = \widehat{y}_t - y_t = C\Delta_t - \xi_{2,t}$$

hence,

$$\begin{array}{c} C^T e_t = C^T \left(C\Delta_t - \xi_{2,t}\right) = \left(C^T C + \delta I\right)\Delta_t - \delta I \Delta_t - C^T \xi_{2,t} \\ \Delta_t = C_\delta^+ e_t + \delta N_\delta \Delta_t + C_\delta^+ \xi_{2,t} \end{array} \tag{6.39}$$

where

$$C_\delta^+ = \left(C^T C + \delta I\right)^{-1} C^T, N_\delta = \left(C^T C + \delta I\right)^{-1}$$

and δ is a small positive scalar.

It is clear that all sigmoid functions $\sigma(\cdot)$ and $\phi(\cdot)$, commonly used in NN, satisfy Lipschitz condition. So, it is natural to assume that

A6.4:

$$\begin{array}{c} \widetilde{\sigma}_t^T \Lambda_1 \widetilde{\sigma}_t \leq \Delta_t^T \Lambda_\sigma \Delta_t \\ \left(\widetilde{\phi}_t u_t\right)^T \Lambda_2 \left(\widetilde{\phi}_t u_t\right) = u_t^T \widetilde{\phi}_t^T \Lambda_2 \widetilde{\phi}_t u_t \\ \leqslant \lambda_{\max}\left(\Lambda_2\right) \overline{\overline{\phi}}^2 \left(v_0 + v_1 \left\|\widehat{x}_t\right\|^2\right) \\ \left\|\widetilde{\phi}_t\right\|^2 \leqslant \overline{\overline{\phi}}^2 \\ \widetilde{\sigma}_t := \sigma(V_1^* \hat{x}_t) - \sigma(V_1^* x_t), \quad \widetilde{\phi}_t := \phi(V_2^* \hat{x}_t) - \phi(V_2^* x_t) \\ \sigma(0) = 0, \phi(0) = 0 \end{array}$$

where, based on Lemma 12.5 (see Appendix A), the introduced variables satisfy :

$$\begin{array}{c} \widetilde{\sigma}_t' := \sigma(V_{1,t}\widehat{x}_t) - \sigma(V_1^*\widehat{x}_t), \quad \widetilde{\phi}_t' := \phi(V_{2,t}\widehat{x}_t) - \phi(V_2^*\widehat{x}_t) \\ \sigma_t := \sigma(V_{1,t}\widehat{x}_t), \ \phi_t := \phi(V_{2,t}\widehat{x}_t) \\ \widetilde{\sigma}_t' = D_\sigma \widetilde{V}_{1,t}\widehat{x}_t + \nu_\sigma, \quad \left(\widetilde{\phi}_t' u_t\right)_i = \sum_{j=1}^{q} \nabla^\intercal \phi_{ij}(V_2^*\widehat{x}_t) u_j \widetilde{V}_{2,t}\widehat{x}_t + \nu_\phi \\ D_{\phi u} = \left\| \sum_{j=1}^{q} \nabla \phi_{1j}(V_2^*\widehat{x}_t) u_j \mid \cdots \mid \sum_{j=1}^{q} \nabla \phi_{qj}(V_2^*\widehat{x}_t) u_j \right\| \\ D_\sigma = \frac{\partial \sigma^T(Z)}{\partial Z} \mid_{Z=V_1^*\widehat{x}_t} \in \mathbb{R}^{m\times m}, \quad \left\|\nu_\sigma\right\|_{\Lambda_1}^2 \leq l_1 \left\|\widetilde{V}_{1,t}\widehat{x}_t\right\|_{\Lambda_1}^2, \quad l_1 > 0 \\ \left\|\nu_\phi\right\|_{\Lambda_2}^2 \leq l_2 \left\|V_{2,t}\widehat{x}_t\right\|_{\Lambda_2}^2, \quad l_2 > 0 \\ \widetilde{V}_{1,t} := V_{1,t} - V_1^*, \quad \widetilde{V}_{2,t} := V_{2,t} - V_2^* \\ \widetilde{W}_{1,t} := W_{1,t} - W_1^*, \quad \widetilde{W}_{2,t} := W_{2,t} - W_2^* \end{array} \tag{6.40}$$

Λ_1, Λ_2, Λ_σ and Λ_ϕ are positive define matrices.

For the general case, when the neural network

$$\dot{\widehat{x}}_t = A\widehat{x}_t + W_{1,t}\sigma(V_{1,t}\widehat{x}_t) + W_{2,t}\phi(V_{2,t}\widehat{x}_t)u_t$$

can not exactly match the given nonlinear system (6.29), this system can be represented as

$$\dot{x}_t = Ax_t + W_1^*\sigma(V_1^*x_t) + W_2^*\phi(V_2^*x_t)u_t + \widetilde{f}_t \tag{6.41}$$

where

$$\widetilde{f}_t = f(x_t, u_t, t) - [Ax_t + W_1^*\sigma(V_1^*x_t) + W_2^*\phi(V_2^*x_t)u_t] + \xi_{1,t}$$

is the unmodeled dynamic term and W_1^*, W_2^*, V_1^* and V_2^* are any known matrices which are selected below as initial conditions for the designed differential learning law.

To guarantee the global existence for the solution of (6.29), the following condition should be satisfied

$$\|f(x_t, u_t, t)\|^2 \le C_1 + C_2\|x_t\|^2 + C_3\|u_t\|^2$$

C_1 and C_2 are positive constants [4]. In view of this and taking into account that the sigmoid functions σ and ϕ are uniformly bounded, the following assumption, concerning the unmodeled dynamics $\widetilde{f}_t$, seems to be natural:

A6.5: There exist positive constants $\overline{\eta}, \overline{\eta}_1$ and $\overline{\eta}_2$ such that

$$\left\|\widetilde{f}_t\right\|_{\Lambda_f}^2 \le \overline{\eta} + \overline{\eta}_1\|x_t\|_{\Lambda_{f_1}}^2 + \overline{\eta}_2\|\hat{x}_t\|_{\Lambda_{f_2}}^2, \quad \Lambda_{f_i} = \Lambda_{f_i}^T > 0 \ (i = 1, 2)$$

The next fact plays a key role in this study. It is well known [3] (see also Appendix A) that if the matrix A is stable, the pair $(A, R^{1/2})$ is controllable, the pair $(Q^{1/2}, A)$ is observable, and the special *local frequency condition* or its matrix equivalent

$$A^TR^{-1}A - Q \ge \frac{1}{4}\left[A^TR^{-1} - R^{-1}A\right]R\left[A^TR^{-1} - R^{-1}A\right]^T \tag{6.42}$$

is fulfilled, then the matrix Riccati equation

$$A^T P + PA + PRP + Q = 0 \tag{6.43}$$

has a positive solution P. In view of this fact, we accept the following additional assumption.

A6.6: There exist a stable matrix A, a positive parameter δ and a positively definite matrix Q_0 such that two matrix Riccati equations in the form (6.43) with

$$\begin{array}{r}
A_1 := A + (L_1 + L_2 h^{-1})\, C \\
R_1 := 2\overline{W}_1 + 2\overline{W}_2 + \Lambda_f^{-1} + \Lambda_{\xi_1}^{-1} \\
+\, (L_1 + L_2 h^{-1})\, \Lambda_{\xi_2}^{-1} (L_1 + L_2 h^{-1})^{\intercal} + 2h^{-1} L_2 \Lambda_{\xi_2}^{-1} L_2^T \\
Q_1 := \Lambda_\sigma + \delta \Lambda_1 + (1+\delta)\, \Lambda_3^{-1} + Q_0
\end{array} \tag{6.44}$$

and

$$\begin{array}{r}
A_2 := A \\
R_2 := \overline{W}_1 + \overline{W}_2 + \\
(L_1 + L_2 h^{-1}) \left(C \Lambda_3^{-1} C^{\intercal} + \Lambda_{\xi_2}^{-1} \right) (L_1 + L_2 h^{-1})^{\intercal} + 2 L_2 \Lambda_{\xi_2}^{-1} L_2^{\intercal} \\
Q_2 := 2 V_1^{*\intercal} \Lambda_\sigma V_1^* + \|\Lambda_2\| \,\overline{\overline{\phi}}^{2}\, v_1 I + \lambda_{\max}(\Lambda_2)\, \overline{\overline{\phi}}^{2}\, v_1 I + Q_0
\end{array}$$

have the positive solutions P and P_2.

This conditions is easily verified if we select A as a stable diagonal matrix. Denote the class of unknown nonlinear systems satisfying **A6.1**-**A6.6** by $\mathcal{H}$.

6.2.3 Learning Algorithm and Neuro Observer Analysis

The main contribution of this study is the new dynamic learning law which can be expressed by the following system of matrix differential equations:

$$\begin{array}{r}
\tilde{L}_{W_1,t} := L_{W_1,t} + 2\sigma(V_{1,t}\widehat{x}_t)\widehat{x}_t^{\intercal} P_2 = 0 \\
\tilde{L}_{W_2,t} := L_{W_2,t} + 2\phi(V_2\widehat{x}_t) u_t \widehat{x}_t^{\intercal} P_2 = 0 \\
\tilde{L}_{V_1,t} := L_{V_1,t} + 2\Lambda_\sigma \tilde{V}_{1,t} \widehat{x}_t \widehat{x}_t^{\intercal} = 0 \\
\tilde{L}_{V_2,t} := L_{V_2,t} = 0
\end{array} \tag{6.45}$$

where

$$
\begin{aligned}
L_{W_{1,t}} &:= 2\,\dot{\widetilde{W}}_{1,t}^{T} K_1^{-1} + \left[2\sigma_t e_t^{\intercal}\left(C_\delta^{+}\right)^{\intercal} P \right. \\
&\quad \left. + \sigma_t\sigma_t^{\intercal}\widetilde{W}_{1,t}^{\intercal} P\left[\delta N_\delta \Lambda_3 \left(N_\delta\right)^{\intercal} + C_\delta^{+}\Lambda_{\xi_2}^{-1}\left(C_\delta^{+}\right)^{\intercal}\right] P\right] \\
L_{W_{2,t}} &:= \left[2\phi_t u_t e_t^{\intercal}\left(C_\delta^{+}\right)^{\intercal} P + \phi_t u_t u_t^{\intercal}\phi_t^{\intercal}\widetilde{W}_{2,t}^{\intercal} P\left[\delta N_\delta \Lambda_3 \left(N_\delta\right)^{\intercal}\right.\right. \\
&\quad \left.\left. + C_\delta^{+}\Lambda_{\xi_2}^{-1}\left(C_\delta^{+}\right)^{\intercal}\right] P + 2\,\dot{\widetilde{W}}_{2,t}^{T} K_2^{-1} \right. \\
L_{V_{1,t}} &:= 2\,\dot{\widetilde{V}}_{1,t}^{\intercal} K_3^{-1} + \left(2\widehat{x}_t e_t^{\intercal}\left(C_\delta^{+}\right)^{\intercal} P W_1^{*} D_\sigma \right. \\
&\quad \left. + \widehat{x}_t\widehat{x}_t^{\intercal}\widetilde{V}_{1,t}^{\intercal}\left[D_\sigma^{\intercal}\left(W_1^{*}\right)^{\intercal} P\left(N_\delta \Lambda_1^{-1} N_\delta^{\intercal} + C_\delta^{+}\Lambda_{\xi_2}^{-1}\left(C_\delta^{+}\right)^{\intercal}\right) P W_1^{*} D_\sigma + l_1\Lambda_1\right]\right) \\
L_{V_{2,t}} &:= \widehat{x}_t\widehat{x}_t^{\intercal}\widetilde{V}_{2,t}^{\intercal}\left[D_{\phi u}\left(W_2^{*}\right)^{\intercal} P\left(C_\delta^{+}\Lambda_{\xi_2}^{-1}\left(C_\delta^{+}\right)^{\intercal} + N_\delta \Lambda_1^{-1} N_\delta^{\intercal}\right) P W_2^{*} D_{\phi u}^{\intercal}\right. \\
&\quad \left. + l_2\Lambda_2\right] + 2\,\dot{\widetilde{V}}_{2,t}^{T} K_4^{-1}
\end{aligned}
$$

where $K_i \in \mathbb{R}^{n\times n}$ $(i = 1\cdots 4)$ are positive defined matrices, P and P_2 are the solutions of the matrix Riccati equations given by (6.43), correspondingly. $D_{\phi u}^{\intercal}$ and D_σ are defined in (6.40). The initial conditions are $W_{1,0} = W_1^{*}$, $W_{2,0} = W_2^{*}$, $V_{1,0} = V_1^{*}$, $V_{2,0} = V_2^{*}$.

Remark 6.6 *It can be seen that the learning law (6.45) of the neuro observer (6.36) consists of several parts: the first term $K_1 P C^{+} e_t \sigma_t^T$ exactly corresponds to the back-propagation scheme as in multilayer networks [19]; the second term $K_1 P C_\delta^{+} e_t \widehat{x}_t^T \widetilde{V}_{1,t}^T D_\sigma$ is intended to assure the robust stable learning law. Even though the proposed learning law looks like the backpropagation algorithm, global asymptotic error stability is guaranteed because it is derived based on the Lyapunov approach (see next Theorem). So, the global convergence problem does not arise in this case.*

Theorem 6.1 *If the gain matrices L_1 and L_2 are selected such a way that the assumption **A6.6** is fulfilled and the weights are adjusted according to (6.45), then under the assumptions **A6.1-A6.5**, for a given class of nonlinear systems given by (6.29), the following properties hold:*

· *(a) the weight matrices remain bounded, that is,*

$$
W_{1,t} \in L^{\infty}, \quad W_{2,t} \in L^{\infty}, \quad V_{1,t} \in L^{\infty}, \quad V_{2,t} \in L^{\infty}, \tag{6.46}
$$

· *(b) for any $T > 0$ the state estimation error fulfills the following*

$$\left[1 - \mu/\sqrt{V_t}\right]_+ \to 0 \tag{6.47}$$

where

$$\begin{gathered} V_t := V_{1,t} + V_{2,t} \\ V_{1,t} = V^0 + \Delta_t^\intercal P \Delta_t + tr\left[\widetilde{W}_1^T K_1^{-1} \widetilde{W}_1\right] \\ +tr\left[\widetilde{W}_2^T K_2^{-1} \widetilde{W}_2\right] + tr\left[\widetilde{V}_1^T K_3^{-1} \widetilde{V}_1\right] + tr\left[\widetilde{V}_2^T K_4^{-1} \widetilde{V}_2\right] \\ V_{2,t} = \hat{x}_t^\intercal P_2 \hat{x}_t + \int\limits_{\tau=t-h}^{t} \Delta_\tau^T P_1 \Delta_\tau d\tau \end{gathered} \tag{6.48}$$

and

$$\begin{gathered} \mu := \sqrt{\beta/\alpha} \\ \beta := \left[\lambda_{\max}(\Lambda_2) + \|\Lambda_2\|\right] \overline{\overline{\phi}}^2 v_0 + \Upsilon_1 + \left(5 + 2h^{-1}\right) \Upsilon_2 + \overline{\eta} \\ \alpha := \min\left\{\lambda_{\min}\left(P^{-1/2} Q_0 P^{-1/2}\right); \lambda_{\min}\left(P_2^{-1/2} Q_0 P_2^{-1/2}\right)\right\} \end{gathered}$$

Remark 6.7 *For a system without any unmodeled dynamics, i.e., neural network matches the given plant exactly ($\overline{\eta} = 0$), without any external disturbances ($\Upsilon_1 = \Upsilon_2 = 0$) and $v_0 = 0$ ($u(0) = 0$), the proposed neuro-observer (6.36) guarantees the "* ***stability*** *" of the state estimation error, that is,*

$$\beta = 0 \textit{ and } V_t \to 0$$

that is equivalent to the fact that

$$\lim_{t \to \infty} \Delta_t = 0$$

Remark 6.8 *Similar to high-gain observers [30], the proved theorem stays only the fact that the estimation error is bounded asymptotically and does not say anything about a bound for a finite time that obligatory demands fulfilling a local uniform observability condition [2]. In our case, some observability properties are contained in* ***A6*** *(for example, if $C = 0$ this condition can not be fulfilled for any matrix A).*

6.2.4 Error Stability Proof

Now we will present the stability proof and tracking error zone-convergence for the class of adaptive controllers based on the suggested neuro observer.

Part 1: Differential inequality for DNN-error

Defining the Lyapunov candidate function as:

$$
\begin{gathered}
V_{1,t} = V^0 + \Delta_t^T P \Delta_t + tr\left[\widetilde{W}_1^T K_1^{-1} \widetilde{W}_1\right] \\
+tr\left[\widetilde{W}_2^T K_2^{-1} \widetilde{W}_2\right] + tr\left[\widetilde{V}_1^T K_3^{-1} \widetilde{V}_1\right] + tr\left[\widetilde{V}_2^T K_4^{-1} \widetilde{V}_2\right]
\end{gathered} \tag{6.49}
$$

with $P = P^T > 0$ and V^0 a positive constant matrix.

In view of **A6.1**, the derivative of the Lyapunov candidate function $V_{1,t}$ can be estimated as

$$
\begin{gathered}
\dot{V}_{1,t} \leq -\lambda \left\|x_t\right\|^2 + 2\Delta_t^T P \dot{\Delta}_t \\
+2tr\left[\dot{\widetilde{W}}_{1,t}^{T} K_1^{-1} \widetilde{W}_{1,t}\right] + 2tr\left[\dot{\widetilde{W}}_{2,t}^{T} K_2^{-1} \widetilde{W}_{2,t}\right] \\
+2tr\left[\dot{\widetilde{V}}_{1,t}^{T} K_3^{-1} \widetilde{V}_{1,t}\right] + 2tr\left[\dot{\widetilde{V}}_{2,t}^{T} K_4^{-1} \widetilde{V}_{2,t}\right]
\end{gathered} \tag{6.50}
$$

In view of **A6.4** and **A6.5,** it follows

$$
\begin{gathered}
\dot{\Delta}_t = A\Delta_t + \left(\widetilde{W}_{1,t}\sigma_t + W_1^*\widetilde{\sigma}_t + W_1^*\widetilde{\sigma}_t'\right) \\
+\left(\widetilde{W}_{2,t}\phi_t + W_{2,t}^*\widetilde{\phi}_t + W_{2,t}^*\widetilde{\phi'}\right) u_t \\
-\widetilde{f}_t - \xi_{1,t} - L_1\left[y_t - \widehat{y}_t\right] - L_2/h\left[(y_t - y_{t-h}) - (\widehat{y}_t - \widehat{y}_{t-h})\right]
\end{gathered} \tag{6.51}
$$

Substituting (6.51) into (6.50) leads to the following relation

$$
\begin{gathered}
2\Delta_t^T P \dot{\Delta}_t = 2\Delta_t^T P A \Delta_t + 2\Delta_t^T P\left(W_1^*\widetilde{\sigma}_t + W_2^*\widetilde{\phi}_t u_t\right) \\
+2\Delta_t^T P\left(\widetilde{W}_{1,t}\sigma_t + \widetilde{W}_{2,t}\phi_t u_t\right) + 2\Delta_t^T P\left(W_1^*\widetilde{\sigma}_t' + W_2^*\widetilde{\phi}_t' u_t\right) \\
-2\Delta_t^T P\widetilde{f}_t - 2\Delta_t^T P\xi_t \\
+2\Delta_t^T P\left\{L_1\left[\widehat{y}_t - y_t\right] + L_2 h^{-1}\left[(\widehat{y}_t - \widehat{y}_{t-h}) - (y_t - y_{t-h})\right]\right\}
\end{gathered} \tag{6.52}
$$

Using the matrix inequality

$$X^T Y + \left(X^T Y\right)^T \leq X^T \Lambda X + Y^T \Lambda^{-1} Y \tag{6.53}$$

valid for any $X, Y \in \mathbb{R}^{n\times k}$ and for any positive defined matrix $0 < \Lambda = \Lambda^T \in \mathbb{R}^{n\times n}$, and in view of **A6.4** and (6.39), the terms in (6.52) can be estimated in the following manner:

1)

$$2\Delta_t^T P A \Delta_t = \Delta_t^T \left(PA + A^\intercal P\right) \Delta_t$$

2)

$$2\Delta_t^T P W_1^* \widetilde{\sigma}_t \leq \Delta_t^T P W_1^* \Lambda_1^{-1} W_1^{*T} P \Delta_t + \widetilde{\sigma}_t^T \Lambda_1 \widetilde{\sigma}_t \leq \Delta_t^T \left(P\overline{W}_1 P + \Lambda_\sigma\right) \Delta_t \tag{6.54}$$

3)

$$2\Delta_t^T P W_2^* \widetilde{\phi}_t u_t \leq \Delta_t^T P \overline{W}_2 P \Delta_t + \lambda_{\max}\left(\Lambda_2\right) \overline{\overline{\phi}}^2 \left(v_0 + v_1 \left\|\widehat{x}_t\right\|^2\right)$$

4)

$$\begin{gathered}
2\Delta_t^T P \widetilde{W}_{1,t} \sigma_t = 2\left(C_\delta^+ e_t + \delta N_\delta \Delta_t + C_\delta^+ \xi_{2,t}\right)^\intercal P \widetilde{W}_{1,t} \sigma_t \\
= 2e_t^\intercal \left(C_\delta^+\right)^\intercal P \widetilde{W}_{1,t} \sigma_t + 2\delta \Delta_t^\intercal \left(N_\delta\right)^\intercal P \widetilde{W}_{1,t} \sigma_t + 2\xi_{2,t}^\intercal \left(C_\delta^+\right)^\intercal P \widetilde{W}_{1,t} \sigma_t \\
\leq tr\left\{\left[2\sigma_t e_t^\intercal \left(C_\delta^+\right)^\intercal P\right] \widetilde{W}_{1,t}\right\} + \delta \Delta_t^\intercal \Lambda_3^{-1} \Delta_t \\
+ tr\left\{\delta \left[\sigma_t \sigma_t^\intercal \widetilde{W}_{1,t}^\intercal P N_\delta \Lambda_3 \left(N_\delta\right)^\intercal P\right] \widetilde{W}_{1,t}\right\} \\
+ \left(\xi_{2,t}^\intercal \Lambda_{\xi_2} \xi_{2,t} + \sigma_t^\intercal \widetilde{W}_{1,t}^\intercal P C_\delta^+ \Lambda_{\xi_2}^{-1} \left(C_\delta^+\right)^\intercal P \widetilde{W}_{1,t} \sigma_t\right) \leq tr\left\{\left[2\sigma_t e_t^\intercal \left(C_\delta^+\right)^\intercal P\right.\right. \\
\left.\left. + \sigma_t \sigma_t^\intercal \widetilde{W}_{1,t}^\intercal P \left[\delta N_\delta \Lambda_3 \left(N_\delta\right)^\intercal + C_\delta^+ \Lambda_{\xi_2}^{-1} \left(C_\delta^+\right)^\intercal\right] P\right] \widetilde{W}_{1,t}\right\} \\
+ \delta \Delta_t^\intercal \Lambda_3^{-1} \Delta_t + \Upsilon_2
\end{gathered}$$

5)

$$\begin{gathered}
2\Delta_t^T P \widetilde{W}_{2,t} \phi_t u_t = 2\left(C_\delta^+ e_t + \delta N_\delta \Delta_t + C_\delta^+ \xi_{2,t}\right)^\intercal P \widetilde{W}_{2,t} \phi_t u_t \\
= 2e_t^\intercal \left(C_\delta^+\right)^\intercal P \widetilde{W}_{2,t} \phi_t u_t + 2\delta \Delta_t^\intercal \left(N_\delta\right)^\intercal P \widetilde{W}_{2,t} \phi_t u_t + 2\xi_{2,t}^\intercal \left(C_\delta^+\right)^\intercal P \widetilde{W}_{2,t} \phi_t u_t \\
\leq tr\left\{\left[2\phi_t u_t e_t^\intercal \left(C_\delta^+\right)^\intercal P\right] \widetilde{W}_{2,t}\right\} + \delta \Delta_t^\intercal \Lambda_3^{-1} \Delta_t \\
+ tr\left\{\left[\delta \phi_t u_t u_t^\intercal \phi_t^\intercal \widetilde{W}_{2,t}^\intercal P N_\delta \Lambda_3 \left(N_\delta\right)^\intercal P\right] \widetilde{W}_{2,t}\right\} \\
+ \left(\xi_{2,t}^\intercal \Lambda_{\xi_2} \xi_{2,t} + u_t^\intercal \phi_t^\intercal \widetilde{W}_{2,t}^\intercal P C_\delta^+ \Lambda_{\xi_2}^{-1} \left(C_\delta^+\right)^\intercal P \widetilde{W}_{2,t} \phi_t u_t\right) \leq \\
tr\left\{\left[2\phi_t u_t e_t^\intercal \left(C_\delta^+\right)^\intercal P + \phi_t u_t u_t^\intercal \phi_t^\intercal \widetilde{W}_{2,t}^\intercal P \left[\delta N_\delta \Lambda_3 \left(N_\delta\right)^\intercal\right.\right.\right. \\
\left.\left.\left. + C_\delta^+ \Lambda_{\xi_2}^{-1} \left(C_\delta^+\right)^\intercal\right] P \widetilde{W}_{2,t}\right\} + \delta \Delta_t^\intercal \Lambda_3^{-1} \Delta_t + \Upsilon_2
\end{gathered}$$

6)

$$\begin{array}{c} 2\Delta_t^T PW_1^*\widetilde{\sigma}_t' = 2\left(C_\delta^+ e_t + \delta N_\delta \Delta_t + C_\delta^+ \xi_{2,t}\right)^\intercal PW_1^* D_\sigma \widetilde{V}_{1,t}\widehat{x}_t \\ +2\Delta_t^T PW_1^* \nu_\sigma \\ = 2e_t^\intercal \left(C_\delta^+\right)^\intercal PW_1^* D_\sigma \widetilde{V}_{1,t}\widehat{x}_t + 2\delta\Delta_t^\intercal N_\delta^\intercal PW_1^* D_\sigma \widetilde{V}_{1,t}\widehat{x}_t \\ +2\xi_{2,t}^\intercal \left(C_\delta^+\right)^\intercal PW_1^* D_\sigma \widetilde{V}_{1,t}\widehat{x}_t + 2\Delta_t^T PW_1^* \nu_\sigma \\ = tr\left\{\left(2\widehat{x}_t e_t^\intercal \left(C_\delta^+\right)^\intercal PW_1^* D_\sigma\right)\widetilde{V}_{1,t}\right\} + \delta\Delta_t^\intercal \Lambda_1 \Delta_t \\ +tr\left\{\left(\widehat{x}_t\widehat{x}_t^\intercal \widetilde{V}_{1,t}^\intercal D_\sigma^\intercal \left(W_1^*\right)^\intercal PN_\delta \Lambda_1^{-1} N_\delta^\intercal PW_1^* D_\sigma\right)\widetilde{V}_{1,t}\right\} \\ +\xi_{2,t}^\intercal \Lambda_{\xi_2}\xi_{2,t} + tr\left\{\left(\widehat{x}_t\widehat{x}_t^\intercal \widetilde{V}_{1,t}^\intercal D_\sigma^\intercal \left(W_1^*\right)^\intercal PC_\delta^+ \Lambda_{\xi_2}^{-1}\left(C_\delta^+\right)^\intercal PW_1^* D_\sigma\right)\widetilde{V}_{1,t}\right\} \\ \Delta_t^T P\overline{W}_1 P\Delta_t + tr\left\{l_1\widehat{x}_t\widehat{x}_t^\intercal \widetilde{V}_{1,t}^\intercal \Lambda_1 \widetilde{V}_{1,t}\right\} \\ = tr\left\{\left(2\widehat{x}_t e_t^\intercal \left(C_\delta^+\right)^\intercal PW_1^* D_\sigma\right.\right. \\ \left.\left. +\widehat{x}_t\widehat{x}_t^\intercal \widetilde{V}_{1,t}^\intercal \left[D_\sigma^\intercal \left(W_1^*\right)^\intercal P\left(N_\delta \Lambda_1^{-1} N_\delta^\intercal + C_\delta^+ \Lambda_{\xi_2}^{-1}\left(C_\delta^+\right)^\intercal\right) PW_1^* D_\sigma + l_1\Lambda_1\right]\right)\widetilde{V}_{1,t}\right\} \\ +\Delta_t^T P\overline{W}_1 P\Delta_t \end{array} \tag{6.55}$$

since the term $2\Delta_t^T PW_1^*\nu_\sigma$ in (6.55) may be also estimated as

$$\begin{array}{c} 2\Delta_t^T PW_1^*\nu_\sigma \leq \Delta_t^T PW_1^{*T}\Lambda_1^{-1}W_1^* P\Delta_t + \nu_\sigma^T \Lambda_1 \nu_\sigma \\ \leq \Delta_t^T P\overline{W}_1 P\Delta_t + l_1\left\|\widetilde{V}_{1,t}\widehat{x}_t\right\|_{\Lambda_1}^2 \end{array} \tag{6.56}$$

7) By the same way we estimate

$$\begin{array}{c} 2\Delta_t^T PW_2^*\widetilde{\phi}_t' u_t = 2\left(C_\delta^+ e_t + \delta N_\delta \Delta_t + C_\delta^+ \xi_{2,t}\right)^\intercal PW_2^* D_{\phi u}^\intercal \widetilde{V}_{2,t}\widehat{x}_t \\ +2\Delta_t^T PW_2^*\nu_\phi \leq tr\left\{\left(2\widehat{x}_t e_t^\intercal \left(C_\delta^+\right)^\intercal PW_2^* D_{\phi u}^\intercal\right)\widetilde{V}_{2,t}\right\} \\ +\delta\Delta_t^\intercal \Lambda_1 \Delta_t + tr\left\{\widehat{x}_t\widehat{x}_t^\intercal \widetilde{V}_{2,t}^\intercal D_{\phi u}\left(W_2^*\right)^\intercal PN_\delta \Lambda_1^{-1} N_\delta^\intercal PW_2^* D_{\phi u}^\intercal \widetilde{V}_{2,t}\right\} \\ +\xi_{2,t}^\intercal \Lambda_{\xi_2}\xi_{2,t} + tr\left\{\widehat{x}_t\widehat{x}_t^\intercal \widetilde{V}_{2,t}^\intercal D_{\phi u}\left(W_2^*\right)^\intercal PC_\delta^+ \Lambda_{\xi_2}^{-1}\left(C_\delta^+\right)^\intercal PW_2^* D_{\phi u}^\intercal \widetilde{V}_{2,t}\right\} \\ +\Delta_t^T P\overline{W}_2 P\Delta_t + tr\left\{\left(l_2\widehat{x}_t\widehat{x}_t^\intercal \widetilde{V}_{2,t}^\intercal \Lambda_2\right)\widetilde{V}_{2,t}\right\} \\ \leq \Delta_t^T\left(P\overline{W}_2 P + \delta\Lambda_1\right)\Delta_t + \Upsilon_2 + \\ tr\left\{\widehat{x}_t\widehat{x}_t^\intercal \widetilde{V}_{2,t}^\intercal \left[D_{\phi u}\left(W_2^*\right)^\intercal P\left(C_\delta^+ \Lambda_{\xi_2}^{-1}\left(C_\delta^+\right)^\intercal + N_\delta \Lambda_1^{-1} N_\delta^\intercal\right) PW_2^* D_{\phi u}^\intercal + l_2\Lambda_2\right]\widetilde{V}_{2,t}\right\} \end{array} \tag{6.57}$$

since

$$2\Delta_t^T PW_2^*\nu_\phi \leq \Delta_t^T P\overline{W}_2 P\Delta_t + l_2\left\|\widetilde{V}_{2,t}\widehat{x}_t\right\|_{\Lambda_2}^2 \tag{6.58}$$

8) So, from **A6.4,** the term $\left(-2\Delta_t^T P\widetilde{f}_t\right)$ can be estimated as

$$-2\Delta_t^T P\widetilde{f}_t \leq \Delta_t^T P\Lambda_f^{-1}P\Delta_t + \widetilde{f}_t^T\Lambda_f\widetilde{f}_t \leq \Delta_t^T P\Lambda_f^{-1}P\Delta_t + \overline{\eta} + \overline{\eta}_1 \|x_t\|_{\Lambda_f}^2 \tag{6.59}$$

9) Using **A6.2,** for the term $2\Delta_t^T P\xi_{1,t}$ in (6.52) we obtain

$$-2\Delta_t^T P\xi_{1,t} \leq \Delta_t^T P\Lambda_{\xi_1}^{-1}P\Delta_t + \xi_{1,t}^T\Lambda_{\xi_1}\xi_{1,t} \leq \Delta_t^T P\Lambda_{\xi_1}^{-1}P\Delta_t + \Upsilon_1 \tag{6.60}$$

10) The last term in (6.52) is

$$\begin{array}{c} 2\Delta_t^T P\left\{L_1\left[\widehat{y}_t - y_t\right] + L_2h^{-1}\left[\left(\widehat{y}_t - \widehat{y}_{t-h}\right) - \left(y_t - y_{t-h}\right)\right]\right\} \\ = 2\Delta_t^T P\left(L_1 + L_2h^{-1}\right)C\Delta_t - 2\Delta_t^T PL_1\xi_{2,t} - 2\Delta_t^T PL_2h^{-1}\xi_{2,t} \\ -2\Delta_t^T PL_2h^{-1}C\Delta_{t-h} + 2\Delta_t^T PL_2h^{-1}\xi_{2,t-h} \\ = \Delta_t^T\left[P\left(L_1 + L_2h^{-1}\right)C + C^{\intercal}\left(L_1 + L_2h^{-1}\right)^{\intercal}P\right]\Delta_t \\ -2\Delta_t^T P\left(L_1 + L_2h^{-1}\right)\xi_{2,t} + 2\Delta_t^T PL_2h^{-1}\xi_{2,t-h} \\ -2\Delta_t^T PL_2h^{-1}C\Delta_{t-h} \end{array} \tag{6.61}$$

Similar to (6.60), the terms $\left(-2\Delta_t^T P\left(L_1 + L_2h^{-1}\right)\xi_{2,t}\right)$, $\left(-2\Delta_{t-h}^T PL_2h^{-1}\xi_{2,t-h}\right)$ and $\left(2\Delta_t^T PL_2/hC\Delta_{t-h}\right)$ in (6.61) can be estimated as

$$\begin{array}{c} -2\Delta_t^T P\left(L_1 + L_2h^{-1}\right)\xi_{2,t} \leq \\ \Delta_t^T P\left(L_1 + L_2h^{-1}\right)\Lambda_{\xi_2}^{-1}\left(L_1 + L_2h^{-1}\right)^{\intercal}P\Delta_t + \Upsilon_2 \\ -2h^{-1}\Delta_t^T PL_2\xi_{2,t-h} \leq h^{-1}\Delta_t^T PL_2\Lambda_{\xi_2}^{-1}L_2^T P\Delta_t + h^{-1}\Upsilon_2 \\ 2h^{-1}\Delta_t^T PL_2C\Delta_{t-h} \leq h^{-1}\Delta_t^T PL_2\Lambda_{\xi_2}^{-1}L_2^T P\Delta_t + h^{-1}\Delta_{t-h}^T\Lambda_{\xi_2}\Delta_{t-h} \end{array} \tag{6.62}$$

Finally, in view of the obtained upper estimates, it follows

$$\begin{array}{c} \dot{V}_{1,t} \leq -\lambda\|x_t\|^2 + \Delta_t^T\left(PA_1 + A_1^{\intercal}P + PR_1P + Q_1\right)\Delta_t \\ +tr\left\{L_{W_1,t}\widetilde{W}_{1,t}\right\} + tr\left\{L_{W_2,t}\widetilde{W}_{2,t}\right\} \\ +tr\left\{L_{V_1,t}\widetilde{V}_{1,t}\right\} + tr\left\{L_{V_2,t}\widetilde{V}_{2,t}\right\} \\ +\lambda_{\max}\left(\Lambda_2\right)\overline{\overline{\phi}}^2\left(v_0 + v_1\|\widehat{x}_t\|^2\right) + \Upsilon_1 + 4\Upsilon_2 \\ +\overline{\eta} + \overline{\eta}_1\|x_t\|_{\Lambda_f}^2 + h^{-1}\Upsilon_2 + \Delta_{t-h}^T h^{-1}\Lambda_{\xi_2}\Delta_{t-h} \end{array} \tag{6.63}$$

where

$$\begin{array}{c} A_1 := A + \left(L_1 + L_2h^{-1}\right)C \\ R_1 := 2\overline{W}_1 + 2\overline{W}_2 + \Lambda_f^{-1} + \Lambda_{\xi_1}^{-1} \\ + \left(L_1 + L_2h^{-1}\right)\Lambda_{\xi_2}^{-1}\left(L_1 + L_2h^{-1}\right)^{\intercal} + 2h^{-1}L_2\Lambda_{\xi_2}^{-1}L_2^T \\ Q_1 := \Lambda_\sigma + \delta\Lambda_1 + (1+\delta)\Lambda_3^{-1} \end{array}$$

and

$$L_{W_1,t} := 2\dot{\widetilde{W}}_{1,t}^{T} K_1^{-1} + \left[2\sigma_t e_t^{\intercal} \left(C_\delta^{+}\right)^{\intercal} P \right.$$
$$\left. + \sigma_t \sigma_t^{\intercal} \widetilde{W}_{1,t}^{\intercal} P \left[\delta N_\delta \Lambda_3 \left(N_\delta\right)^{\intercal} + C_\delta^{+} \Lambda_{\xi_2}^{-1} \left(C_\delta^{+}\right)^{\intercal}\right] P\right]$$
$$L_{W_2,t} := \left[2\phi_t u_t e_t^{\intercal} \left(C_\delta^{+}\right)^{\intercal} P + \phi_t u_t u_t^{\intercal} \phi_t^{\intercal} \widetilde{W}_{2,t}^{\intercal} P \left[\delta N_\delta \Lambda_3 \left(N_\delta\right)^{\intercal}\right.\right.$$
$$\left.\left. + C_\delta^{+} \Lambda_{\xi_2}^{-1} \left(C_\delta^{+}\right)^{\intercal}\right] P + 2\dot{\widetilde{W}}_{2,t}^{T} K_2^{-1}$$
$$L_{V_1,t} := 2\dot{\widetilde{V}}_{1,t}^{\intercal} K_3^{-1} + \left(2\widehat{x}_t e_t^{\intercal} \left(C_\delta^{+}\right)^{\intercal} P W_1^{*} D_\sigma\right.$$
$$\left. + \widehat{x}_t \widehat{x}_t^{\intercal} \widetilde{V}_{1,t}^{\intercal} \left[D_\sigma^{\intercal} \left(W_1^{*}\right)^{\intercal} P \left(N_\delta \Lambda_1^{-1} N_\delta^{\intercal} + C_\delta^{+} \Lambda_{\xi_2}^{-1} \left(C_\delta^{+}\right)^{\intercal}\right) P W_1^{*} D_\sigma + l_1 \Lambda_1\right]\right)$$
$$L_{V_2,t} := \widehat{x}_t \widehat{x}_t^{\intercal} \widetilde{V}_{2,t}^{\intercal} \left[D_{\phi u} \left(W_2^{*}\right)^{\intercal} P \left(C_\delta^{+} \Lambda_{\xi_2}^{-1} \left(C_\delta^{+}\right)^{\intercal} + N_\delta \Lambda_1^{-1} N_\delta^{\intercal}\right) P W_2^{*} D_{\phi u}^{\intercal}\right.$$
$$\left. + l_2 \Lambda_2\right] + 2\dot{\widetilde{V}}_{2,t}^{T} K_4^{-1}$$

Part 2: Differential inequality for DNN-states

Select the second Lyapunov-Krasovski functional as follows:

$$V_{2,t} = \hat{x}_t^{\intercal} P_2 \hat{x}_t + \int_{\tau=t-h}^{t} \Delta_\tau^T P_1 \Delta_\tau d\tau$$

where $P_2 = P_2^{\intercal} > 0$. Then, analogously to the previous calculations, it follows

$$\dot{V}_{2,t} = 2\,\widehat{x}_t P_2\,\dot{\widehat{x}}_t + \Delta_t^{\intercal} P_1 \Delta_t - \Delta_{t-h}^{\intercal} P_1 \Delta_{t-h}$$

$$2\widehat{x}_t^{\intercal} P_2\,\dot{\widehat{x}}_t =$$
$$2\widehat{x}_t^{\intercal} P_2 \left[A\widehat{x}_t + W_{1,t}\sigma(V_{1,t}\widehat{x}_t) + W_{2,t}\phi(V_{2,t}\widehat{x}_t)u_t +\right.$$
$$\left. L_1\left[\widehat{y}_t - y_t\right] + L_2 h^{-1}\left[\left(\widehat{y}_t - \widehat{y}_{t-h}\right) - \left(y_t - y_{t-h}\right)\right]\right]$$
$$\leqslant \widehat{x}_t^{\intercal}\left(P_2 A + A^{\intercal} P_2\right)\widehat{x}_t + tr\left\{2\sigma(V_{1,t}\widehat{x}_t)\widehat{x}_t^{\intercal} P_2 \widetilde{W}_{1,t}\right\} +$$
$$\widehat{x}_t^{\intercal} P_2 W_1^{*} \Lambda_1^{-1} W_1^{*\intercal} P_2 \widehat{x}_t + 2tr\left\{\Lambda_\sigma \tilde{V}_{1,t}\widehat{x}_t\widehat{x}_t^{\intercal}\tilde{V}_{1,t}^{\intercal}\right\} + 2\widehat{x}_t^{\intercal} V_1^{*\intercal}\Lambda_\sigma V_1^{*}\widehat{x}_t$$
$$tr\left\{2\phi(V_2\widehat{x}_t)u_t\widehat{x}_t^{\intercal} P_2 \widetilde{W}_2\right\} + \widehat{x}_t^{\intercal} P_2 W_2^{*}\Lambda_2^{-1} W_2^{*\intercal} P_2\widehat{x}_t +$$
$$\left\|\Lambda_2\right\| \overline{\phi}^2 \left(v_0 + v_1\left\|\widehat{x}_t\right\|^2\right) +$$
$$\widehat{x}_t^{\intercal} P_2\left(L_1 + L_2 h^{-1}\right) C\Lambda_3^{-1} C^{\intercal}\left(L_1 + L_2 h^{-1}\right)^{\intercal} P_2\widehat{x}_t + \Delta_t^{\intercal}\Lambda_3\Delta_t +$$
$$\widehat{x}_t^{\intercal} P_2\left(L_1 + L_2 h^{-1}\right)\Lambda_{\xi_2}^{-1}\left(L_1 + L_2 h^{-1}\right)^{\intercal} P_2\widehat{x}_t + \Upsilon_2 +$$
$$h^{-1}\widehat{x}_t^{\intercal} P_2 L_2 \Lambda_{\xi_2}^{-1} L_2^{\intercal} P_2 \widehat{x}_t + h^{-1}\Upsilon_2 +$$
$$h^{-1}\widehat{x}_t^{\intercal} P_2 L_2 \Lambda_{\xi_2}^{-1} L_2^{\intercal} P_2 \widehat{x}_t + h^{-1}\Delta_{t-h}^{\intercal}\Lambda_{\xi_2}\Delta_{t-h}$$

since

$$\begin{gathered}2\widehat{x}_t^{\intercal}P_2\left(L_1\left[\widehat{y}_t-y_t\right]+L_2h^{-1}\left[\left(\widehat{y}_t-\widehat{y}_{t-h}\right)-\left(y_t-y_{t-h}\right)\right]\right)=\\ 2\widehat{x}_t^{\intercal}P_2\left(L_1+L_2h^{-1}\right)C\Delta_t-2\widehat{x}_t^{\intercal}P_2\left(L_1+L_2h^{-1}\right)\xi_{2,t}+\\ 2\widehat{x}_t^{\intercal}P_2L_2h^{-1}\xi_{2,t-h}-2\widehat{x}_t^{\intercal}P_2L_2h^{-1}C\Delta_{t-h}\leq\\ \widehat{x}_t^{\intercal}P_2\left(L_1+L_2h^{-1}\right)C\Lambda_3^{-1}C^{\intercal}\left(L_1+L_2h^{-1}\right)^{\intercal}P_2\widehat{x}_t+\Delta_t^{\intercal}\Lambda_3\Delta_t+\\ \widehat{x}_t^{\intercal}P_2\left(L_1+L_2h^{-1}\right)\Lambda_{\xi_2}^{-1}\left(L_1+L_2h^{-1}\right)^{\intercal}P_2\widehat{x}_t+\Upsilon_2+\\ h^{-1}\widehat{x}_t^{\intercal}P_2L_2\Lambda_{\xi_2}^{-1}L_2^{\intercal}P_2\widehat{x}_t+h^{-1}\Upsilon_2+\\ h^{-1}\widehat{x}_t^{\intercal}P_2L_2\Lambda_{\xi_2}^{-1}L_2^{\intercal}P_2\widehat{x}_t+h^{-1}\Delta_{t-h}^{\intercal}\Lambda_{\xi_2}\Delta_{t-h}\end{gathered}$$

As a result, we obtain

$$\begin{gathered}\dot{V}_{2,t}\leq\widehat{x}_t^{\intercal}\left(P_2A+A^{\intercal}P_2+P_2R_2P_2+Q_2\right)\widehat{x}_t\\ +\Delta_t^{\intercal}\left(P_1+\Lambda_3\right)\Delta_t-\Delta_{t-h}^{\intercal}\left(P_1-h^{-1}\Lambda_{\xi_2}\right)\Delta_{t-h}+H\\ +tr\left\{2\sigma(V_{1,t}\widehat{x}_t)\widehat{x}_t^{\intercal}P_2\widetilde{W}_{1,t}\right\}+2tr\left\{\Lambda_\sigma\tilde{V}_{1,t}\widehat{x}_t\widehat{x}_t^{\intercal}\tilde{V}_{1,t}^{\intercal}\right\}\\ +tr\left\{2\phi(V_2\widehat{x}_t)u_t\widehat{x}_t^{\intercal}P_2\widetilde{W}_2\right\}\end{gathered}\tag{6.64}$$

where

$$\begin{gathered}R_2:=\overline{W}_1+\overline{W}_2+\\ \left(L_1+L_2h^{-1}\right)\left(C\Lambda_3^{-1}C^{\intercal}+\Lambda_{\xi_2}^{-1}\right)\left(L_1+L_2h^{-1}\right)^{\intercal}+2L_2\Lambda_{\xi_2}^{-1}L_2^{\intercal}\\ Q_2:=2V_1^{*\intercal}\Lambda_\sigma V_1^*+\left\|\Lambda_2\right\|\overline{\overline{\phi}}^2v_1I\\ H:=\left\|\Lambda_2\right\|\overline{\overline{\phi}}^2v_0+\left(1+h^{-1}\right)\Upsilon_2\end{gathered}$$

Part 3: Joint differential inequality

The use of (6.63) and (6.64) implies

$$\begin{gathered}\dot{V}_{1,t}+\dot{V}_{2,t}\leq-\left(\lambda-\left\|\Lambda_f\right\|\overline{\eta}_1\right)\left\|x_t\right\|^2-\Delta_t^TQ_0\Delta_t\\ +\Delta_t^T\left(PA_1+A_1^{\intercal}P+PR_1P+Q_1+P_1+\Lambda_3+Q_0\right)\Delta_t\\ +tr\left\{\left(L_{W_{1,t}}+2\sigma(V_{1,t}\widehat{x}_t)\widehat{x}_t^{\intercal}P_2\right)\widetilde{W}_{1,t}\right\}+\\ tr\left\{\left(L_{W_{2,t}}+2\phi(V_2\widehat{x}_t)u_t\widehat{x}_t^{\intercal}P_2\right)\widetilde{W}_{2,t}\right\}+\\ tr\left\{\left(L_{V_{1,t}}+2\Lambda_\sigma\tilde{V}_{1,t}\widehat{x}_t\widehat{x}_t^{\intercal}\right)\widetilde{V}_{1,t}\right\}+tr\left\{L_{V_{2,t}}\widetilde{V}_{2,t}\right\}+\\ +\Delta_{t-h}^T\left(2h^{-1}\Lambda_{\xi_2}-P_1\right)\Delta_{t-h}+\\ \widehat{x}_t^{\intercal}\left(P_2A+A^{\intercal}P_2+P_2R_2P_2+Q_2+\lambda_{\max}\left(\Lambda_2\right)\overline{\overline{\phi}}^2v_1I+Q_0\right)\widehat{x}_t+\\ -\widehat{x}_t^{\intercal}Q_0\widehat{x}_t+H+\lambda_{\max}\left(\Lambda_2\right)\overline{\overline{\phi}}^2v_0+\Upsilon_1+4\Upsilon_2+\overline{\eta}+h^{-1}\Upsilon_2\end{gathered}\tag{6.65}$$

Finally, in view of the applied learning law (6.45)

$$\begin{aligned}
\tilde{L}_{W_1,t} &:= L_{W_1,t} + 2\sigma(V_{1,t}\widehat{x}_t)\widehat{x}_t^{\intercal} P_2 = 0 \\
\tilde{L}_{W_2,t} &:= L_{W_2,t} + 2\phi(V_2\widehat{x}_t)u_t\widehat{x}_t^{\intercal} P_2 = 0 \\
\tilde{L}_{V_1,t} &:= L_{V_1,t} + 2\Lambda_\sigma \tilde{V}_{1,t}\widehat{x}_t\widehat{x}_t^{\intercal} = 0 \\
\tilde{L}_{V_2,t} &:= L_{V_2,t} = 0
\end{aligned}$$

selecting $P_1 = 2h^{-1}\Lambda_{\xi_2}$, $\|\Lambda_f\| \leq \lambda/\overline{\eta}_1$ and by A6, from (6.65) for $V_t := V_{1,t} + V_{2,t}$ it follows

$$\begin{aligned}
\tfrac{d}{dt}V_t = \tfrac{d}{dt}\left(V_{1,t} + V_{2,t}\right) &\leq -\Delta_t^T Q_0 \Delta_t - \widehat{x}_t^{\intercal} Q_0 \widehat{x}_t \\
H + \lambda_{\max}\left(\Lambda_2\right) \overline{\overline{\phi}}^2 v_0 &+ \Upsilon_1 + 4\Upsilon_2 + \overline{\eta} + h^{-1}\Upsilon_2 \\
&= -\Delta_t^T P^{1/2}\left(P^{-1/2}Q_0P^{-1/2}\right)P^{1/2}\Delta_t \\
&\quad -\widehat{x}_t^{\intercal} P_2^{1/2}\left(P_2^{-1/2}Q_0P_2^{-1/2}\right)P_2^{1/2}\widehat{x}_t + \beta \\
&\leq -\lambda_{\min}\left(P^{-1/2}Q_0P^{-1/2}\right)\Delta_t^{\intercal} P\Delta_t \\
&\quad -\lambda_{\min}\left(P_2^{-1/2}Q_0P_2^{-1/2}\right)\widehat{x}_t^{\intercal} P_2\widehat{x}_t + \beta \\
&\leq -\alpha\left(\Delta_t^{\intercal} P\Delta_t + \widehat{x}_t^{\intercal} P_2\widehat{x}_t\right) + \beta
\end{aligned} \tag{6.66}$$

where

$$\begin{aligned}
\beta &= \left[\lambda_{\max}\left(\Lambda_2\right) + \|\Lambda_2\|\right]\overline{\overline{\phi}}^2 v_0 + \Upsilon_1 + \left(5 + 2h^{-1}\right)\Upsilon_2 + \overline{\eta} \\
\alpha &:= \min\left\{\lambda_{\min}\left(P^{-1/2}Q_0P^{-1/2}\right); \lambda_{\min}\left(P_2^{-1/2}Q_0P_2^{-1/2}\right)\right\}
\end{aligned}$$

Introduce the function

$$G_t := \left[\sqrt{V_t} - \mu\right]_+^2 = V_t\left[1 - \mu/\sqrt{V_t}\right]_+^2$$

where the function $[.]_+$ is defined as

$$[z]_+ := \begin{cases} z & \text{if } z \geq 0 \\ 0 & \text{if } z < 0 \end{cases}$$

which is a "*cutting function*" or a "*dead zone*". For the derivative of this function we obtain

$$\begin{array}{c} \dot{G}_t := \frac{1}{2}\left[\sqrt{V_t} - \mu\right]_+ / \sqrt{V_t}\dot{V}_t = \frac{1}{2}\left[1 - \mu/\sqrt{V_t}\right]_+ \dot{V}_t \\ \leq \frac{1}{2}\left[1 - \mu/\sqrt{V_t}\right]_+ \left(-\alpha\left(\Delta_t^\intercal P\Delta_t + \widehat{x}_t^\intercal P_2\widehat{x}_t\right) + \beta\right) \\ = -\frac{1}{2}\alpha\left(\Delta_t^\intercal P\Delta_t + \widehat{x}_t^\intercal P_2\widehat{x}_t\right)\left[1 - \mu/\sqrt{V_t}\right]_+ \left(1 - \beta\left[\alpha\left(\Delta_t^\intercal P\Delta_t + \widehat{x}_t^\intercal P_2\widehat{x}_t\right)\right]^{-1}\right) \\ \leq -\frac{1}{2}\alpha\left(\Delta_t^\intercal P\Delta_t + \widehat{x}_t^\intercal P_2\widehat{x}_t\right)\left[1 - \mu/\sqrt{V_t}\right]_+ \left(1 - \beta\left(\alpha V_t\right)^{-1}\right) \\ \leq -\frac{1}{2}\alpha\left(\Delta_t^\intercal P\Delta_t + \widehat{x}_t^\intercal P_2\widehat{x}_t\right)\left[1 - \mu/\sqrt{V_t}\right]_+ \left(1 - \mu^2/V_t\right) \leq 0 \end{array} \tag{6.67}$$

if take

$$\mu := \sqrt{\frac{\beta}{\alpha}}$$

The last inequality implies that

$$\begin{array}{c} G_t = \left[\sqrt{V_t} - \mu\right]_+^2 \leq G_0 < \infty \\ 0 \leq V_t \leq \left(\mu + \sqrt{G_0}\right)^2 < \infty \end{array}$$

and, hence, $x_t, \widehat{x}_t, \Delta_t, W_{i,t}, V_{i,t} \in L_\infty$, that is, they are bounded. So, *a)* is proven.

The integration of (6.67) from 0 to T yields

$$\begin{array}{c} G_T - G_0 \leq \\ -\frac{1}{2}\alpha\int_0^T \left(\Delta_t^\intercal P\Delta_t + \widehat{x}_t^\intercal P_2\widehat{x}_t\right)\left[1 - \mu/\sqrt{V_t}\right]_+ \left(1 - \mu^2/V_t\right) dt \end{array}$$

that leads to the following inequality

$$\begin{array}{c} \frac{1}{2}\alpha\int_0^T \left(\Delta_t^\intercal P\Delta_t + \widehat{x}_t^\intercal P_2\widehat{x}_t\right)\left[1 - \mu/\sqrt{V_t}\right]_+ \left(1 - \mu^2/V_t\right) dt \\ \leq G_0 - G_T \leq G_0 \end{array} \tag{6.68}$$

Dividing by T and taking the upper limits of both sides, we finally obtain:

$$\overline{\lim_{T\to\infty}}\frac{1}{T}\int_0^T \left(\Delta_t^\intercal P\Delta_t + \widehat{x}_t^\intercal P_2\widehat{x}_t\right)\left[1 - \mu/\sqrt{V_t}\right]_+ \left(1 - \mu^2/V_t\right) dt \leq 0$$

and, hence,

$$\left(\Delta_t^\intercal P\Delta_t + \widehat{x}_t^\intercal P_2\widehat{x}_t\right) V_t\left[1 - \mu/\sqrt{V_t}\right]_+ \left(1 - \mu^2/V_t\right) \to 0$$

or,

$$\left[1 - \mu/\sqrt{V_t}\right]_+ \to 0$$

That is *b)*. Theorem is proven. ■

6.2.5 TRACKING ERROR ANALYSIS

The control system behavior expected is to move the states to track a signal response generated by a nonlinear reference model given by:

$$\dot{x}_m = f_m(x_m, t) \tag{6.69}$$

Define the following seminomas:

$$|z|_Q^2 = \overline{\lim_{T\to\infty}} \frac{1}{T} \int_0^T z^T(t) Q z(t) dt \tag{6.70}$$

Here $Q = Q^T > 0$. The state trajectory tracking can be formulated as:

$$J_{\min} = \min_u J \quad J = |x_t - x_m|_{Q_c}^2 + |u_t|_{\widetilde{R}_c}^2 \tag{6.71}$$

So, for any $\eta > 0$ we have

$$J \leqslant (1+\eta) \left|\widehat{x}_t - x_m\right|_{Q_c}^2 + |u_t|_{\widetilde{R}_c}^2 \tag{6.72}$$

We will minimize the term $|x_t - x_m|_{Q_c}^2$, selecting

$$\widetilde{R}_c = \left(1 + \eta^{-1}\right) R_c$$

so we can reformulate the control goal as follows: minimize the term

$$|\widehat{x}_t - x_m|_{Q_c}^2 + |u_t|_{\widetilde{R}_c}^2$$

For this purpose, we define the state trajectory error as:

$$\Delta_m = \widehat{x}_t - x_t \tag{6.73}$$

and the *energetic function* $\Psi_t(u)$ as:

$$\Psi_t(u) = \gamma^T(u) W_{2,t} P_L \Delta_m + u^T R_c u \tag{6.74}$$

where P_L is the solution of the following differential Riccati equation:

$$P_L A + A^T P_L + P_L \left(W_{1,t}^T Z^{-1} W_{1,t} + \Lambda_\xi^{-1} + W_{2,t}^T W_{2,t} \right) P_L + 2\Lambda_\sigma + Q = -\dot{P}_L \tag{6.75}$$

With the initial condition $P_L(0)$ equal to the positive solution of the algebraic Riccati equation corresponding to (some equation) at time $t = 0$ with zero right hand side.

Proposition 1 *We will select the control action $u(t)$ such a way to minimize the energetic function $\Psi_t(u)$ at each time t, i.e.,*

$$u_t^* = \min_u \Psi_t(u) \tag{6.76}$$

To calculate the control action $u(t)$, which minimize $\Psi_t(u)$, we have to fulfill

$$d\Psi_t(u)/du = 0$$

To perform this minimization, we assume that, at the given positive t (positive), $x_m(t)$, and $\widehat{x}(t)$ are known and do not depend on $u(t)$.

Remark 6.9 *We name the u_t^* as locally optimal control because it is calculated based on local information available at time t.*

To solve this optimization problem, let us consider the following recursive gradient scheme:

$$u_k(t) = u_{k-1}(t) - \tau_k \frac{d\Psi_t(u_{k-1}(t))}{du}, \qquad u_0(t) = 0 \tag{6.77}$$

where the gradient $d\Psi_t(u)/du$ is calculated as:

$$\frac{d\Psi_t(u)}{du} = 2\frac{d\gamma^T(u)}{du} W_{2,t} P_L(t) \Delta_m(t) + 2R_c u \tag{6.78}$$

and the sequence of the scalar parameter τ_k satisfies the condition

$$\tau_k \geqslant 0, \quad \sum_{k=0}^{\infty} \tau_k = \infty, \quad \tau_k \to 0$$

For example, we can select $\tau_k = (1/(1+k)^{\tau}), \tau \in (0,1]$. Concerning $u^*(t)$, we state the following lemma.

Lemma 6.1 *The $u^*(t)$ can be calculated as the limit of the sequence $\{u_k(t)\}$, i.e.,*

$$u_k(t) \to u^*(t), \qquad k \to \infty \tag{6.79a}$$

Proof. it directly follows from the properties of gradient method [23], taking into account(6.69), and (6.79a) ■

Corollary 6.1 *If nonlinear input function to the DNN depends linearly on $u(t)$, we can select $d\gamma^T(u)/du = \Gamma$, and we can compensate the measurable signal $\xi^*(t)$ by the modified control law*

$$u(t) = u_{comp}(t) + u^*(t) \tag{6.80}$$

Where $u_{comp}(t)$ satisfies the relation

$$\begin{gathered} W_{2,t}^T u_{comp}(t) + \xi^*(t) = 0 \\ \xi^*(t) = Ax_m - f_m(x_m, t) \end{gathered}$$

And u^* is selected according to the *linear squares optimal control law* [3]

$$u^*(t) = -R_c^{-1}\Gamma^{-1}W_{2,t}^T P_c(t)\Delta_m(t) \tag{6.81}$$

At this point, we establish another contribution

Theorem 6.2 *For the nonlinear system (6.29), the given neural network (6.36), the nonlinear reference model (6.69) and the control law (6.81), the following property holds:*

$$|\Delta_m|_Q^2 + |u^*|_Q^2 \leqslant 2\,|x_m|_{\Lambda_\sigma}^2 + \overline{\lim_{T\to\infty}}\frac{1}{T}\int_0^T \Psi_t(u^*(t))dt \tag{6.82}$$

Remark 6.10 *Equation (6.82) fixes a tolerance level for the trajectory tracking error.*

On the final structure of the DNN the weights are learned on line.

6.3 Simulation Results

Below we present simulation results which illustrate the applicability of the proposed neuro-observer.

Example 6.1 *We consider the same example as Example 2.1 in Chapter 2. We implement the control law given by equation (6.8) and (6.28). It constitutes a feedback control with an on-line adaptive gain. Figure 6.2 and Figure 6.3 present the respective response, where the solid lines correspond to reference singles* x_t^* *and the dashed lines are the nonlinear system responses* x_t*.The time evolution for the weight of the selected neural network and the solution of differential Riccati equation are shown in Figure 6.4 and Figure 6.5.The performance index is selected as*

$$J_T^\Delta := \frac{1}{T}\int_0^T \Delta^{*T} Q_c \Delta^{*T} dt$$

can be seen in Figure 6.6.

Example 6.2 *We consider the same example as Example 3.2 of Chapter 3. We implement the control law given by equation (6.3). It constitutes a feedback control with an on-line adaptive gain. Figure 6.7 and Figure 6.8 present the respective response,*

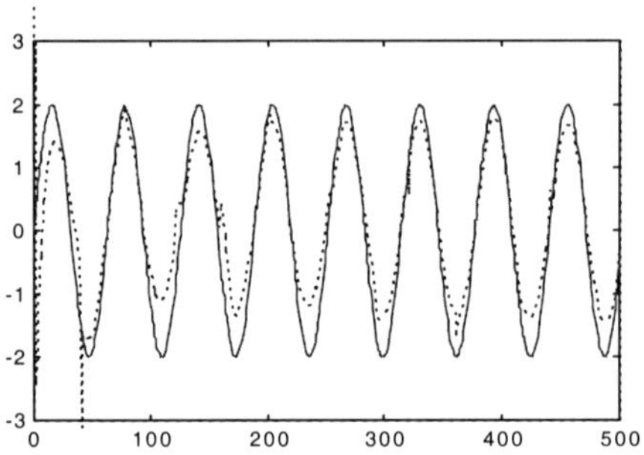

FIGURE 6.2. Response with feedback control for x.

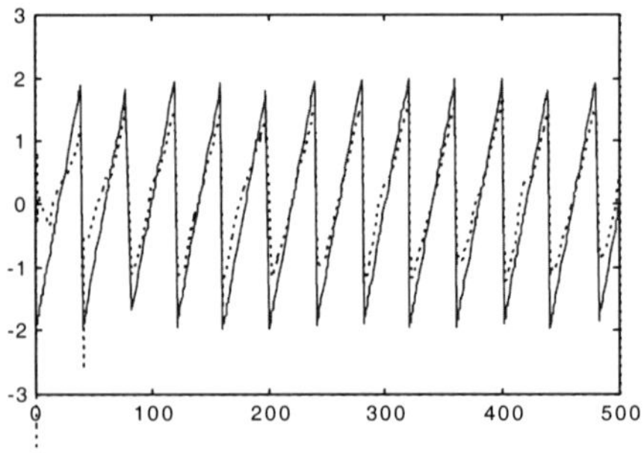

FIGURE 6.3. Rewsponse with feedback control of x_2.

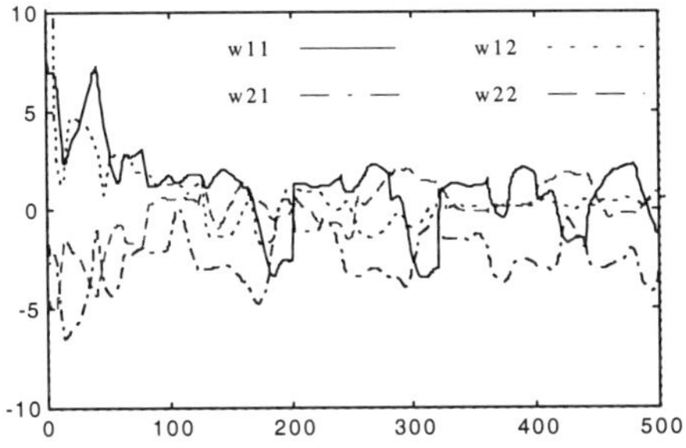

FIGURE 6.4. Time evolution of $W_{1,t}$ matrix entries.

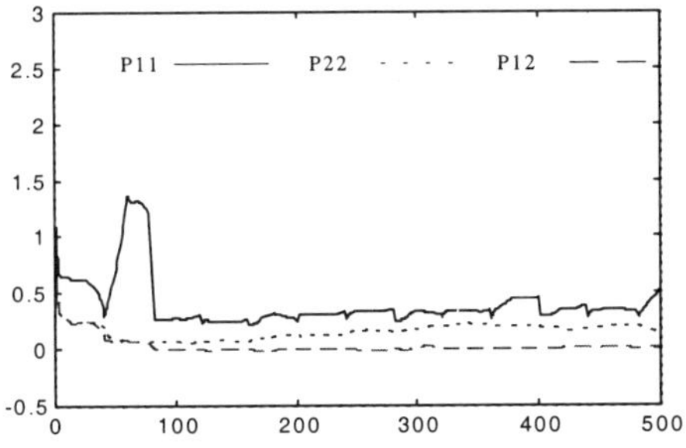

FIGURE 6.5. Time evolution of P_c matrix entries.

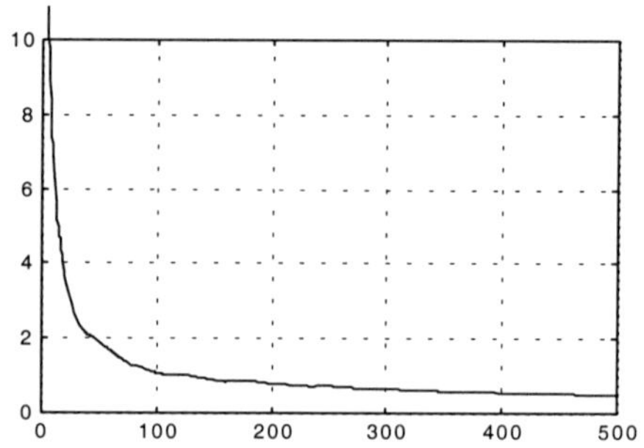

FIGURE 6.6. Tracking error J_t^{Δ}.

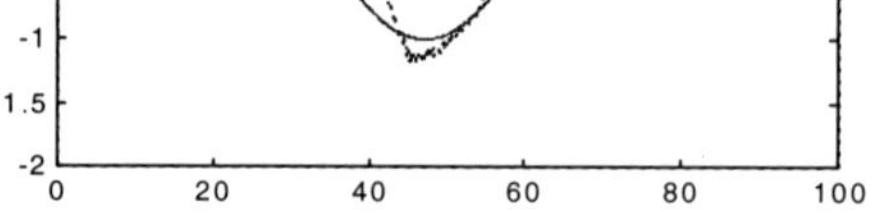

FIGURE 6.7. Trajectory tracking for x_1.

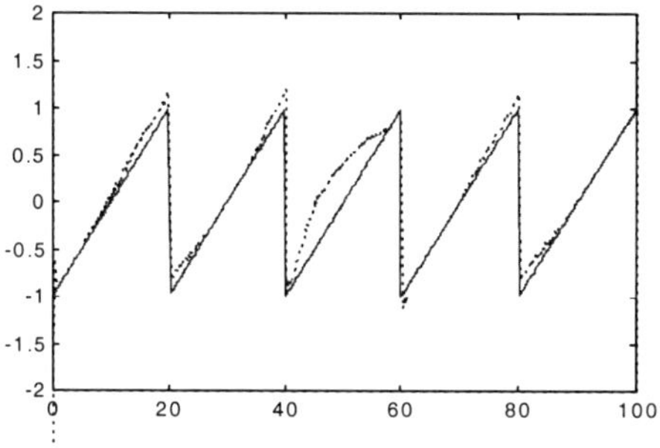

FIGURE 6.8. Trajectory tracking for x_2.

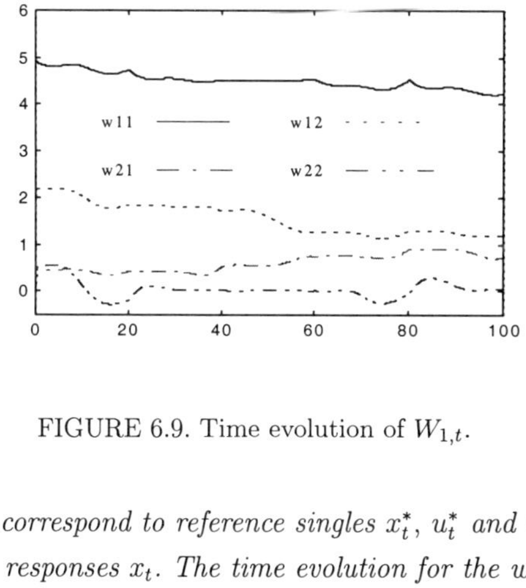

FIGURE 6.9. Time evolution of $W_{1,t}$.

where the solid lines correspond to reference singles x_t^*, u_t^* and the dashed lines are the nonlinear system responses x_t. The time evolution for the weight of the selected neural network is shown in Figure 6.9. The time evolution of two performance indexes

$$J_\tau^\Delta := \frac{1}{\tau} \int_0^\tau \Delta^{*T} Q_c \Delta^{*T} dt, \; J_\tau^u := \frac{1}{\tau} \int_0^\tau u^{*T} R_c u^* dt$$

can be seen in Figure 6.10 and Figure 6.11.

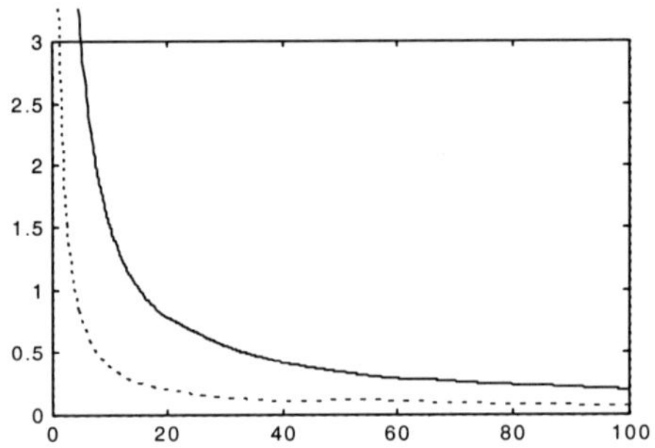

FIGURE 6.10. Performance indexes error $J_t^{\Delta_1}$, $J_t^{\Delta_2}$.

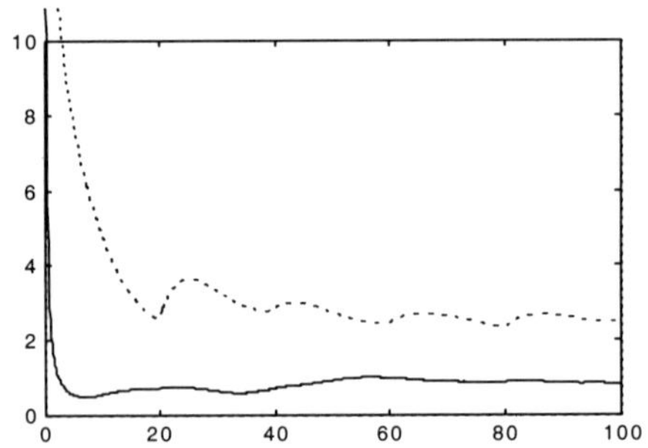

FIGURE 6.11. Performance indexes of inputs $J_t^{u_1}$, $J_t^{u_2}$.

6.4 Conclusions

In this chapter we have shown that the use of neuro-observers, with Luneburger structure and with a new learning law for the gain and weight matrices, provides a good enough estimation process, for a wide class of nonlinear systems in presence of external perturbations on the state and the outputs.

The gain matrix, guaranteeing the robustness property, is constructed solving a differential matrix Riccati equation with time-varying parameters which are dependent on on-line measurements. An important feature of the proposed neuro-observer is the use of the pseudoinverse operation applied to calculate the gain of observer. A new learning law is used to guarantee the boundness of the dynamic neural network weights.

As a continuation of the previous chapters, we are able to develop and implement a new trajectory tracking controller based on a new neuro-observer. The proposed scheme is composed of two parts: the neuro-observer and tracking controller. As our main contribution, we establish a theorem on the trajectory tracking error for closed-loop system based on the adaptive neuro-observer described above.

We test the proposed scheme with an interesting system: it has multiple equilibrium and associated vector field is not smooth. As the results show, the performances of the scheme is good enough.

The analogous approach can be successfully implemented to more complete nonlinear systems, such as saturation, friction, hysteresis and systems with nonlinear output functions.

6.5 References

[1] A.Albert, "Regression and the Moore-Penrose Pseudoinverse", Academic Press, 1972.

[2] G.Ciccarella, M.Dalla Mora and A.Germani, A Luenberger-Like Observer for Nonlinear System, *Int. J. Control*, Vol.57, 537-556, 1993.

[3] C.A.Desoer ann M.Vidyasagar, *Feedback Systems: Input-Output Properties*, New York: Academic, 1975.

[4] E.A.Coddington and N.Levinson. *Theory of Ordinary Differential Equations.* Malabar, Fla: Krieger Publishing Company, New York, 1984.

[5] F.Esfandiari and H.K.Khalil, Output Feedback Stabilization of Fully Linearizable Systems, *Int. J. Control*, Vol.56, 1007-1037, 1992.

[6] K.Funahashi, On the approximation Realization of Continuous Mappings by the Neural Networks, *Neural Networks*, Vol.2, 181-192, 1989

[7] J.P.Gauthier, H.Hammouri and S.Othman,"A simple observer for nonlinear systems: applications to bioreactors", *IEEE Trans. Automat. Contr.*, vol.37, 875-880, 1992.

[8] W.Hahn, *Stability of Motion,* Springer-Verlag: New York, 1976.

[9] K.J.Hunt and D.Sbarbaro, Neural Networks for Nonlinear Internal Model Control, *Proc. IEEE Pt.D*, Vol.138, 431-438, 1991

[10] K.J.Hunt, D.Sbarbaro, R.Zbikowski and P.J.Gawthrop, Neural Networks for Control Systems-A Survey, *Automatica*, Vol.28, 1083-1112, 1992

[11] P.A.Ioannou and J.Sun, *Robust Adaptive Control*, Prentice-Hall, Inc, Upper Saddle River: NJ, 1996

[12] L.Jin, P.N.Nikiforuk and M.M.Gupta, Adaptive Control of Discrete-Time Nonlinear Systems Using Recurrent Neural Networks, *IEE Proc.-Control Theory Appl.*, Vol.141, 169-176, 1994

[13] Y.H.Kim, F.L.Lewis and C.T.Abdallah,"Nonlinear observer design using dynamic recurrent neural networks", *Proc. 35th Conf. Decision Contr.*, 1996.

[14] E.B.Kosmatopoulos, M.M.Polycarpou, M.A.Christodoulou and P.A.Ioannpu, "High-Order Neural Network Structures for Identification of Dynamical Systems", *IEEE Trans. on Neural Networks*, Vol.6, No.2, 442-431, 1995.

[15] E.B.Kosmatpoulos, M.A.Christodoulou and P.A.Ioannou, Dynamical Neural Networks that Ensure Exponential Identification Error Convergence, *IEEE Trans. on Neural Networks*, Vol.10, 299-314,1997.

[16] R.Marino and P.Tomei, "Adaptive observer with arbitrary exponential rate of convergence for nonlinear system", *IEEE Trans. Automat. Contr.*, vol.40, 1300-1304, 1995.

[17] F.L.Lewis, A.Yesildirek and K.Liu,"Neural net robot controller with guaranteed tracking performance", *IEEE Trans. Neural Network*, Vol.6, 703-715, 1995.

[18] D.G.Luenberger, Observing the State of Linear System, *IEEE Trans. Military Electron*, Vol.8, 74-90, 1964.

[19] W.T.Miller, S.A.Sutton and P.J.Werbos, *Neural Networks for Control*, MIT Press, Cambridge, MA, 1990.

[20] K.S.Narendra and K.Parthasarathy, "Identification and Control of Dynamical Systems Using Neural Networks", *IEEE Trans. on Neural Networks*, Vol.1,4-27, 1989.

[21] S.Nicosia and A.Tornambe, High-Gain Observers in the State and Parameter Estimation of Robots Having Elastic Joins, *System & Control Letter*, Vol.13, 331-337, 1989.

[22] M.M.Polycarpou, Stable Adaptive Neural Control Scheme for Nonlinear Systems, *IEEE Trans. Automat. Contr.*, vol.41, 447-451, 1996.

[23] B.T. Polyak, *Introduction to Optimization* New York, Optimization Software, 1987.

[24] A.S. Poznyak, Learning for Dynamic Neural Networks, *10th Yale Workshop on Adaptive and Learning System*, 38-47, 1998.

[25] A.S.Poznyak, Wen Yu , Hebertt Sira Ramirez and Edgar N. Sanchez, Robust Identification by Dynamic Neural Networks Using Sliding Mode Learning, *Applied Mathematics and Computer Sciences*, Vol.8, No.1, 101-110, 1998.

[26] A.S.Poznyak, W.Yu, E. N. Sanchez and J. Perez, 1999, "Nonlinear Adaptive Trajectory Tracking Using Dynamic Neural Networks", IEEE Trans. on Neur. Netw. Vol. 10 No. 6 November, 1402-1411.

[27] A.S.Poznyak and W.Yu, 2000, "Robust Asymptotic Neuro-Observer with Time Delay Term", Int.Journal of Robust and Nonlinear Control. Vol. 10, 535–559.

[28] G.A.Rovithakis and M.A.Christodoulou, " Adaptive Control of Unknown Plants Using Dynamical Neural Networks", *IEEE Trans. on Syst., Man and Cybern.*, Vol. 24, 400-412, 1994.

[29] G.A.Rovithakis and M.A.Christodoulou, "Direct Adaptive Regulation of Unknown Nonlinear Dynamical System via Dynamical Neural Networks", *IEEE Trans. on Syst., Man and Cybern.*, Vol. 25, 1578-1594, 1994.

[30] A.Tornambe, Use of Asymptotic Observer Having High-Gains in the State and Parameter Estimations, *Proc. 28th Conf. Dec. Control*, 1791-1794, 1989.

[31] A.Tornambe, High-Gains Observer for Nonlinear Systems, *Int. J. Systems Science*, Vol.23, 1475-1489, 1992.

[32] Wen Yu and Alexander S.Poznyak, Indirect Adaptive Control via Parallel Dynamic Neural Networks, *IEE Proceedings - Control Theory and Applications*, Vol.146, No.1, 25-30, 1999.

[33] B.Widrow and S.D.Steans, *Adaptive Signal Processing*, Prentice-Hall, Englewood Cliffs, NJ, 1985.

[34] H.K.Wimmer, Monotonicity of Maximal Solutions of Algebraic Riccati Equations, *System and Control Letters*, Vol.5, pp317-319, 1985.

[35] J.C.Willems,"Least squares optimal control and algebraic Riccati equations", *IEEE Trans. Automat. Contr.*, vol.16, 621-634, 1971.

[36] A.Yesildirek and F.L.Lewis, Feedback Linearization Using Neural Networks, *Automatica*, Vol.31, 1659-1664, 1995.

Part II

Neurocontrol Applications

7

Neural Control for Chaos

In this chapter we consider identification and control of unknown chaotic dynamical systems. Our aim is to regulate the unknown chaos to a fixed points or a stable periodic orbits. This is realized by following two contributions: first, a dynamic neural network is used as identifier. The weights of the neural networks are updated by the sliding mode technique. This neuro-identifier guarantees the boundness of identification error. Secondly, we derive a local optimal controller via the neuro-identifier to remove the chaos in a system. This on-linc tracking controller guarantees a bound for the trajectory error. The controller proposed in this paper is shown to be highly effective for many chaotic systems including Lorenz system, Duffing equation and Chua's circuit.

7.1 Introduction

Control chaos is one of the topics acquiring big importance and attention in physics and engineering publications. Although the model description of some chaotic systems are simple, nevertheless the dynamic behaviors are complex (see Figures 7.1, 7.9, 7.14 and 7.19).

Recently many researchers manage to use modern elegant theories to control chaotic systems, most of them are based on the chaotic model (differential equations). *Linear state feedback* is very simple and easily implemented for the nonlinear chaotic systems [1, 14]. *Lyapunov-type method* is a more general synthesis approach for nonlinear controller design [7]. *Feedback linearization technique* is an effective nonlinear geometric theory for nonlinear chaos control [3]. If the chaotic system is partly known, for example, the differential equation is known but some the parameters are unknown, *adaptive control methods* are required [17].

In general, *the unknown chaos* is a black box belonging to a given class of nonlinearities. So, a non-model-based method is suitable. The PID-type controller have been applied to control Lorenz model [4]. The neuro-controller is also popular for control unknown chaotic system. Yeap and Ahmed [16] used multilayer perceptrons to control chaotic systems. Chen and Dong suggested direct and indirect neuro controller for chaos [2]. Both of them were based on inverse modelling, i.e., neural networks are applied to learn the inverse dynamics of the chaotic systems. There are some drawbacks to this kind of technique: lack of robustness, the demand of persistent excitation for the input signal and a not one-to-one mapping of the inverse model [7].

There exists another approach to control such unknown systems: first, construct some sort of identifier or observer, then, using this model, to the generate a control in order to guarantee "a good behavior" of unknown systems. When we have no a priori information on the structure of the chaotic system, neural networks are very effective to approach the behavior of chaos. Two types of neural networks can be applied to identify dynamic systems with chaotic trajectories:

- the *static* neural network connected with a dynamic linear model is used to approximate a chaotic system [2], but the computing time is very long and some priori knowledge of chaotic systems are need;

- the *dynamic* neural networks can minimize the approximation error of the chaotic behavior [12]. However, the number of neurons and the value of their weights are not determined. Because the dynamics of chaos are much faster, they can only realize off-line identifier (need more time for convergence). From a practical point of view, the existing results are not satisfied for controller design.

One main point of this chapter is to apply the sliding mode technique to the weights learning of dynamic neural networks. This approach can overcome the shortages of chaos identification. To the best of our knowledge sliding mode technique has been scarcely used in neural network weights learning [9]. We will proof that the identification error converges to a bounded zone by means of a Lyapunov function technique. A *local optimal controller* [6] which is based on the neural network identifier is then implemented. The controller uses a solution of a corresponding differential Riccati equation. Lyapunov-like analysis is also implemented as a basic mathematical instrument to prove the convergence of the performance index. The effectiveness are illustrated by several chaotic system such as Lorenz system, Duffing equation and Chua's circuit.

The chapter is organized as follows. First, identification and trajectory tracking for most Lorenz system is demonstrated. Then, Duffing equation is analyzed. After that Chua Circuit is studied. Finally, the relevant conclusions are established.

7.2 Lorenz System

Lorenz model is used for the fluid conviction description especially for some feature of atmospheric dynamic [14]. The uncontrolled model is given by

$$\begin{aligned} \dot{x}_1 &= \sigma(x_2 - x_1) \\ \dot{x}_2 &= \rho x_1 - x_2 - x_1 x_3 \\ \dot{x}_3 &= -\beta x_3 + x_1 x_2 \end{aligned} \tag{7.1}$$

where x_1, x_2 and x_3 represent measures of fluid velocity, horizontal and vertical temperature variations, correspondingly. The parameters σ, ρ and β are positive parameters that represent the Prandtl number, Rayleigh number and geometric factor, correspondingly.

If

$$\rho < 1$$

the origin is a stable equilibrium.

If

$$1 < \rho \leq \rho^*(\sigma, \beta)$$

the system has two stable equilibrium points with components

$$(\pm\sqrt{\beta(\rho - 1)}, \pm\sqrt{\beta(\rho - 1)}, (\rho - 1))$$

and one unstable equilibrium (the origin).

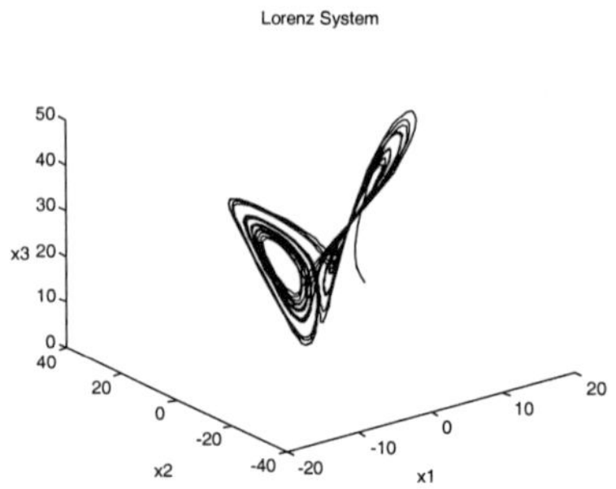

FIGURE 7.1. Phase space trajectory of Lorenz system.

If

$$\rho^*(\sigma,\beta) < \rho,$$

all three equilibrium points become unstable.

As in the commonly studied case, we select

$$\sigma = 10 \text{ and } \beta = 8/3$$

that leads to

$$\rho^*(\sigma,\beta) = 24.74.$$

In this example we will consider the system with

$$\rho = 28.$$

Figure 7.1 shows the chaotic behavior on the initial uncontrolled system.

Experiment 1.1 (*Identification of original uncontrolled chaotic via Neural Network*).

We design a on-line neuro identifier as in Chapter 2, but more simple structure:

$$\dot{\widehat{x}}_t = A\widehat{x}_t + W_{1,t}\sigma(\widehat{x}_t) + u_t \tag{7.2}$$

where

$$A = diag(-8, -8, -8).$$

The initial conditions for $\widehat{x}_t$ can by any small values. Here we select

$$\widehat{x}_0 = [1, -5, 0]^T$$

The weights $W_{1,t}$ are the elements of the 3×3 matrix. The elements $\sigma(\cdot)$ are selected as

$$\sigma(x_i) = \frac{2}{1 + e^{-2x_i}} - 0.5 \tag{7.3}$$

and

$$P = diag(20, 20, 20), \quad \tau = 0.01$$

Here we will use the sliding mode learning as in Chapter 3. The identification results are shown in Figure 7.2, 7.3 and 7.4. The solid lines correspond to the state of the original uncontrolled Lorenz system, the dashed lines are the states of neural network identifier. Because we adopt the sliding mode learning, this dynamic neural network can follow the fast system very well. We know that most of the existed updating laws of neural networks cannot give so quick response. So, based on those neuro models, it is difficult to design a on-line controller which can guarantee a respectively good trajectory behavior. The drawback of these neuro identifier is that its weights change very quickly and as, a fact, it is not easy to use them in any control loop. If apply the derived sliding mode identifier to local optimal controller, we can avoid these big deviations. Figure 7.5 shows the time evaluation of the element $(w_{11})_t$ of the weight matrix $W_{1,t}$.

Experiment 1.2 (*Regulation of the controlled chaotic via Neural Network*).

Based on the neural network model (7.2), a local optimal controller (see Appendix C) is taken to force the Lorenz system into a stable periodic orbits of a fixed point. The Lorenz system subjected to a control can be expressed as [16]

$$\begin{aligned} \dot{x}_1 &= \sigma(x_2 - x_1) + u_1 \\ \dot{x}_2 &= \rho x_1 - x_2 - x_1 x_3 + u_2 \\ \dot{x}_3 &= -\beta x_3 + x_1 x_2 + u_3 \end{aligned} \tag{7.4}$$

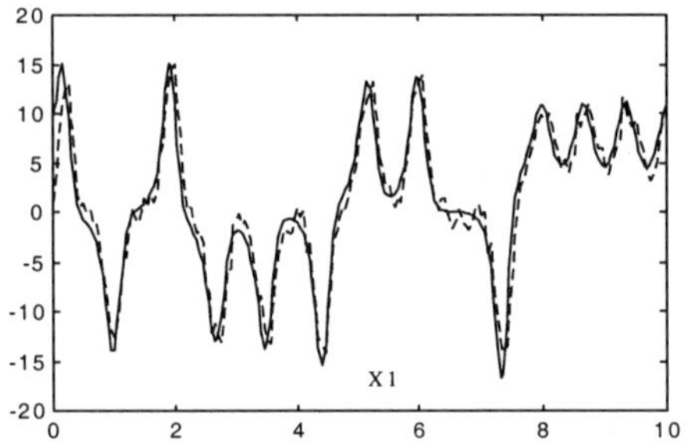

FIGURE 7.2. Identification results for x_1.

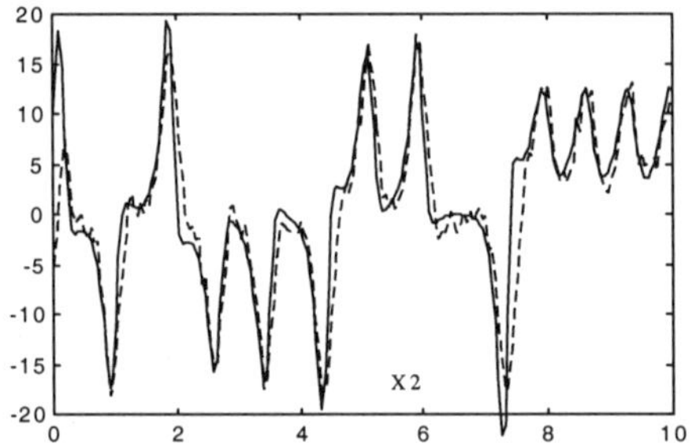

FIGURE 7.3. Identification results for x_2.

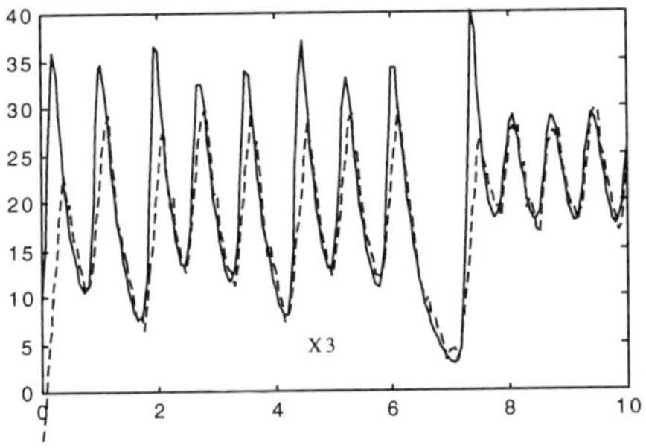

FIGURE 7.4. Identification results for x_3.

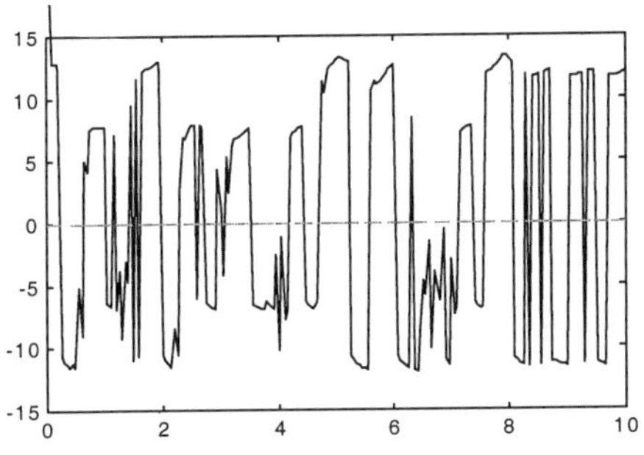

FIGURE 7.5. The time evaluation of w_{11} of $W_{1,t}$.

The controller is selected as

$$\begin{aligned} u_t &= [W_{2,t}\phi(\widehat{x}_t)]^+ (u_{1,t} + u_{2,t}) \\ u_{1,t} &= \varphi(x_t^*, t) - Ax_t^* - W_{1,t}\sigma(\widehat{x}_t) \\ u_{2,t} &= -2R^{-1}P\Delta_t \end{aligned} \tag{7.5}$$

with $W_{2,t} = I$. The matrix P_c is the solution of the matrix Riccati equation (6.25), where

$$\begin{aligned} P_c &= \begin{bmatrix} 0.84 & 0 & 0 \\ 0 & 0.74 & 0 \\ 0 & 0 & 0.84 \end{bmatrix} \\ Q_c &= diag(1,1,1) \\ R_c &= diag(0.3, 0.3, 0.3) \end{aligned}$$

and

$$Z = I$$

Using the controller as in (7.5) we may regulate the Lorenz system (7.4) to set points. The Lorenz system start from

$$x_0 = [10, 10, 10]^T$$

First, we control the system to a set point

$$X_1 = [11, 11, 45]^T$$

and let it stay in this point until

$$t = 4.8s.$$

Then, we force the system to the other new set point

$$X_2 = [8.5, 8.5, 28]^T.$$

Figures 7.6, 7.7 and 7.8 give the regulation results of these three states and Figure 7.9 shows the corresponding phase space trajectory.

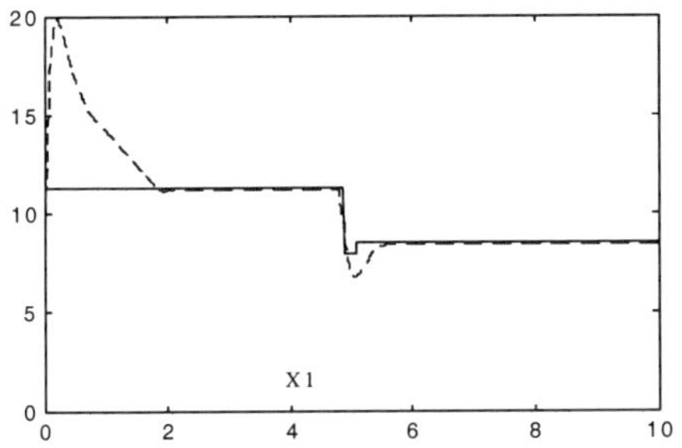

FIGURE 7.6. Regulation of state x_1.

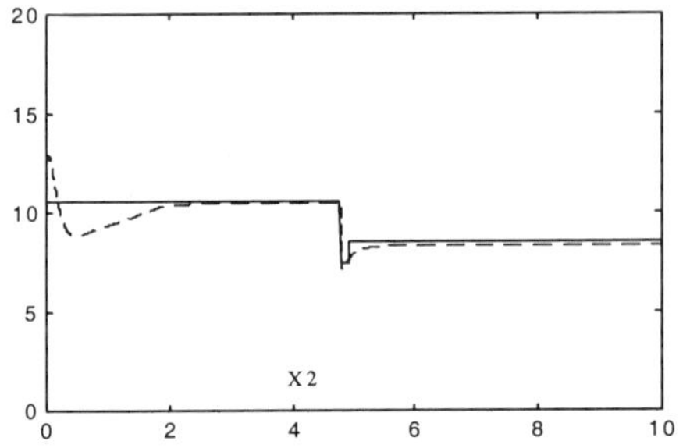

FIGURE 7.7. Regulation of state x_2.

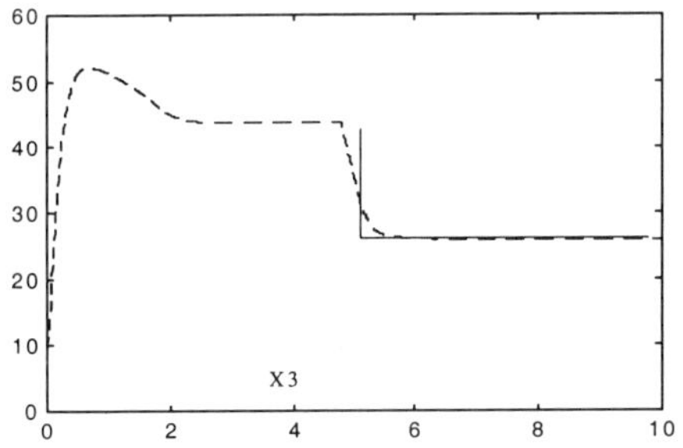

FIGURE 7.8. Regulation of state x_3.

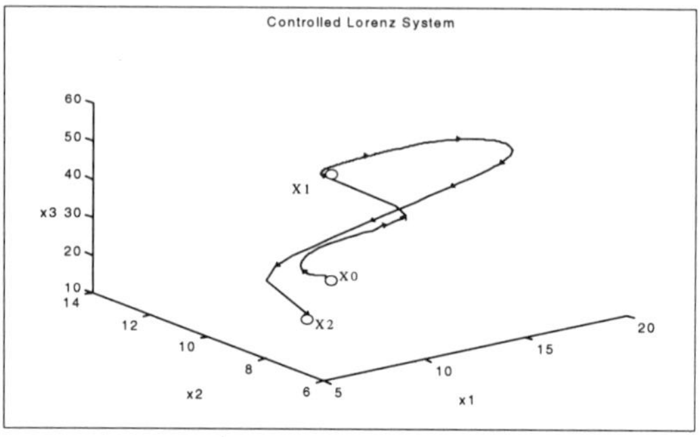

FIGURE 7.9. Phase space trajectory.

We note that the close loop system by the use of the suggested technique is free of chaotic transients. Each of states subjected to local optimal control reaches a constant value in short time and stay there for a long period.

Experiment 1.3 (*Trajectory tracking of the controlled chaotic via Neural Network*).

Now we will manage to force this system into a desirable periodic trajectory. It is more difficult problem than the regulating to set points. The nonlinear reference model to be followed is selected as a circle:

$$\begin{aligned} \dot{x}_1^* &= x_2^* \\ \dot{x}_2^* &= \sin(x_1^*) \\ x_3^* &= 50 \end{aligned} \tag{7.6}$$

with initial conditions equal to

$$x_1^*(0) = 1, x_2^*(0) = 0$$

The trajectory tracking results are shown in Figures 7.10 and 7.11. The control inputs are shown in Figure 7.12.

We observe that the control input is not changed quickly as the weights, since the control u_t is proportional to the "slow" solution of the differential Riccati equation which time evaluation is shown at Figure 7.13.

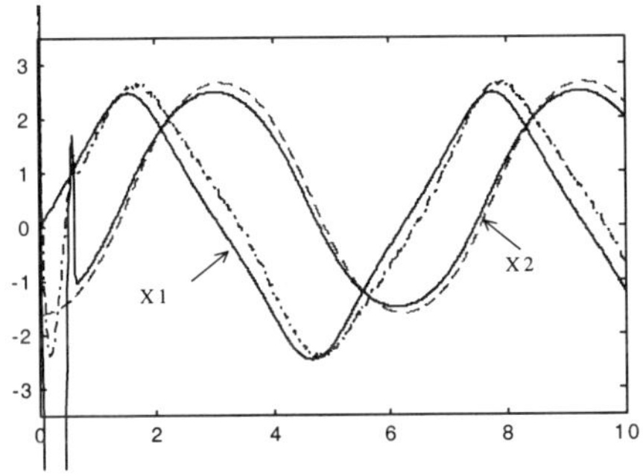

FIGURE 7.10. States tracking.

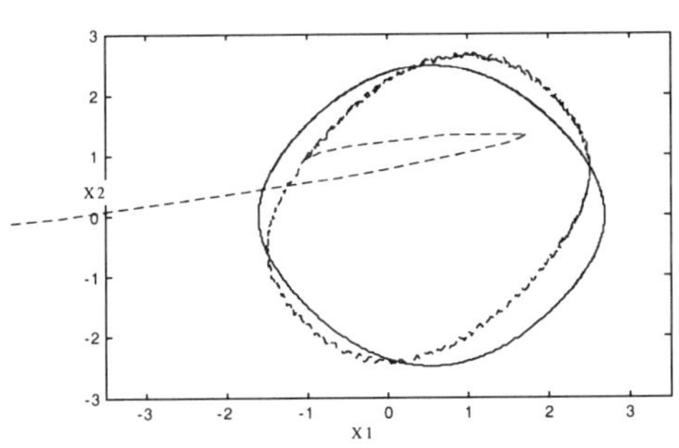

FIGURE 7.11. Phase space.

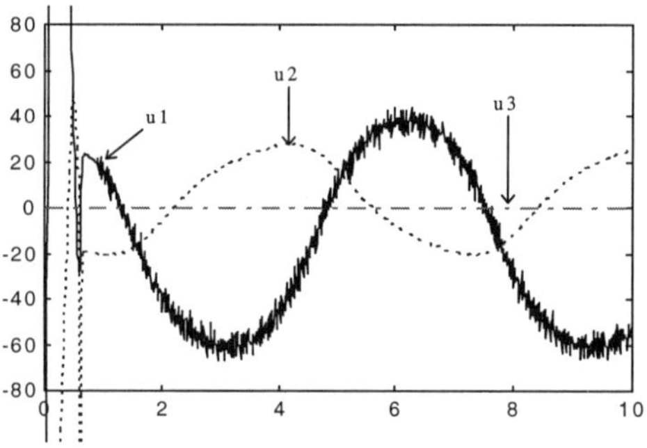

FIGURE 7.12. Control inputs.

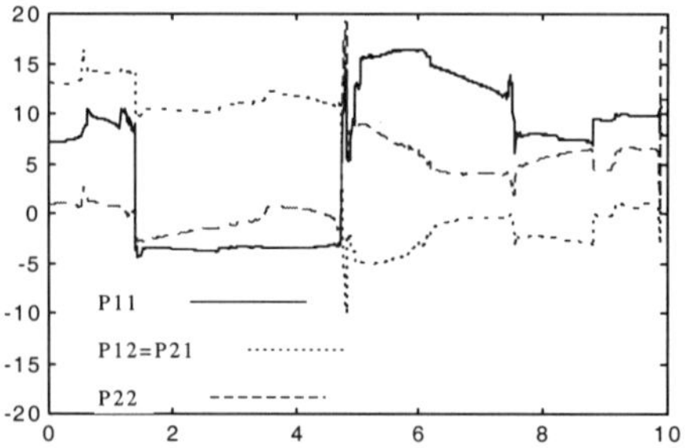

FIGURE 7.13. The time evaluation of $P(t)$.

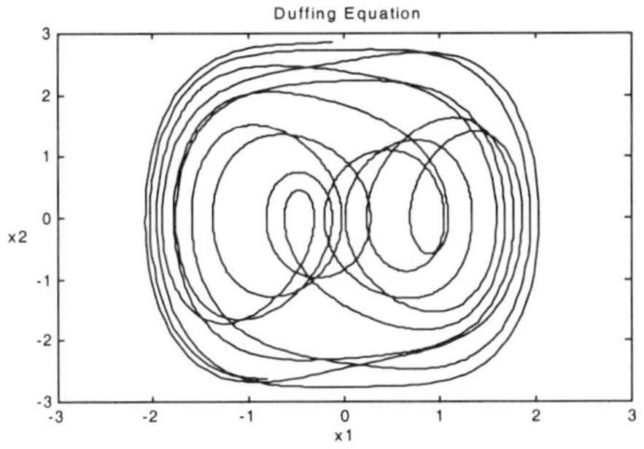

FIGURE 7.14. Phase space trajectory of Duffing equation.

7.3 Duffing Equation

Duffing equation describes a specific nonlinear circuit or, so named, *"the hardening spring effect"* observed in many mechanical problems [1]. It can be written as

$$\begin{aligned} \dot{x}_1 &= x_2 \\ \dot{x}_2 &= -p_1 x_1 - p_2 x_1^3 - p x_2 + q\cos(\omega t) + u_t \end{aligned} \tag{7.7}$$

where p, p_1, p_2, q and ω are constants and u_t is a control input. It is known that the solution of (7.7) exhibits almost periodic and chaotic behavior. In uncontrolled case ($u_t = 0$), if we select

$$p_1 = 1.1, p_2 = 1, p = 0.4, q = 2.1, \omega = 1.8,$$

the Duffing oscillator has a chaotic response as in Figure 7.14.

Experiment 2.1 (*Identification of original uncontrolled chaotic via Neural Network*).

Since Duffing oscillator is a two dimension dynamics, to identify this system we use the same neural network as in (7.2), but with two dimension state space, i.e.,

$$A = diag(-8, -8), \quad \widehat{x}_0 = [1, -5]^T,$$

$W_{1,t}$ is 2×2 matrixes. The elements of $\phi(\cdot)$ is selected as in (7.3),

$$P = diag(20, 20) \text{ and } \tau = 0.01.$$

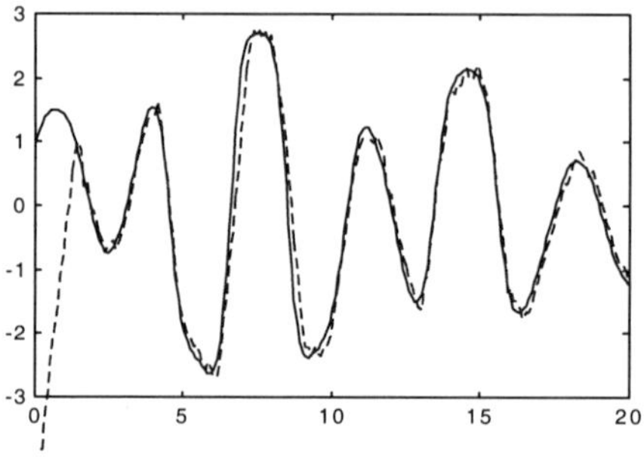

FIGURE 7.15. Identification of x_1.

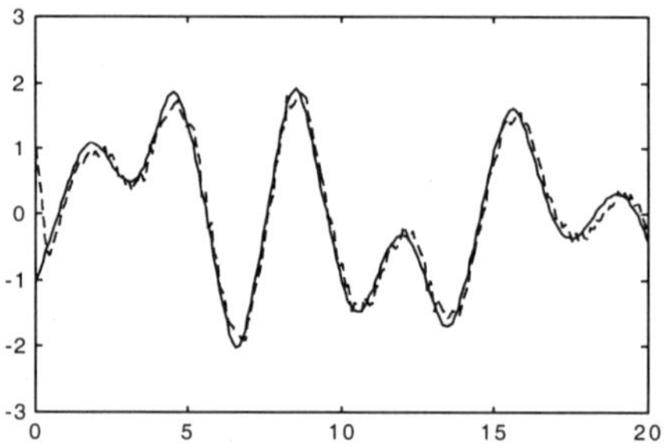

FIGURE 7.16. Identification of x_2.

Sliding mode learning as in Chapter 3 is used. The identification results are shown in Figures 7.15 and 7.16.

Experiment 2.2 (*Trajectory tracking of the controlled chaotic via Neural Network*).

Controlled Duffing equation differs from Lorenz system because we have only one control input. We also force the Duffing equation to the periodic orbits as in (7.6). The corresponding results are shown in Figures 7.17 and 7.18.

We note that the local optimal controller, which we are applying here, is independent of the chaotic systems, because this controller is based on only the neuro identifier data. Numerical simulations show that good identification results provide

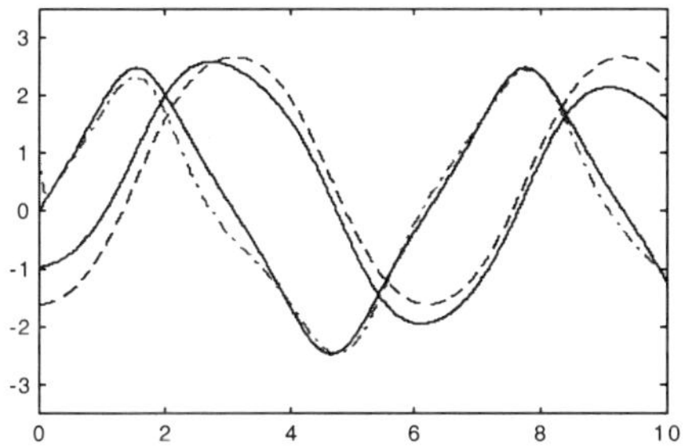

FIGURE 7.17. States tracking.

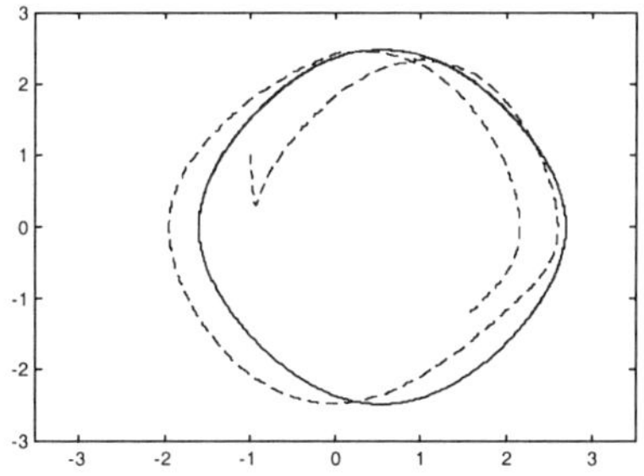

FIGURE 7.18. Phase space.

a small enough tracking error.

7.4 Chua's Circuit

Chua's circuit is a interesting electronics system that display rich and typical bifurcation and chaotic phenomena such as *double scroll* and *double hook* [2]. To study the controlled circuit, we introduce its differential equation in the following form:

$$\left\{\begin{array}{r} C_1 \dot{x}_1 = G(x_2 - x_1) - g(x_1) + u_1 \\ C_2 \dot{x}_2 = G(x_1 - x_2) + x_3 + u_2 \\ L \dot{x}_3 = -x_2 \end{array}\right.$$
$$g(x_1) = m_0 x_1 + \tfrac{1}{2}(m_1 - m_0)\left[|x_1 + B_p| + |x_1 - B_p|\right]$$

where x_1, x_2, x_3 denote, respectively, the voltages across the capacities C_1 and C_2 and the current through the induction L. It is known (see [1]) that with

$$\frac{1}{C_1} = 9, \frac{1}{C_2} = 1, \frac{1}{L} = 7$$

and

$$G = 0.7, m_0 = -\frac{1}{2}, m_1 = -\frac{4}{5}, B_p = 1$$

the circuit displays double scroll. The chaos of Chua's circuit is shown in Figure 7.19.The circuit displays as a double scroll.

Experiment 3.1 (*Identification of original uncontrolled chaotic via Neural Network*).

To demonstrate the effectiveness of the approach suggested in this book, we also use the same neural network as in (7.2). The identification results are shown in Figures 7.20 and 7.21.

Experiment 3.2 (*Trajectory tracking of the controlled chaotic via Neural Network*)

The controlled tracking behavior is shown in Figure 7.22 and 7.23.

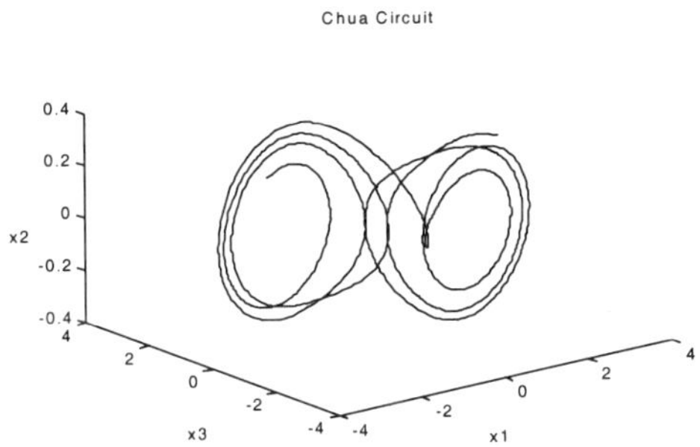

FIGURE 7.19. The chaos of Chua's Circuit.

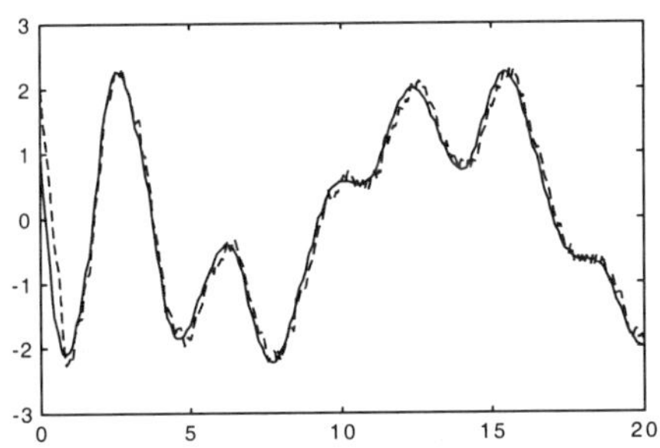

FIGURE 7.20. Identification of x_1.

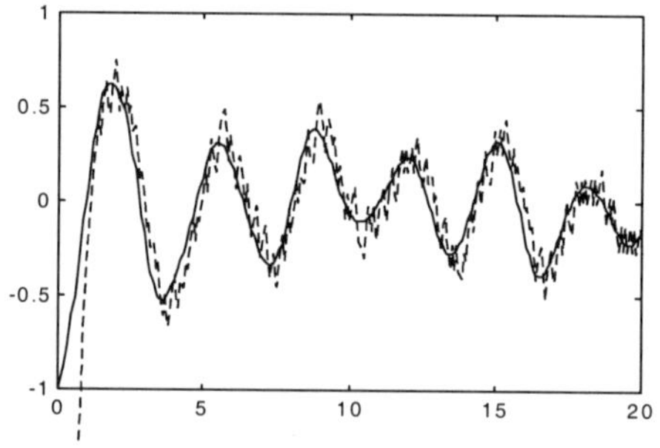

FIGURE 7.21. Identification of x_2.

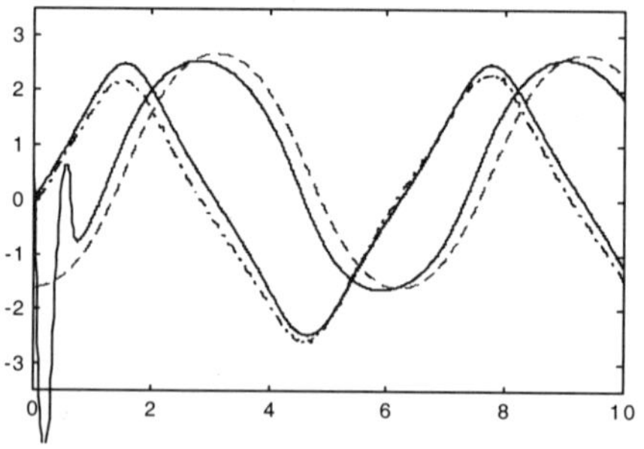

FIGURE 7.22. State Tracking of Chua' s Circuit.

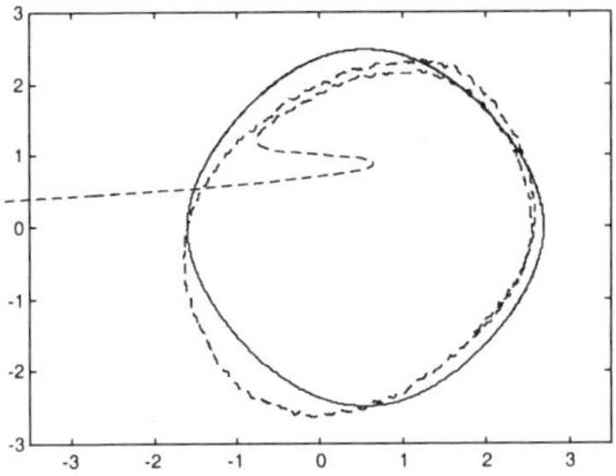

FIGURE 7.23. Phase space.

7.5 Conclusion

In this chapter we present a new method for designing a control for the chaotic systems. The suggested controller is independent of the chaotic models. We assume that the states of chaos are observable, the dynamic equations are unknown. Our approach does not use any inverse model.

The proposed controller is composed by two parts [21]:

- neuro identifier

- and tracking controller.

The identifier uses the sliding mode technique to increase learning speed of neural network weights. It is shown that for different chaotic dynamic the same neural network identifier can work very well practically without corrections of the algorithm.

The implemented controller uses the local optimal method to avoid inversion of the weight matrices.

Lyapunov-like analysis and the differential Riccati equation are used to guarantee the corresponding bounds for the tracking errors. Simulation results show that for different chaotic systems, the derived control via neuro identifier turns out to be very effective.

7.6 References

[1] G.Chen and X.Dong, "On feedback control of chaotic continuous-time systems", *IEEE Trans. Circuits Syst.*, Vol.40, pp.591-601, 1993.

[2] G.Chen and X.Dong, "Identification and Control of chaotic systems", *Proc. of IEEE Int'l Symposium on Circuits and Systems*, Seattle, WA, 1995.

[3] J.A.Gallegos, "Nonlinear Regulation of a Lorenz System by Feedback Linearization Techniques", *Dynamic and Control*, Vol. 4, 277-298, 1994.

[4] T.T.Hartley and F.Mossayebi, Classical Control of a Chaotic System, *IEEE Conference on Control Application*, Dayton USA, 522-526,1992

[5] K.J.Hunt, D.Sbarbaro, R.Zbikowski and P.J.Gawthrop, "Neural Networks for Control Systems-A Survey", *Automatica*, Vol.28, pp.1083-1112, 1992.

[6] G.K.Kel'mans, A.S.Poznyak and A.V.Cherniser, Adaptive Locally Optimal Control, *Int. J. System Sci.*, Vol.12, pp.235-254, 1981.

[7] H.Nijmeijer and H.Berghuis, "On Lyapunov Control of the Duffing Equation", *IEEE Trans. Circuits Syst.*, Vol.42, pp.473-477, 1995.

[8] A.S.Poznyak and E.N.Sanchez, "Nonlinear System Approximation by Neural Networks: Error Stability Analysis", *Intl. Journ. of Intell. Autom. and Soft Comput.*, Vol. 1, pp 247-258, 1995.

[9] Alexander S.Poznyak, Wen Yu , Hebertt Sira Ramirez and Edgar N. Sanchez, "Robust Identification by Dynamic Neural Networks Using Sliding Mode Learning", *Applied Mathematics and Computer Sciences*, Vol.8, 101-110, 1998.

[10] Alexander S.Poznyak, Wen Yu and Edgar N. Sanchez, Identification and Control of Unknown Chaotic Systems via Dynamic Neural Networks, *IEEE Trans. Circuits and Systems, Part I*, Vol.46, No.12, 1999.

[11] G.A.Rovithakis and M.A.Christodoulou, "Adaptive Control of Unknown Plants Using Dynamical Neural Networks", *IEEE Trans. Syst., Man and Cybern.*, vol. 24, pp 400-412, 1994.

[12] J.A.K.Suykens and J.Vandewalle, "Learning a Simple Recurrent Neural State Space Model to Behave Like Chua's Double Scroll", *IEEE Trans. Circuits Syst.*, Vol.42, pp.499-502, 1995.

[13] J.A.K.Suykens and J.Vandewalle, "Control of a Recurrent Neural Network Emulator for Double Scroll", *IEEE Trans. Circuits Syst.*, Vol.43, pp.511-514, 1996.

[14] T.L.Vincent and J.Yu, "Control of a Chaotic System", *Dynamic and Control*, Vol.1, 35-52, 1991.

[15] B.Widrow and S.D.Steans, *Adaptive Signal Processing*, Prentice-Hall, Englewood Cliffs, NJ, 1985.

[16] T.H.Yeap and N.U.Ahmed, Feedback Control of Chaotic Systems, *Dynamic and Control*, Vol.4, 97-114, 1994.

[17] Y.Zeng and S.N.Singh, "Adaptive Control of Chaos in Lorenz System", *Dynamic and Control*, Vol.7, 143-154, 1997.

8

Neuro Control for Robot Manipulators

In this chapter the neuro tracking problem for a robot manipulator with two degrees of mobility and with unknown load, friction and the parameters of the mechanical system , subject to variations within a given interval, is tackled. The design of the neuro robust nonlinear controller is proposed such a way that a certain accuracy of the tracking is achieved. The suggested neuro controller has a direct linearization part and a locally optimal compensator. Compared with sliding mode type and linear state feedback controllers, the numerical simulations of this robust controller illustrate its effectiveness.

8.1 Introduction

Based on *Lagrange Equalities Approach,* the most of mechanical systems turn out to be considered as a class of *nonlinear systems* containing known as well as unknown parameters in its model description [30]. *Robot manipulators* can be also considered as a class of nonlinear systems with a *friction coefficient* and *load* as unknown parameters which assumed to be a priory within a given region and may be varying in time.

Friction models are not yet completely understood. Some friction phenomena such as a hysteresis, Daih's effect (nonlinear dynamic friction properties) and Stribeck's effect (positive damping at low velocities) require further investigation. The comprehensive survey on this topic can be found in [2].

State feedback control is one of the topic acquiring big importance and attention in engineering publications that in the last two decades, guarantees the desired performance of a nonlinear dynamic system containing uncertain elements was discussed in [3, 10]. In this direction there exists already some results which can be classified in five large groups:

- **Adaptive Control** (see [22] and [31]) is a popular and powerful approach to control systems with unknown parameters. So, in [36] virtual decomposition-based adaptive motion/force control scheme is presented to deal with the control problem of coordinated multiple manipulators with flexible joints holding a common object in contact with the environment. The main feature is that the developed technique can successfully work only if the corresponding unknown parameters are assumed to be constant.

- **Sliding Mode Control** [8] consists in the selection a hypersurface switching surface such a way which leads to the asymptotic trajectory convergence to this sliding surface. In spite of the fact that this control is robust with respect to external disturbances, its implementation is never perfect because of *"chattering effect"* (state oscillation around sliding surface).

- **Robust Feedback Control** [9] is usually designed to guarantee the stability and some quality of control in the presence of parametric or unparametric uncertainties. Robust control of flexible joint manipulators with unmodeled parameters and unknown disturbances has recently been reported in [27]. Global uniform ultimate boundness was discussed in [4]. The most of publications deal with linear models in the presence of L_2-bounded disturbances.

- **Robust Adaptive Control.** Since the time derivative of the Lyapunov function is only negative semidefinite under adaptive control, any of un-parametrizable dynamics (such as frictions) can potentially destabilize the system. This observation leads to the following two ways:

  - by adding *minimax control* or *saturation-type control* to the existing adaptive control [23],

  - or by changing the adaptation law so there is a negative defined term (*leakage-like adaptation*) [20].

- **Adaptive-Robust Control** (see [8] and [29]) estimates on-line the size of the uncertainties and uses these estimates in the traditional robust procedures [8]. Unfortunately, the corresponding theoretical study is still not completed.

It is well known that most of the industrial manipulators are equipped with the simplest proportional and derivative (PD) controller. Various modified PD control schemes and their successful experimental tests have been published [30], [22]. But there exist two main weaknesses in PD control:

1. PD control required the measurements both joint position and joint velocity. It is necessary to implement position and velocity sensors at each joint. The joint position measurement can be obtained by means of encoder, which gives very accurate measurement. The joint velocity is usually measured by velocity tachometer, which is expensive and often contaminated by noise [10].

2. Due to the existence of friction and gravity forces, the PD-control cannot guarantee that the steady state error becomes zero [15].

It is very important to realize the PD control scheme with only joint position measurement. One of possible method is to use a velocity observer. Many papers have been published devoted to the theory and practice implementation of velocity observers of manipulators. Two kinks of observer may be used: model-based observer and model- free observer. The model-based observer assumes that the dynamics of the robot are complete known or partial known. In the case of only inertia matrix of robotic dynamic being known, the sliding model observer was proposed in [5]. The adaptive observer was proposed in [6]. The passivity method was developed to design the velocity observer in [1]. The model-free observer means that no exact knowledge of robot dynamics is required. Most popular model-free observers are high-gain observers, they can estimate the derivative of the output [28]. Recently, neural networks observer was presented in [10], only the inertia matrix is assumed known, the nonlinearities of manipulator were estimated by static neural networks.

Since friction and gravity may influence the steady and dynamic properties of PD control, two kinds of compensation can be used. The global asymptotic stability PD control was realized by pulsing gravity compensation in [28]. If the parameter in the gravitational torque vector are unknown, the adaptive version of PD control with gravity compensation was introduced in [26]. PID control does not require any component of robot dynamics into its control law, but it lacks a global asymptotic stability proof [16]. By adding integral actions or computed feedforward, the global asymptotic stability PD control were proposed in [15] and [32].

In this chapter we consider the robust tracking problem of a robot manipulator with two degrees of mobility and with unknown friction parameter, subject to variations within a given interval.

The main result consists in the proposition of a robust nonlinear controller which can guarantee a certain accuracy of a tracking process. The suggested robust controller has the same structure as in Chapter 6.

We also propose a new modified algorithm which may overcome the two drawbacks of PD control at same time. First, the high-gain observer is joined with a PD control which achieves stability with the knowledge of friction and gravity. Unlike the other papers which used singular perturbation method [27], we give the upper bound of observer error by means of Lyapunov analysis. Second, a RBF neural network is used to estimated the nonlinear terms of friction and gravity. The learning rules obtained for the neural networks are very closed to the backpropagation rules but with some additional terms. No off-line learning phase is required. We show that the closed-loop system with high-gain observer and neuro compensator is stable. Some experimental tests are carried out in order to validate the modified PD control with high-gain observer and neural networks compensator .

Experimental results and numerical simulations illustrate its effectiveness in comparison with the sliding mode type and linear state feedback controllers.

8.2 Manipulator Dynamics

First, derive the dynamic model for a Robot Manipulator with two degrees of freedom and containing an internal uncertainty connected with an unknown (and, may be, time-varying) friction parameter. The scheme of a two-links robot manipulator is shown in Figure 8.1.

he corresponding Lagrange dynamic equation can be expressed as follows [30]:

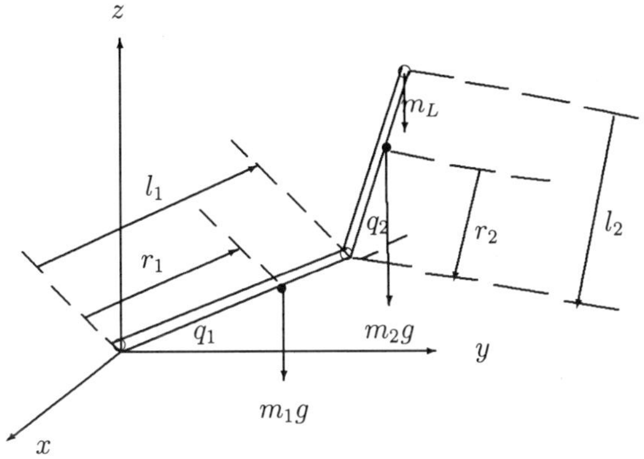

FIGURE 8.1. A scheme of two-links manipulator.

$$\boxed{M(\theta)\ddot{\theta} + W\left(\theta, \dot{\theta}\right) = u \ \left(\theta, u \in R^2\right)} \tag{8.1}$$

where $M(\theta)$ represents the positive defined inertia matrix

$$M(\theta) = M^T(\theta) = \begin{bmatrix} M_{11} & M_{12} \\ M_{21} & M_{22} \end{bmatrix} > 0$$

with the elements

$$\begin{aligned} M_{11} &= (m_1 + m_2)\, a_1^2 + m_2 a_2^2 + 2m_2 a_1 a_2 c_2 \\ M_{12} &= m_2 a_2^2 + m_2 a_1 a_2 c_2, \;\; M_{22} = m_2 a_2^2 \\ M_{21} &= M_{12}, \; a_i = l_i, \; c_i = \cos\theta_i, \; s_i = \sin\theta_i \\ c_{12} &= \cos(\theta_1 + \theta_2) \end{aligned}$$

Here $m_i, l_i \ (i = 1, 2)$ are the mass and length of the corresponding links and $W\left(\theta, \dot{\theta}\right)$ is the Coriolis matrix representing the centrifugal and friction effects (with the uncertain parameters). It can be described as follows:

$$W\left(\theta, \dot{\theta}\right) = W_1\left(\theta, \dot{\theta}\right) + W_2\left(\dot{\theta}\right)$$

where $W_1\left(\theta,\dot{\theta}\right)$ corresponds to the Coriolis and centrifugal components:

$$W_1\left(\theta,\dot{\theta}\right)=\begin{pmatrix} W_{1a} \\ W_{1b} \end{pmatrix}$$

$$W_{1a}=-m_1a_1a_2(2\,\dot{\theta}_1\dot{\theta}_2+\dot{\theta}_2^2)s_2+(m_1+m_2)\,ga_1c_1+m_2ga_2c_{12}$$

$$W_{1b}=m_2a_1a_2\,\dot{\theta}_1^2\,s_2+m_2ga_2c_{12}$$

and $W_2\left(\dot{\theta}\right)$ corresponds to the friction component:

$$W_2\left(\dot{\theta}\right)=\kappa\upsilon$$

where

$$\kappa:=\begin{pmatrix} \nu_1 & \kappa_1 & 0 & 0 \\ 0 & 0 & \nu_2 & \kappa_2 \end{pmatrix}$$

$$\upsilon^T=\begin{pmatrix} \dot{\theta}_1 & sign\,\dot{\theta}_1 & \dot{\theta}_2 & sign\,\dot{\theta}_2 \end{pmatrix}$$

In (8.1) the input vector u is a joint torque vector which is assumed to be given. We don't consider any external perturbations in this concrete context, but as it follows from the theory presented above, we can do it.

This robot model (8.1) has the following structural properties which will be used in the design of velocity observer and nonlinearities compensation.

Property 1. The inertia matrix is symmetric and positive definite [30], i. e.

$$m_1\left\|x\right\|^2\leq x^TM(x_1)x\leq m_2\left\|x\right\|^2;\forall x\in R^n$$

where m_1, m_2 are known positive scalar constant, and $\|\circ\|$ denotes the euclidean vector norm.

Property 2. The centripetal and Coriolis matrix is skew-symmetric, i.e., satisfies the following relationship:

$$x^T \left[\dot{M}(q) - 2C(q,\dot{q})\right] x = 0; \forall x \in R^n$$
$$C(q,x)y = C(q,y)x; \forall x, y \in R^n$$
$$C(q,\dot{q}) = \sum_{k=1}^{n} C_k(q)\, \dot{q}_k$$
$$\left\| C(q,\dot{q}) \right\| \leq k_c \left\| \dot{q} \right\|$$
$$C(q,\dot{q})\, \dot{q} = \dot{q}^T\, C_0(q)\, \dot{q}$$

where

$$C_{k,ij}(q) = \left(\frac{\partial B_{ij}}{\partial q_k} + \frac{\partial B_{ik}}{\partial q_j} - \frac{\partial B_{jk}}{\partial q_i} \right)$$

$$k_c = \frac{1}{2} \max_{q \in R^n} \sum_{k=1}^{n} \|C_k(q)\|$$

$C_0(q)$ is a bounded matrix.

Let us now represent this system in the standard form which will be in force throughout of this paper. To do this, we introduces the extended vector

$$\boxed{x^T = \left(\theta_1,\ \theta_2; \dot{\theta}_1, \dot{\theta}_2 \right)}$$

and in view of this definition we can rewrite the dynamic equation (8.1) as follows:

$$\dot{x} = \begin{pmatrix} \dot{x}_1 \\ \dot{x}_2 \\ \dot{x}_3 \\ \dot{x}_4 \end{pmatrix} = \begin{pmatrix} x_3 \\ x_4 \\ \left(-M^{-1}(x)W_1(x) - M^{-1}(x)\kappa v(x) + M^{-1}(x)u\right)_1 \\ \left(-M^{-1}(x)W_1(x) - M^{-1}(x)\kappa v(x) + M^{-1}(x)u\right)_2 \end{pmatrix} \tag{8.2}$$

Let assume also that the matrix κ can be expressed in the following form

$$\kappa := \kappa_0 + \Delta\kappa_t \tag{8.3}$$

where the internal uncertainty $\Delta\kappa$ satisfies

$$\forall t: \qquad \Delta\kappa_t^T \Delta\kappa_t \leq \Delta \tag{8.4}$$

Here the matrix Δ is assumed to be a priory known. In view of the notations accepted above we can represent our system (8.2) in the following standard form:

$$\dot{x}_t = F_0(x_t, t) + \Delta F(x_t, t) + F_1(x_t, t) u_t \tag{8.5}$$

where

$$F_0(x_t, t) = \begin{pmatrix} x_3 \\ x_4 \\ \mathbf{f}_0(x_t) \end{pmatrix} = Ex_t + \overline{F_0(x_t, t)}, \ \Delta F(x_t, t) = \begin{pmatrix} 0 \\ 0 \\ \Delta\mathbf{f}(x_t) \end{pmatrix}$$

$$F_1(x_t, t) = \begin{pmatrix} 0 \ \ 0 \\ 0 \ \ 0 \\ B(x_t) \end{pmatrix}$$

and

$$\mathbf{f}_0(x_t) := -M^{-1}(x_t)\left[W_1(x_t) + \kappa_0 v(x_t)\right] \in R^2$$

$$\Delta\mathbf{f}(x_t) := -M^{-1}(x_t)\, \Delta\kappa_t \cdot v(x_t) \in R^2$$
$$B(x_t) = M^{-1}(x_t) \in R^{2\times 2}$$

$$E = \begin{pmatrix} 0_{2x2} & I_{2x2} \\ 0_{2x2} & 0_{2x2} \end{pmatrix}, \ \overline{F_0(x_t, t)} = \begin{pmatrix} 0 \\ 0 \\ \mathbf{f}_0(x_t) \end{pmatrix} \tag{8.6}$$

Taking into account the restrictions (8.4), we can estimate the corresponding nonlinear term, containing the uncertainty mentioned above, as follows:

$$\|\Delta F(x_t,t)\|^2_{\Lambda_0} = \Delta F^T(x_t,t)\,\Lambda_0 \Delta F(x_t,t) = \Delta \mathbf{f}^T(x_t)\,\Lambda_{02}\Delta \mathbf{f}(x_t) =$$

$$= v^T(x_t)\,\Delta\kappa_t^T M^{-1}(x_t)\Lambda_{02}M^{-1}(x_t)\Delta\kappa_t v(x_t) \leq$$

$$\leq \lambda_{max}(S(x_t))\;\; v^T(x_t)\,\Delta\kappa_t^T\Delta\kappa_t v(x_t) \leq \mu_t \tag{8.7}$$

where

$$\mu_t := \lambda_{max}(S(x_t))\;\; v^T(x_t)\,\Delta v(x_t) \tag{8.8}$$

$$S(x_t) := M^{-1}(x_t)\Lambda_{02}M^{-1}(x_t) \tag{8.9}$$

and Λ_0 is the weight matrix selected for the simplicity in the block-diagonal form:

$$\Lambda_0 := \begin{bmatrix} \Lambda_{01} & 0 \\ 0 & \Lambda_{02} \end{bmatrix} \in R^{4\times 4}.$$

To apply the ideas described above (see Chapter 5) we do not need to represent this system in the standard form (8.5) to design a controller. Only input-output signal should be available to construct a neuro-observer and then, based on its model, to design a locally optimal controller. We will follow this line.

8.3 Robot Joint Velocity Observer and RBF Compensator

The motion equations of the serial $n-$link rigid robot manipulator (8.1) can be rewritten as state space form [27]:

$$\begin{aligned} \dot{x}_1 &= x_2 \\ \dot{x}_2 &= H(x,u) \\ y &= x_1 \end{aligned} \tag{8.10}$$

where $x_1 = q = [q_1 \cdots q_n]$ is joint position of $n-$link, $x_2 = \dot{q}$ is joint velocity of the link, $x = [x_1^T, x_2^T]^T$, $u = \tau$ is control input.

The system (8.10) has solution for any $t \in [0, T]$. The output is the position which is measurable.

$$H(x,u) := f(x) + g(x)u$$

and

$$\begin{array}{c} f(x) := -M(x_1)^{-1}[C(x_1,x_2)\dot{x}_1 + G(x_1) + F\dot{x}_1] \\ g(x_1) := M(x_1)^{-1} \end{array} \tag{8.11}$$

Now we use high-gain observer to get the estimates of the joint velocity

$$\begin{array}{c} \dot{\widehat{x}}_1 = \widehat{x}_2 + \varepsilon^{-1}K_1(x_1 - \widehat{x}_1) \\ \dot{\widehat{x}}_2 = \varepsilon^{-2}K_2(x_1 - \widehat{x}_1) \end{array} \tag{8.12}$$

where $\widehat{x}_1 \in \Re^n$, $\widehat{x}_2 \in \Re^n$ denote the estimated values of x_1, x_2 respectively; ε is chosen as a small positive parameter; and K_1, K_2 are positive definite matrices chosen such that the matrix $\begin{bmatrix} -K_1 & I \\ -K_2 & 0 \end{bmatrix}$ is stable.

Let define the observer error as

$$\widetilde{x} := x - \widehat{x} \tag{8.13}$$

where $\widehat{x} = [\widehat{x}_1^T, \widehat{x}_2^T]^T$. From (8.10) and (8.12) the observer error equation can be written as

$$\begin{array}{c} \dot{\widetilde{x}}_1 = \widetilde{x}_2 - \varepsilon^{-1}K_1\widetilde{x}_1 \\ \dot{\widetilde{x}}_2 = -\varepsilon^{-2}K_2\widetilde{x}_1 + H(x,u) \end{array} \tag{8.14}$$

If we define the new pair of variables as

$$\begin{array}{c} \widetilde{z}_1 := \widetilde{x}_1 \\ \widetilde{z}_2 := \varepsilon\widetilde{x}_2 \end{array} \tag{8.15}$$

(8.14) can be rewritten as:

$$\begin{aligned} \varepsilon \dot{\tilde{z}}_1 &= \tilde{z}_2 - K_1 \tilde{z}_1 \\ \varepsilon \dot{\tilde{z}}_2 &= -K_2 \tilde{z}_1 + \varepsilon^2 H(x,u) \end{aligned} \tag{8.16}$$

or, in the matrix form:

$$\varepsilon \dot{\tilde{z}} = A\tilde{z} + \varepsilon^2 BH(x,u) \tag{8.17}$$

where

$$\tilde{z} := [\tilde{z}_1^T, \tilde{z}_2^T]^T, \quad A := \begin{bmatrix} -K_1 & I \\ -K_2 & 0 \end{bmatrix}, \quad B := \begin{bmatrix} 0 \\ I \end{bmatrix} \tag{8.18}$$

Next theorem gives the upper bound of the joint velocity estimation.

Theorem 8.1 *If we use the high gain observer (8.12) to estimate the velocity of the robot dynamic (8.10), the observer error $\tilde{x}$ converges to the residual set*

$$D_\varepsilon = \left\{ \tilde{x} \mid \|\tilde{x}\| \le 2\varepsilon^2 C_{i,T} \right\}$$

where

$$C_{i,T} := \sup_{t\in[0,T]} \|BH(x,u)\|^T \|P\| \tag{8.19}$$

Proof. Due to the fact that the spectra of K_1 and K_2 are in the left half plane, there exists a constant positive definite matrix P such that it satisfies the Lyapunov equation:

$$A^T P + PA = -I \tag{8.20}$$

where A is defined as in (8.18). Consider the following candidate Lyapunov function:

$$V(\tilde{z}) = \varepsilon \tilde{z}^T P \tilde{z}$$

The derivation along the solutions of (8.16) is:

$$\begin{aligned} \dot{V} &= \varepsilon \dot{\tilde{z}}^T P\tilde{z} + \varepsilon \tilde{z}^T P \dot{\tilde{z}} \\ &= \left(A\tilde{z} + \varepsilon^2 BH(x,u)\right)^T P\tilde{z} + \tilde{z}^T P \left(A\tilde{z} + \varepsilon^2 BH(x,u)\right) \\ &= \tilde{z}^T \left(A^T P + PA\right) \tilde{z} + 2\varepsilon^2 \left(BH(x,u)\right)^T P\tilde{z} \\ &\le \tilde{z}^T \left(A^T P + PA\right) \tilde{z} + 2\varepsilon^2 \|BH(x,u)\|^T \|P\| \, |\tilde{z}| \end{aligned} \tag{8.21}$$

Because the control u can make (8.10) have solution for any $t \in [0,T]$, $\|H(x,u)\|$ is bounded for any finite time T [27]. We may conclude that $\|BH(x,u)\|^T \|P\|$ is bounded. From(8.20) we have

$$\dot{V} \leq -\|\widetilde{z}\|^2 + \overline{K}(\varepsilon)\|\widetilde{z}\|$$

where

$$\overline{K}(\varepsilon) := 2\varepsilon^2 C_{i,T}.$$

It is noted that if

$$\|\widetilde{z}(t)\| > \overline{K}(\varepsilon) \tag{8.22}$$

then $\dot{V}_i < 0$, $\forall t \in [0,T]$. So the total time during which $\|\widetilde{z}(t)\| > \overline{K}(\varepsilon)$ is finite. Let T_k denotes the time interval during which $\|\widetilde{z}(t)\| > \overline{K}(\varepsilon)$

· If only finite times that $\widetilde{z}(t)$ stay outside the ball of radius $\overline{K}(\varepsilon)$ (and then reenter), $\widetilde{z}(t)$ will eventually stay inside of this ball.

· If $\widetilde{z}(t)$ leave the ball infinite times, since the total time $\widetilde{z}(t)$ leave the ball is finite, $\sum_{k=1}^{\infty} T_k < \infty$,

$$\lim_{k\to\infty} T_k = 0 \tag{8.23}$$

So $\widetilde{z}(t)$ is bounded via an invariant set argument. From (8.17) it follows that $\dot{\widetilde{z}}(t)$ is also bounded.

Denote by $\|\widetilde{z}_k(t)\|$ the largest tracking error during the T_k interval. Then (8.23) and boundness of $\dot{\widetilde{z}}(t)$ imply

$$\lim_{k\to\infty} \left[\|\widetilde{z}_k(t)\| - \overline{K}(\varepsilon)\right] = 0$$

So $\|\widetilde{z}_k(t)\|$ converges to $\overline{K}(\varepsilon)$. Because

$$\widetilde{x} = \begin{bmatrix} I & 0 \\ 0 & \frac{1}{\varepsilon}I \end{bmatrix} \widetilde{z}$$

and $\varepsilon < 1$, it follows that $\|\widetilde{x}\|$ converges to the ball of radius $\overline{K}(\varepsilon)$. ■

Remark 8.1 *Since $C_{i,T}$ is bounded, we can select ε arbitrary small (the gain of observer (8.12) becomes bigger) in order to make the observer error small enough.*

Remark 8.2 *The high-gain observer in this paper has a similar structure as in [27], but the proof is different. [27] used singular perturbation which assumed that $\varepsilon \to 0$. It is difficult to applied their results on neuro compensation. In next section we will shown that Lyapunov method can provide a good condition for PD control.*

It is well known that PD control with friction and gravity compensation may reach asymptotic stability [15]. Using neural networks to compensate the nonlinearities of robot dynamic may be found in [19] and [10]. In [19] the authors use neural networks to approximate the whole nonlinearities of robot dynamic. With this neuro feedforward compensator and a PD control, they can guarantee a good track performance.

The friction and gravity in (8.1) can be approximated by a RBF neural network as follows,

$$P(q, \dot{q}) = W^*\Phi(V^*x) + \widetilde{P} \tag{8.24}$$

where $P(q, \dot{q}) := G(q) + F\dot{q}$, where W^*, V^* are fixed bounded weights, $\widetilde{P}$ is the approximated error, whose magnitude depends on the values of W^* and V^*.

The estimation of $P(q, \dot{q})$ may be defined as $\widehat{P}(q, \dot{q})$

$$\widehat{P}(q, \dot{q}) = \widehat{W}\Phi(\widehat{V}x) \tag{8.25}$$

In order to implement neural networks, the following assumption for $\widetilde{P}$ in (8.24) is needed.

A1:

$$\widetilde{P}^T\Lambda_1\widetilde{P} \leq \overline{\eta}, \quad \overline{\eta} > 0 \tag{8.26}$$

It is clear that the Gaussian function, commonly used in RBF neural networks, satisfy Lipschitz condition. We may conclude that:

Property 3

$$\begin{array}{c} \widetilde{\Phi}_t := \Phi(V^{*T}x) - \Phi(\widehat{V}_t^Tx) = D_\sigma\widetilde{V}_t^Tx + \nu_\sigma \\ D_\sigma = \frac{\partial\Phi^T(Z)}{\partial Z}\big|_{Z=\widehat{V}_t^Tx}, \quad \|\nu_\sigma\|_\Lambda^2 \leq l\left\|\widetilde{V}_t^Tx\right\|_\Lambda^2, \quad l > 0 \end{array} \tag{8.27}$$

where Λ is a positive definite matrix,

$$\widetilde{W} = W^* - \widehat{W}, \quad \widetilde{V} = V^* - \widehat{V} \tag{8.28}$$

Remark 8.3 *One can see that this condition is similar with [19] (Taylor series). The upper bound found for ν_σ will be essential for proving stability of the PD control with high gain observer and neuro compensator.*

8.4 PD Control with Velocity Estimation and Neuro Compensator

First, let us study the PD control with a neuro compensation. In this case, we assume the velocities are measurable, so PD control is as follows

$$u = -K_p(x_1 - x_1^d) - K_d(x_2 - x_2^d) + \widehat{W}\Phi(\widehat{V}x) \tag{8.29}$$

where $x_1^d \in \Re^n$ is the desired position. x_2^d is the desired velocity bounded as

$$\left\|x_2^d\right\| \le \overline{x}_2^d, \quad \left\|\dot{x}_2^d\right\| \le \overline{\dot{x}}_2^d$$

The input control vector $\tau = u$. K_p and K_d are positive defined matrices corresponding to proportional and derivative coefficients.

Let define the tracking error as:

$$\overline{x}_1 := x_1 - x_1^d, \quad \overline{x}_2 := x_2 - x_2^d$$

$$s = \left(x_1^T, \widehat{x}_2^T\right)^T$$

Theorem 8.2 *If following learning laws for the weights of neural networks (8.25) are used*

$$\begin{aligned} \dot{\widehat{W}}_t &= -2d_t K_w \sigma(\widehat{V}_t^T s)\overline{x}_2^T - 2d_t K_w D_\sigma \widetilde{V}_t^T s \overline{x}_2^T \\ \dot{\widehat{V}}_t &= -2d_t K_v s \overline{x}_2^T \widehat{W}_t^T D_\sigma + 2d_t l K_v s s^T \widetilde{V}_t \Lambda_3 \end{aligned} \tag{8.30}$$

where $0 < \Lambda_1 = \Lambda_1^T \in \Re^{n \times n}$,

$$P_0 = \begin{bmatrix} M(x_1) & 0 \\ 0 & K_p \end{bmatrix}$$
$$d_t := \left[\frac{1}{2} - \mu \left\| P_0^{\frac{1}{2}} \begin{bmatrix} \overline{x}_2 \\ \overline{x}_1 \end{bmatrix} \right\|^{-1} \right]_+$$
$$\mu = \frac{\chi^2}{4\lambda_{\min}(\Gamma)} \lambda_{\min} \left(P_0^{-\frac{1}{2}} \begin{bmatrix} R & 0 \\ 0 & 0 \end{bmatrix} P_0^{-\frac{1}{2}} \right)$$

$$[z]_+ = \begin{cases} z & si \quad z \geq 0 \\ 0 & si \quad z < 0 \end{cases}$$

$$\Gamma := K_d - \tfrac{1}{4}\Lambda_1^{-1} - k_c \left\| \overline{x}_2^d \right\| I - R, \; R = R^T > 0$$
$$\chi := \overline{\eta} + k_c \left\| \overline{x}_2^d \right\|^2 + \lambda_{\max}(M) \left\| \dot{\overline{x}}_2^d \right\|$$

then

(I) The weights of neural networks $\widehat{W}_t$, $\widehat{V}_t$ *and tracking error* $\overline{x}_2$ *are bounded.*

(II) For any $T \in (0, \infty)$ *the tracking error* $\overline{x}_2$ *satisfies*

$$\limsup_{T \to \infty} \frac{1}{T} \int_0^T d_t \overline{x}_2^T R \overline{x}_2 dt \leq \frac{\chi^2}{8\lambda_{\min}(\Gamma)} \tag{8.31}$$

Proof. From (8.29) the closed-loop system is:

$$M(x_1) \dot{x}_2 + C(x_1, \dot{x}_1) x_2 + K_p \overline{x}_1 + K_d \overline{x}_2 - \widehat{W}_t \Phi(\widehat{V}_t x) + W^* \Phi(V^* x) + \widetilde{P} = 0 \tag{8.32}$$

The proposed candidate Lyapunov function is

$$V_1 = \left[\frac{1}{2} \left\| \begin{bmatrix} M(x_1) & 0 \\ 0 & K_p \end{bmatrix}^{\frac{1}{2}} \begin{bmatrix} \overline{x}_2 \\ \overline{x}_1 \end{bmatrix} \right\| - \mu \right]_+^2 + \tfrac{1}{2} tr \left(\widetilde{W}_t^T K_w^{-1} \widetilde{W}_t \right) + \tfrac{1}{2} tr \left(\widetilde{V}_t^T K_v^{-1} \widetilde{V}_t \right) \tag{8.33}$$

where K_w and K_v are any positive definite constant matrices. The derivative of (8.33) is

$$\begin{array}{l}\dot{V}_1 = 2\left[\frac{1}{2} - \mu \left\| \begin{bmatrix} M & 0 \\ 0 & K_p \end{bmatrix}^{\frac{1}{2}} \begin{bmatrix} \overline{x}_2 \\ \overline{x}_1 \end{bmatrix} \right\|^{-1}\right]_+ \\ \left[\overline{x}_2^T M(x_1)\, \dot{\overline{x}}_2 + \frac{1}{2}\overline{x}_2^T \dot{M}(x)\overline{x}_2 + \overline{x}_2^T K_p \overline{x}_1\right] \\ +tr\left(\widetilde{W}_t^T K_w^{-1} \dot{\widetilde{W}}_t\right) + tr\left(\widetilde{V}_t^T K_v^{-1} \dot{\widetilde{V}}_t\right)\end{array} \tag{8.34}$$

Using (8.32) we obtain

$$\begin{array}{l}\overline{x}_2^T M \dot{\overline{x}}_2 = -\overline{x}_2^T M \dot{x}_2^d - \overline{x}_2^T \left[Cx_2 + K_p\overline{x}_1 + K_d\overline{x}_2 - \widehat{W}_t\Phi(\widehat{V}_t x) + W^*\Phi(V^*x) + \widetilde{P}\right] \\ = -\overline{x}_2^T M \dot{x}_2^d - \overline{x}_2^T C\overline{x}_2 - \overline{x}_2^T C x_2^d - \overline{x}_2^T K_p \overline{x}_1 - \overline{x}_2^T K_d \overline{x}_2 \\ -\overline{x}_2^T \left[-\widehat{W}_t\Phi(\widehat{V}_t x) + W^*\Phi(V^*x) + \widetilde{P}\right]\end{array} \tag{8.35}$$

Using Property 2 and (8.35), (8.34) becomes

$$\begin{array}{l}\dot{V}_1 = 2d_t\left[-\overline{x}_2^T M \dot{x}_2^d - \overline{x}_2^T C x_2^d - \overline{x}_2^T K_d \overline{x}_2\right] \\ -2d_t\overline{x}_2^T\left[-\widehat{W}_t\Phi(\widehat{V}_t s) + W^*\Phi(V^*s) + \widetilde{P}\right] \\ +tr\left(\widetilde{W}_t^T K_w^{-1} \dot{\widetilde{W}}_t\right) + tr\left(\widetilde{V}_t^T K_v^{-1} \dot{\widetilde{V}}_t\right)\end{array} \tag{8.36}$$

The term

$$-\widehat{W}_t\Phi(\widehat{V}_t x) + W^*\Phi(V^*x)$$

can be expressed as

$$\begin{array}{l}W^*\Phi(V^*x) - W^*\Phi(\widehat{V}_t x) + W^*\Phi(\widehat{V}_t x) - \widehat{W}_t\Phi(\widehat{V}_t x) \\ = \widetilde{W}_t^T\Phi(\widehat{V}_t^T x) + W^{*T}D_\sigma\widetilde{V}_t^T x + W^{*T}\nu_\sigma = \widetilde{W}_t^T\Phi(\widehat{V}_t^T x) + \widehat{W}^T D_\sigma \widetilde{V}_t^T x \\ +\widetilde{W}^T D_\sigma \widetilde{V}_t^T x + W^{*T}\nu_\sigma\end{array} \tag{8.37}$$

In view of the matrix inequality,

$$X^TY+\left(X^TY\right)^T\le X^T\Lambda^{-1}X+Y^T\Lambda Y \tag{8.38}$$

which is valid for any $X,Y\in\Re^{n\times k}$ and for any positive defined matrix $0<\Lambda=\Lambda^T\in\Re^{n\times n}$ [35], $-\overline{x}_2^T\left[\widetilde{P}+W^{*T}\nu_\sigma\right]$ can be estimated as

$$\begin{array}{l}-\overline{x}_2^T\left[\widetilde{P}+W^{*T}\nu_\sigma\right]\le\|\overline{x}_2\|\left\|\widetilde{P}\right\|+\frac{1}{4}\overline{x}_2^T\Lambda_1^{-1}\overline{x}_2+\left[W^{*T}\nu_\sigma\right]^T\Lambda_1\left[W^{*T}\nu_\sigma\right]\\ \le\|\overline{x}_2\|\overline{\eta}+\frac{1}{4}\overline{x}_2^T\Lambda_1^{-1}\overline{x}_2+l\left\|\widetilde{V}_t^Tx\right\|_{\Lambda_3}^2\end{array} \tag{8.39}$$

where $\Lambda_3:=W^*\Lambda_1W^{*T}$. Using Property 2, it follows

$$\begin{array}{l}-\overline{x}_2^TC(x_1,x_2)x_2^d\le\|\overline{x}_2\|\,\|C(x_1,x_2)\|\,\left\|x_2^d\right\|\le\\ \le\left\|\overline{x}_2^d\right\|\overline{x}_2^Tk_cI\overline{x}_2+k_c\left\|\overline{x}_2^d\right\|^2\|\overline{x}_2\|\\ -\overline{x}_2^TM(x_1)\dot{x}_2^d\le\lambda_{\max}(M)\left\|\dot{\overline{x}}_2^d\right\|\|\overline{x}_2\|\end{array} \tag{8.40}$$

So

$$\begin{array}{l}\dot{V}_1\le-2d_t\overline{x}_2^T\Gamma\overline{x}_2+2d_t\chi\|\overline{x}_2\|+L_w+L_v-2d_t\overline{x}_2^TR\overline{x}_2\\ \le-2d_t\lambda_{\min}(\Gamma)\left(\|\overline{x}_2\|-\frac{\chi}{2\lambda_{\min}(\Gamma)}\right)^2+\frac{2d_t\chi^2}{4\lambda_{\min}(\Gamma)}-2d_t\overline{x}_2^TR\overline{x}_2+L_w+L_v\\ \le\frac{2d_t\chi^2}{4\lambda_{\min}(\Gamma)}-2d_t\overline{x}_2^TR\overline{x}_2+L_w+L_v\end{array} \tag{8.41}$$

where

$$\begin{array}{c}L_w:=tr\left[\left(K_w{}^{-1}\dot{\widetilde{W}}_t-2d_t\sigma(\widehat{V}_t^Ts)\overline{x}_2^T-2d_tD_\sigma\widetilde{V}_t^Ts\overline{x}_2^T\right)\widetilde{W}_t^T\right]\\ L_v:=tr\left[\left(K_v^{-1}\dot{\widetilde{V}}_t-2d_ts\overline{x}_2^T\widehat{W}_t^TD_\sigma+2d_tlss^T\widetilde{V}_t\Lambda_3\right)\widetilde{V}_t^T\right]\end{array}$$

Using adaptive law (8.30),

$$\dot{V}_1\le-2d_t\left[\overline{x}_2^TR\overline{x}_2-\frac{\chi^2}{4\lambda_{\min}(\Gamma)}\right]$$

So

$$\dot{V}_1\le-4d_t\lambda_{\min}\left(P_0^{-\frac{1}{2}}RP_0^{-\frac{1}{2}}\right)\left[\frac{1}{2}\left\|P_0^{\frac{1}{2}}\left[\begin{array}{c}\overline{x}_2\\ \overline{x}_1\end{array}\right]\right\|^2-\mu\right]\le0$$

V_1 is bounded, so (I) is proven.

Integrating it from 0 to $T > 0$, we get

$$V_{1,T} - V_{1,0} \leq -2 \int_0^T d_t \left[\overline{x}_2^T R \overline{x}_2 - \frac{\chi^2}{4\lambda_{\min}(\Gamma)} \right] dt$$

That is,

$$2 \int_0^T d_t \overline{x}_2^T R \overline{x}_2 dt \leq V_{1,0} - V_{1,T} + \left[\frac{\chi^2}{4\lambda_{\min}(\Gamma)} \right] T \leq V_{1,0} + \left[\frac{\chi^2}{4\lambda_{\min}(\Gamma)} \right] T$$

where $V_{1,0}$ is correspond to $\widetilde{W}_t = W^*$ and $\widetilde{V}_t = V^*$, (8.31) is proven ■

Now, let us study PD control with velocities estimation and neuro compensation. We select a new PD control with velocity estimation and neuro compensator as

$$u = -K_p(x_1 - x_1^d) - K_d(\widehat{x}_2 - x_2^d) + \widehat{W}\Phi(\widehat{V}s) \tag{8.42}$$

If the joint velocities are measurable and gravity and friction are unknown, we only need to change $\widehat{x}_2$ to x_2. From (8.10) and (8.42) the tracking error equation can be expressed as:

$$\begin{aligned} \dot{\overline{x}}_1 &= \overline{x}_2 \\ \dot{\overline{x}}_2 &= \dot{x}_2 - \dot{x}_2^d = H(\overline{x}_1, \overline{x}_2, x_1^d, x_2^d, \widetilde{x}_2) - \dot{x}_2^d \end{aligned} \tag{8.43}$$

where

$$\begin{aligned} H(\overline{x}_1, \overline{x}_2, x_1^d, x_2^d, \widetilde{x}_2) = M(\overline{x}_1 + x_1^d)^{-1}[-K_p\overline{x}_1 + K_d\widetilde{x}_2 - K_d\overline{x}_2 + \widehat{W}\Phi(\widehat{V}s) \\ -W^*\Phi(V^*s) - \widetilde{P} - C\left(\overline{x}_2 + x_2^d\right)] \end{aligned} \tag{8.44}$$

Substituting PD control (8.42) into high-gain observer (8.16), we get

$$\begin{aligned} \varepsilon \dot{\widetilde{z}}_1 &= \widetilde{z}_2 - K_1\widetilde{z}_1 \\ \varepsilon \dot{\widetilde{z}}_2 &= -K_2\widetilde{z}_1 + \varepsilon^2 H \end{aligned}$$

The closed-loop system with observer is

$$\begin{aligned} \dot{\overline{x}}_1 &= \overline{x}_2 \\ \dot{\overline{x}}_2 &= H(\overline{x}_1, \overline{x}_2, x_1^d, x_2^d, \widetilde{x}_2) - \dot{x}_2^d \\ \varepsilon \dot{\widetilde{z}}_1 &= \widetilde{z}_2 - K_1\widetilde{z}_1 \\ \varepsilon \dot{\widetilde{z}}_2 &= -K_2\widetilde{z}_1 + \varepsilon^2 H \end{aligned} \tag{8.45}$$

The equilibrium point of (8.45) is $(\overline{x}_1, \overline{x}_2, \widetilde{z}_1, \widetilde{z}_2) = (0,0,0,0)$. Clearly (8.45) has the singularly perturbed term form (8.16). If we put $\varepsilon = 0$, then

$$0 = \widetilde{z}_2 - K_1 \widetilde{z}_1, \quad 0 = -K_2 \widetilde{z}_1$$

It implies that the vector $\widetilde{z}$ has zero components

$$\widetilde{z}_1 = x_1 - \widehat{x}_1 = 0, \quad \widetilde{z}_2 = \varepsilon \left(x_2 - \widehat{x}_2\right) = 0$$

and has an equilibrium point $(x_1, x_2) = (\widehat{x}_1, \widehat{x}_2)$. The system (8.45) therefore is in the standard form. Although on the singularly perturbed analysis it's assume $\varepsilon = 0$, the equilibrium point $(x_1, x_2) = (\widehat{x}_1, \widehat{x}_2)$ for the case $0 < \varepsilon < 1$ is unique.

Substituting the equilibrium point into (8.45), we obtain the quasi-steady-state model

$$\begin{aligned} \dot{\overline{x}}_1 &= \overline{x}_2 \\ \dot{\overline{x}}_2 &= H(\overline{x}_1, \overline{x}_2, x_1^d, x_2^d, 0) - \dot{x}_2^d \\ &= M(\overline{x}_1 + x_1^d)^{-1}[-K_p \overline{x}_1 - K_d \overline{x}_2 - C\left(\overline{x}_2 + x_2^d\right) \\ &\quad + \widehat{W}\Phi(\widehat{V}s) - W^*\Phi(V^*s) - \widetilde{P}] - \dot{x}_2^d \end{aligned} \tag{8.46}$$

The boundary layer system of (8.45) is

$$\begin{aligned} \dot{\widetilde{z}}_1(\tau) &= \widetilde{z}_2 - K_1 \widetilde{z}_1(\tau) \\ \dot{\widetilde{z}}_2(\tau) &= -K_2 \widetilde{z}_1(\tau) \end{aligned} \tag{8.47}$$

where $\tau = t/\varepsilon$. (8.47) can be written as:

$$\dot{\widetilde{z}}(\tau) = A\widetilde{z}(\tau) \tag{8.48}$$

If the velocities x_2 are assumed to be measurable, following theorem shows the stability properties of the slow system $((\overline{x}_1, \overline{x}_2) = (0,0))$

Theorem 8.3 *The equilibrium point $(\widetilde{z}_1, \widetilde{z}_2) = (0,0)$ of (8.48) is asymptotically stable.*

Proof. Since A is a Hurwitz matrix, there exist a positive definite matrix P such that

$$A^T P + PA = -Q \tag{8.49}$$

where Q is a positive definite matrix. Consider the candidate Lyapunov function

$$V_2(\widetilde{z}_1, \widetilde{z}_2) = \widetilde{z}^T P \widetilde{z}$$

with its derivative with respect to time and along with $\dot{\widetilde{z}}(\tau) = A\widetilde{z}(\tau)$:

$$\dot{V}_2 = \widetilde{z}^T \left(A^T P + PA\right) \widetilde{z} = -\widetilde{z}^T Q \widetilde{z} < 0$$

that implies the asymptotical stability ■

Remark 8.4 *Singular perturbation technique is used to analysis the whole system: high-gain observer(fast system) and PD controller with neuro compensation (slow system). The advantage of this approach is that it may divide the original problem in two subsystems: the slow subsystem or quasi-steady state system and the fast subsystem or boundary layer system. Both systems can be studied independently. From the point of the singular perturbation analysis, one can see that high-gain observer (8.12) has a faster dynamic than the robot (8.10) and the PD control (8.42). Under the assumption that* $\varepsilon = 0$, *the observer error and the tracking error of PD control are asymptotic stable if the joint velocities are measurable.*

Remark 8.5 *Defining* $P = \begin{bmatrix} P_{11} & P_{12} \\ P_{21} & P_{22} \end{bmatrix}$, *since* $A = \begin{bmatrix} -K_1 & I \\ -K_2 & 0 \end{bmatrix}$, $P_{22} \in \Re^{n \times n}$, *we may free to select to satisfy Lyapunov equation (8.49). We only need to match the positive define condition of P.*

The another main contribution of this chapter is that we give a new on-line learning law for RBF neuro compensator

$$\begin{aligned} \dot{\widehat{W}}_t &= -\tfrac{2}{(1-d)} d_t K_w \sigma(\widehat{V}_t^T s) \Psi^T - \tfrac{2}{(1-d)} d_t K_w D_\sigma \widetilde{V}_t^T s \Psi^T \\ \dot{\widehat{V}}_t &= -\tfrac{2}{(1-d)} d_t K_v s \Psi^T \widehat{W}_t^T D_\sigma + \tfrac{2}{(1-d)} d_t l K_v s s^T \widetilde{V}_t \Lambda_3 \end{aligned} \tag{8.50}$$

where

$$\Psi = 2d\varepsilon\eta_{M} (x_1 - \widehat{x}_1)^T P_{12} - 2d\varepsilon^2\eta_{M}\widehat{x}_2^T P_{22} - (1-d)\, x_2^{dT} \tag{8.51}$$

$$\begin{aligned}
P_1 &:= \begin{bmatrix} \frac{1}{2}(1-d)K_p & 0 & 0 \\ 0 & \frac{1}{2}(1-d)M & 0 \\ 0 & 0 & dP \end{bmatrix} \\
d_t &:= \left[1 - \mu \left\| P_1^{\frac{1}{2}} \begin{bmatrix} \overline{x}_1 \\ \overline{x}_2 \\ \widetilde{z} \end{bmatrix} \right\|^{-1} \right]_+ , \quad R_1 = R_1^T > 0 \\
\mu &- (a - (1-d)2d_t \left\|\overline{x}_2\right\| (\lambda_{\min}(K_d) \left\|\overline{x}_2\right\| - b)) / \lambda_{\min}\left(P_1^{-\frac{1}{2}} R_1 P_1^{-\frac{1}{2}}\right) \\
y &:= \begin{bmatrix} \|\overline{x}\| \\ \|\widetilde{z}\| \end{bmatrix}, \; a := \overline{\Psi}\overline{\eta} + \frac{\lambda_{\max}(\Lambda_1^{-1})}{4}\overline{\Psi}^2, \; b := k_c^2 \overline{x}_2^{d2} + \lambda_{\max}(M)\dot{\overline{x}}_2^{d} \\
\overline{\Psi} &\geq 2d\varepsilon\eta_{M} \left\|\widetilde{x}_1\right\| \left\|P_{12}\right\| + 2d\varepsilon^2\eta_{M} \left\|\widehat{x}_2\right\| \left\|P_{22}\right\| + (1-d)\, \overline{x}_2^d \\
\eta_{M} &= \sup \left\|M^{-1}(x_1)\right\|, \; 0 < d < 1
\end{aligned} \tag{8.52}$$

Remark 8.6 *The structure of the new updating law (8.50) is similar to (8.30). Since $\overline{x}_2$ and x_2 are not available in this case, they are replaced by Ψ and s. But we need one more condition: M should be known. This requirement is necessary if both velocity and friction are unknown (see [10]).*

When we realize the high-gain observer (8.12), it is impossible to make $\varepsilon \to 0$ (singular perturbation). One can see that the observer error is less than $2\varepsilon^2 C_{i,T}$.

Can we find a largest value of ε which can assure the whole system is stable? For this purpose we proposed a modified version of [33]. The following theorem may answer to this question. It states the fact that if the velocity, friction and gravity are unknown, the learning law, suggested above, turns out to be stable.

Theorem 8.4 *If P_{22} is selected as*

$$P_{22} = -\frac{2(1-d)}{4d\varepsilon^2\eta_{M}} I \tag{8.53}$$

where η_{M} is upper bound of M^{-1}, the learning laws for the weights of neural networks (8.25) are used as in (8.50), $\varepsilon \in [0, \overline{\varepsilon}]$, $\overline{\varepsilon}$ is the solution of following inequality

$$\begin{array}{c} d^2 K_2^2 \varepsilon^4/2 + ((1-d)d\beta_1 K_2)\varepsilon^2 + ((1-d)\alpha_1 K_1)\varepsilon \\ +(1-d)^2\beta_1^2/2 - (1-d)d\alpha_1\alpha_2 < 0 \end{array} \tag{8.54}$$

then

(I) The weights of neural networks $\widehat{W}_t$, $\widehat{V}_t$, observer and tracking errors $\|\widetilde{z}\|$ and $\|\overline{x}\|$ are bounded.

(II) For any $T \in (0, \infty)$ the tracking and observer error satisfies

$$\begin{array}{c} \sup \lim\limits_{T\to\infty} \frac{1}{T}\int_0^T y^T R_1 y dt \\ \leq a - (1-d)2d_t \|\overline{x}_2\| \left(\lambda_{\min}(K_d)\|\overline{x}_2\| - b\right) \end{array}$$

where a and b are defined in (8.52).

Proof. Let us select the following candidate Lyapunov function for (8.45):

$$\begin{array}{c} V_3 = (1-d)V_1(\overline{x}_1, \overline{x}_2) + dV_2(\widetilde{z}_1, \widetilde{z}_2) \\ = (1-d)\left[\frac{1}{2}\overline{x}_2^T M(x_1)\overline{x}_2 + \frac{1}{2}\overline{x}_1^T K_p \overline{x}_1 + \frac{1}{2}tr\left(\widetilde{W}_t^T K_w^{-1}\widetilde{W}_t\right) + \frac{1}{2}tr\left(\widetilde{V}_t^T K_v^{-1}\widetilde{V}_t\right)\right] \\ +d\widetilde{z}^T P\widetilde{z} \end{array} \tag{8.55}$$

where V_1 and V_2 are defined before. So,

$$\begin{array}{c} V_3 = \left[\left\| \begin{bmatrix} \frac{1}{2}(1-d)K_p & 0 & 0 \\ 0 & \frac{1}{2}(1-d)M & 0 \\ 0 & 0 & dP \end{bmatrix}^{\frac{1}{2}} \begin{bmatrix} \overline{x}_1 \\ \overline{x}_2 \\ \widetilde{z} \end{bmatrix} \right\| - \mu \right]_+^2 \\ +(1-d)\left[\frac{1}{2}tr\left(\widetilde{W}_t^T K_w^{-1}\widetilde{W}_t\right) + \frac{1}{2}tr\left(\widetilde{V}_t^T K_v^{-1}\widetilde{V}_t\right)\right] \end{array}$$

where $0 < d < 1$. Since the control (8.42) is different from (8.29), we cannot apply the result of $\dot{V}_1$ (8.34) directly. The derivative of V_3 is

$$\dot{V}_3 = 2\left[\left\|\begin{bmatrix} \frac{1}{2}(1-d)K_p & 0 & 0 \\ 0 & \frac{1}{2}(1-d)M & 0 \\ 0 & 0 & dP \end{bmatrix}^{\frac{1}{2}} \begin{bmatrix} \overline{x}_1 \\ \overline{x}_2 \\ \widetilde{z} \end{bmatrix}\right\| - \mu\right]_+ \\ \left[(1-d)\left(\overline{x}_2^T M \dot{\overline{x}}_2 + \tfrac{1}{2}\overline{x}_2^T \dot{M}\,\overline{x}_2 + \overline{x}_2^T K_p \overline{x}_1^T\right) + 2d\widetilde{z}^T P \dot{\widetilde{z}}\right] \\ +(1-d)\left[tr\left(\widetilde{W}_t^T K_w^{-1} \dot{\widetilde{W}}_t\right) + tr\left(\widetilde{V}_t^T K_v^{-1} \dot{\widetilde{V}}_t\right)\right]$$

From (8.17) we have

$$\dot{\widetilde{z}} = \frac{1}{\varepsilon} A\widetilde{z} + \varepsilon BH(\overline{x}_1, \overline{x}_2, x_1^d, x_2^d, \widetilde{x}_2)$$

where $H(\overline{x}_1, \overline{x}_2, x_1^d, x_2^d, \widetilde{x}_2)$ is defined as in (8.44). From (8.45) it follows

$$\dot{\overline{x}}_2 = \left[H(\overline{x}_1, \overline{x}_2, x_1^d, x_2^d, 0) - \dot{x}_2^d\right] \\ + \left[H(\overline{x}_1, \overline{x}_2, x_1^d, x_2^d, \widetilde{x}_2) - H(\overline{x}_1, \overline{x}_2, x_1^d, x_2^d, 0)\right]$$

and

$$H(\overline{x}_1, \overline{x}_2, x_1^d, x_2^d, \widetilde{x}_2) - H(\overline{x}_1, \overline{x}_2, x_1^d, x_2^d, 0) = \frac{1}{\varepsilon} M^{-1} K_d \widetilde{z}_2 \tag{8.56}$$

So, the derivative of (8.55) with respect to time and along (8.45) is:

$$(1-d)2d_t\left[\overline{x}_2^T MH(\overline{x}_1, \overline{x}_2, x_1^d, x_2^d, 0) + \overline{x}_2^T \dot{M}\,\overline{x}_2 + \overline{x}_2^T K_p \overline{x}_1^T\right] \\ +(1-d)\left[tr\left(\widetilde{W}_t^T K_w^{-1} \dot{\widetilde{W}}_t\right) + tr\left(\widetilde{V}_t^T K_v^{-1} \dot{\widetilde{V}}_t\right)\right] \\ +(1-d)\,2d_t x_2^T M^{-1} K_d \widetilde{z}_2/\epsilon \\ +2d_t\left(d/\varepsilon\left[\widetilde{z}^T\left(PA + A^T P\right)\widetilde{z}\right] + \left[2d\varepsilon\widetilde{z}^T PBH(\overline{x}_1, \overline{x}_2, x_1^d, x_2^d, \widetilde{x}_2)\right]\right)$$

Compare (8.46) and (8.32) the first term of right hand side is the same as (8.34).

Consider (8.49):

$$\begin{array}{c}\dot{V_3}=(1-d)2d_t\left[-\overline{x}_2^TM\overset{\cdot}{x}_2^d-\overline{x}_2^TCx_2^d-\overline{x}_2^TK_d\overline{x}_2\right]\\ -(1-d)2d_t\overline{x}_2^T\left[-\widehat{W}_t\Phi(\widehat{V}_ts)+W^*\Phi(V^*s)+\widetilde{P}\right]\\ +(1-d)\left[tr\left(\widetilde{W}_t^TK_w^{-1}\dot{\widetilde{W}}_t\right)+tr\left(\widetilde{V}_t^TK_v^{-1}\dot{\widetilde{V}}_t\right)\right]\\ +(1-d)\,2d_t\frac{1}{\epsilon}\overline{x}_2^TM^{-1}K_d\widetilde{z}_2\\ +2d_t\left(-d/\varepsilon\left[\widetilde{z}^TQ\widetilde{z}\right]+\left[2d\varepsilon\widetilde{z}^TPBH(\overline{x}_1,\overline{x}_2,x_1^d,x_2^d,\widetilde{x}_2)\right]\right)\end{array}$$

where $\dot{V}_1$ is same as in (8.36). The relation (8.44) leads to

$$\begin{array}{c}2d\varepsilon\widetilde{z}^TPBH=-2d\varepsilon\widetilde{z}^TP_1\left(Cx_2+K_p\overline{x}_1+K_d\overline{x}_2-K_d\widetilde{x}_2\right)\\ -2d\varepsilon\widetilde{z}^TPBM^{-1}\left(-\widehat{W}\Phi(\widehat{V}s)+W^*\Phi(V^*s)+\widetilde{P}\right)\end{array}\tag{8.57}$$

The last term of (8.57) has the same structure as the last term of (8.35), so

$$\begin{array}{c}\dot{V}_3=\left[(1-d)\overline{x}_2^T+2d\varepsilon\widetilde{z}^TPBM^{-1}\right]\left[\widehat{W}\Phi(\widehat{V}s)-W^*\Phi(V^*s)-\widetilde{P}\right]\\ +2d\varepsilon\widetilde{z}^TP_1\left(-Cx_2-K_p\overline{x}_1-K_d\overline{x}_2+K_d\widetilde{x}_2\right)\\ +(1-d)\left[-\overline{x}_2^TM\overset{\cdot}{x}_2^d-\overline{x}_2^TCx_2^d-\overline{x}_2^TK_d\overline{x}_2\right]\\ +(1-d)\overline{x}_2^TK_d\widetilde{x}_2-d/\varepsilon\widetilde{z}^TQ\widetilde{z}+tr\left(\widetilde{W}_t^TK_w^{-1}\dot{\widetilde{W}}_t\right)+tr\left(\widetilde{V}_t^TK_v^{-1}\dot{\widetilde{V}}_t\right)\end{array}\tag{8.58}$$

since

$$\begin{array}{l}(1-d)\,\overline{x}_2^T+2d\varepsilon\widetilde{z}^TPBM^{-1}\\ =(1-d)\left(x_2-x_2^d\right)^T+2d\varepsilon\eta_{_M}\left(x_1-\widehat{x}_1\right)^TP_{12}M^{-1}\\ +2d\varepsilon^2\eta_{_M}\left(x_2-\widehat{x}_2\right)^TP_{22}M^{-1}\\ =x_2^T\left[(1-d)\,I+2d\varepsilon^2\eta_{_M}P_{22}M^{-1}\right]+2d\varepsilon\eta_{_M}\left(x_1-\widehat{x}_1\right)^TP_{12}M^{-1}\\ -2d\varepsilon^2\eta_{_M}\widehat{x}_2^TP_{22}M^{-1}-(1-d)\,x_2^{dT}\end{array}$$

Using (8.53), we get

$$(1-d)\,I+2d\varepsilon^2\eta_{_M}P_{22}M^{-1}=0$$

From (8.37), we derive

$$\widehat{W}\Phi(\widehat{V}s) - W^*\Phi(V^*s) - \widetilde{P} = \widetilde{W}_t^T\Phi(\widehat{V}_t^T s)$$
$$+\widehat{W}^T D_\sigma \widetilde{V}_t^T s + \widetilde{W}^T D_\sigma \widetilde{V}_t^T s + W^*\nu_\sigma - \widetilde{P}$$

The first term and the last term of (8.58) can be rewritten as

$$L_{w1} := tr\left[\left(K_w{}^{-1}\dot{\widetilde{W}}_t - 2d_t\sigma(\widehat{V}_t^T s)\Psi^T/(1-d) - 2d_t D_\sigma \widetilde{V}_t^T s\Psi^T/(1-d)\right)\widetilde{W}^T\right]$$
$$L_{v1} := tr\left[\left(K_v^{-1}\dot{\widetilde{V}} - 2d_t s\Psi^T \widehat{W}_t^T D_\sigma/(1-d)\right)\widetilde{V}^T\right]$$
$$L_{w1} + L_{v1} + \Psi^T\left(W^*\nu_\sigma - \widetilde{P}\right) \tag{8.59}$$

where Ψ is defined as in (8.51). Similar as (8.39), $\Psi^T\left(W^*\nu_\sigma - \widetilde{P}\right)$ may be estimated as

$$\Psi^T\left(W^*\nu_\sigma - \widetilde{P}\right) \leq \|\Psi\|\,\overline{\eta} + \frac{1}{4}\Psi^T\Lambda_1^{-1}\Psi + l\left\|\widetilde{V}_t^T x\right\|_{\Lambda_3}^2 \tag{8.60}$$

The last term of (8.60) can be joined into (8.59). With the learning law (8.50), (8.59) and the last term of (8.60) is zero. So

$$\begin{aligned}&\dot{V}_3 \leq \|\Psi\|\,\overline{\eta} + \tfrac{1}{4}\Psi^T\Lambda_1^{-1}\Psi - \\ &4d_t d\varepsilon\widetilde{z}^T PBM^{-1}\left(C\left(x_1,x_2\right)x_2 + K_p\overline{x}_1 + K_d\overline{x}_2 - K_d\widetilde{x}_2\right) \\ &+(1-d)2d_t\left[-\overline{x}_2^T M\,\dot{x}_2^{\,d} - \overline{x}_2^T C x_2^d\right] + \\ &(1-d)2d_t\overline{x}_2^T\eta_M K_d\widetilde{z}_2/\epsilon - (1-d)2d_t\overline{x}_2^T K_d\overline{x}_2 - 2d_t d\tfrac{1}{\varepsilon}\widetilde{z}^T Q\widetilde{z}\end{aligned} \tag{8.61}$$

$\left[-\overline{x}_2^T M\,\dot{x}_2^{\,d} - \overline{x}_2^T C x_2^d\right]$ can be estimated as in (8.40)

$$-\overline{x}_2^T M\,\dot{x}_2^{\,d} - \overline{x}_2^T C x_2^d \leq \left\|x_2^d\right\|\overline{x}_2^T k_c\overline{x}_2$$
$$+k_c\left\|x_2^d\right\|^2\|\overline{x}_2\| + \lambda_{\max}(M)\left\|\dot{x}_2^{\,d}\right\|\left\|\overline{x}_2^T\right\|$$

Using *Property 2*, $d\varepsilon\widetilde{z}^T P_1\left(Cx_2\right)$ becomes

$$d\varepsilon\widetilde{z}^T P_1\left[C\left(x_1,x_2\right)x_2\right] \leq 2d\varepsilon\widetilde{z}^T P_1 C_0\left(x_1\right)\overline{x}_2$$

So

$$\begin{aligned} & 2d\varepsilon\widetilde{z}^T P_1 \left[C(x_1,x_2)x_2 + K_p\overline{x}_1 + K_d\overline{x}_2 - K_d\widetilde{x}_2\right] \\ & \leq 2d\varepsilon\widetilde{z}^T P_1 \begin{bmatrix} 0 & 0 \\ K_p & K_d + 2C_0 \end{bmatrix} \overline{x} - 2d\widetilde{z}^T P_1 \begin{bmatrix} 0 & 0 \\ 0 & K_d \end{bmatrix} \widetilde{z} \end{aligned}$$

(8.61) is

$$\begin{aligned} \dot{V}_3 &\leq -y^T T y + a + (1-d)2d_t \left(b\left\|\overline{x}_2\right\| - \lambda_{\min}(K_d)\left\|\overline{x}_2\right\|^2\right) \\ &= -2d_t y^T T_0 y + a - (1-d)2d_t\left\|\overline{x}_2\right\|\left(\lambda_{\min}(K_d)\left\|\overline{x}_2\right\| - b\right) - y^T R_1 y \end{aligned} \tag{8.62}$$

where

$$T = 2d_t T_0 - R_1 = 2d_t \begin{bmatrix} (1-d)\alpha_1 & -\frac{(1-d)\beta_1}{2\varepsilon} - \frac{\varepsilon d K_2}{2} \\ -\frac{(1-d)\beta_1}{2\varepsilon} - \frac{\varepsilon d K_2}{2} & d\left[\frac{\alpha_2}{\varepsilon} - K_1\right] \end{bmatrix} - R_1$$

$$y = \begin{bmatrix} \|\overline{x}\| \\ \|\widetilde{z}\| \end{bmatrix}$$

$$\alpha_1 = \left\| \begin{bmatrix} 0 & 0 \\ 0 & Q_q \end{bmatrix} \right\|, \; \alpha_2 = \|Q\|, \; \beta_1 = \left\| \eta_M \begin{bmatrix} 0 & 0 \\ 0 & K_d \end{bmatrix} \right\|$$

$$K_1 = \left\| 2\eta_M P \begin{bmatrix} 0 & 0 \\ 0 & K_d \end{bmatrix} \right\|, \; K_2 = \left\| 2\eta_M P \begin{bmatrix} 0 & 0 \\ K_p & K_d + 2C_0 \end{bmatrix} \right\|$$

And T is positive definite if there exists a continuous interval $\Gamma = (0, \overline{\varepsilon})$ such that for all $\varepsilon \in \Gamma$ (8.54) is satisfied. So $\overline{\varepsilon}$ is an upper bound of ε. (8.62) is as follows

$$\dot{V}_3 \leq \left(a - (1-d)2d_t\left\|\overline{x}_2\right\|\left(\lambda_{\min}(K_d)\left\|\overline{x}_2\right\| - b\right)\right) - y^T R_1 y \tag{8.63}$$

where $y = \begin{bmatrix} \|\overline{x}\| \\ \|\widetilde{z}\| \end{bmatrix} = \begin{bmatrix} \left\| \begin{bmatrix} \overline{x}_1 \\ \overline{x}_2 \end{bmatrix} \right\| \\ \|\widetilde{z}\| \end{bmatrix}$. Since (8.63) has the same structure as (8.41), the similar proof to (I) and (II) can be established. ■

Remark 8.7 *Since $y = \left[\|\overline{x}\|, \|\widetilde{z}\|\right]^T$ is not measurable, the dead zone d_t in (8.52) cannot be realized. We will use the available date $\overline{y} = \widehat{x} - x^d$ to determine the dead zone. The dead zone can be represented as*

$$\|\overline{y}\|_{R_1}^2 > a + \frac{\lambda_{\min}^2(K_d)}{4b^2} + \|\overline{y}\|_{R_1}^2 - \|y\|_{R_1}^2$$

Because $\overline{y} = \overline{x} - \widetilde{x}$,

$$\left\|\overline{y}\right\|_{R_1}^2 - \left\|y\right\|_{R_1}^2 \leq \left\|\overline{x}\right\|_{R_1}^2 + \left\|\widetilde{x}\right\|_{R_1}^2 - \left\|\overline{x}\right\|_{R_1}^2$$
$$+ \left\|\widetilde{z}\right\|_{R_1}^2 = \left\|\widetilde{x}\right\|_{R_1}^2 + \left\|\widetilde{z}\right\|_{R_1}^2 \leq 4\varepsilon^2\lambda_{\max}\left(R_1\right) C_{i,T}$$

So, the new dead zone is

$$d_t = \begin{cases} 0 & if \quad \left\|\overline{y}\right\|_{R_1}^2 \leq a + \left[\lambda_{\min}^2\left(K_d\right) + 4\varepsilon^2\lambda_{\max}\left(R_1\right) C_{i,T}\right]/4b^2 \\ 1 & if \quad \left\|\overline{y}\right\|_{R_1}^2 > a + \lambda_{\min}^2\left(K_d\right)/4b^2 + 4\varepsilon^2\lambda_{\max}\left(R_1\right) C_{i,T} \end{cases}$$

Remark 8.8 *Since (8.54) has 4 possible solutions, the theorem will be valid if there exist a positive real root such that in the interval* $[0, \overline{\epsilon}]$ *(8.54) is negative. The condition (8.54) is only necessary. The main differences between [33] and this material are:*

· *we neglect the assumption 3-a in [33], because our Lyapunov function does not depend on* $\overline{x}$.

· *the assumption 3-b of [33] does not depend on* ϵ.

·*the assumption 3-c of [33] includes the constant* K_2 *with* ϵ, *not* ϵ^2 *as our result.*

· *the condition for* $\overline{\epsilon}$ *found in [33] has a more simple formula than ours.*

8.5 Simulation Results

The values of the manipulator parameters in (8.1) are listed below

$$m_1 = m_2 = 1.53\ kg$$
$$l_1 = l_2 = 0.365\ m$$
$$r_1 = r_2 = 0.1$$
$$v_1 = v_2 = 0.4$$
$$g = 9.81$$

and the time-varying uncertainty is as follows:

$$k_0 = \begin{pmatrix} 0.8 & 0.8 & 0 & 0 \\ 0 & 0 & 0.8 & 0.8 \end{pmatrix}$$

$$\Delta k_t = \begin{pmatrix} 0.5\omega\sin(\omega t) & 0.9\omega\cos(\omega t) & 0 & 0 \\ 0 & 0 & 0.2\omega\sin(\omega t) & 0.6\omega\cos(\omega t) \end{pmatrix}$$

with $\omega = 2$.

8.5.1 Robot's Dynamic Identification based on Neural Network

We assume the parameters in (8.1) are known, only the position and the velocity of θ are available. We test two independent neural networks to identify the position θ_1 and θ_2 , and the velocity $\dot{\theta}_1$ and $\dot{\theta}_2$.

The first NN is given by

$$\begin{aligned} \dot{\widehat{\theta}}_1 &= -2\widehat{\theta}_1 + w_{11}\sigma\left(\widehat{\theta}_1\right) + w_{12}\sigma\left(\widehat{\theta}_2\right) + w'_{11}\phi\left(\widehat{\theta}_1\right)\tau_1 + w'_{12}\phi\left(\widehat{\theta}_1\right)\tau_2 \\ \dot{\widehat{\theta}}_2 &= -2\widehat{\theta}_2 + w_{21}\sigma\left(\widehat{\theta}_1\right) + w_{22}\sigma\left(\widehat{\theta}_2\right) + w'_{21}\phi\left(\widehat{\theta}_1\right)\tau_1 + w'_{22}\phi\left(\widehat{\theta}_2\right)\tau_2 \end{aligned}$$

The second NN is as

$$\begin{aligned} \ddot{\widehat{\theta}}_1 &= -2\,\dot{\widehat{\theta}}_1 + \overline{w}_{11}\sigma\left(\dot{\widehat{\theta}}_1\right) + \overline{w}_{12}\sigma\left(\dot{\widehat{\theta}}_2\right) + \overline{w}'_{11}\phi\left(\dot{\widehat{\theta}}_1\right)\tau_1 + \overline{w}'_{12}\phi\left(\dot{\widehat{\theta}}_1\right)\tau_2 \\ \ddot{\widehat{\theta}}_2 &= -2\,\dot{\widehat{\theta}}_2 + \overline{w}_{21}\sigma\left(\dot{\widehat{\theta}}_1\right) + \overline{w}_{22}\sigma\left(\dot{\widehat{\theta}}_2\right) + \overline{w}'_{21}\phi\left(\dot{\widehat{\theta}}_1\right)\tau_1 + \overline{w}'_{22}\phi\left(\dot{\widehat{\theta}}_2\right)\tau_2 \end{aligned}$$

Here

$$\begin{aligned} \sigma(x) &= \frac{2}{(1+e^{-2x})} - \frac{1}{2} \\ \phi(x) &= \frac{0.2}{(1+e^{-0.2x})} - 0.05 \end{aligned}$$

The initial conditions are selected as

$$\begin{aligned} W_0 &= \overline{W}_0 = \begin{bmatrix} w_{11}(0) & w_{12}(0) \\ w_{21}(0) & w_{22}(0) \end{bmatrix} = \begin{bmatrix} 1 & 10 \\ 10 & 1 \end{bmatrix} \\ W'_0 &= \overline{W}'_0 = \begin{bmatrix} w'_{11}(0) & w'_{12}(0) \\ w'_{21}(0) & w'_{22}(0) \end{bmatrix} = \begin{bmatrix} 0.1 & 0 \\ 0 & 0.1 \end{bmatrix} \end{aligned}$$

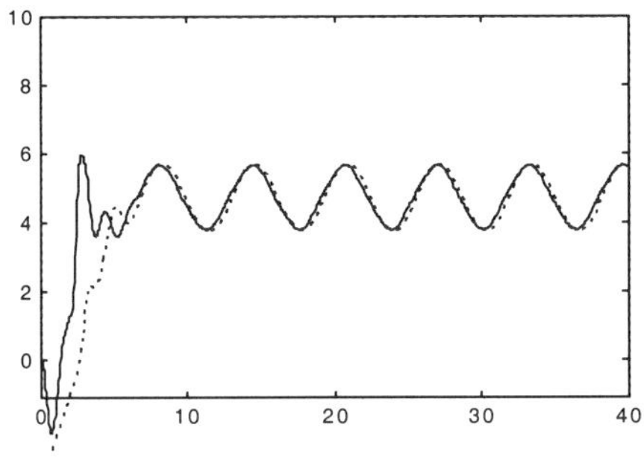

FIGURE 8.2. Identification results for θ_1.

The update laws are the same as in (8.30), we select

$$A = \begin{bmatrix} -2 & 0 \\ 0 & -2 \end{bmatrix}, \quad P = \begin{bmatrix} 0.2 & 0 \\ 0 & 0.2 \end{bmatrix}$$

$$k = \begin{bmatrix} 5 & 0 \\ 0 & 2 \end{bmatrix}$$

$$\dot{W}_t = -s_t k P \begin{bmatrix} \sigma(\widehat{\theta}_1) \\ \sigma(\widehat{\theta}_2) \end{bmatrix} \begin{bmatrix} \Delta_1 & \Delta_2 \end{bmatrix}$$

$$\Delta = \theta - \widehat{\theta}, \quad s_t = \begin{cases} 1 & \text{if} \quad \|\Delta_t\| > 0.1 \\ 0 & \text{if} \quad \|\Delta_t\| \le 0.1 \end{cases}$$

For the generalized forces as

$$\tau_1 = 7 \sin t$$
$$\tau_2 = 0$$

the identification results for the state vector θ are shown in Figure 8.2 and Figure 8.3. The time evaluation of the weights W_t are shown in Figure 8.4. The identification results for $\dot{\theta}$ are shown in Figure 8.5 - Figure 8.7. The identification errors exists in this experiments because we use the second order neural network to realize the modelling of the dynamic of two links robot. So, there are unmodeled dynamics.

On the other hand, if we use sliding mode learning (as in Chapter 3) for the identification of this robot, we can obtain much better results shown in Figure 8.8 - Figure 8.11.

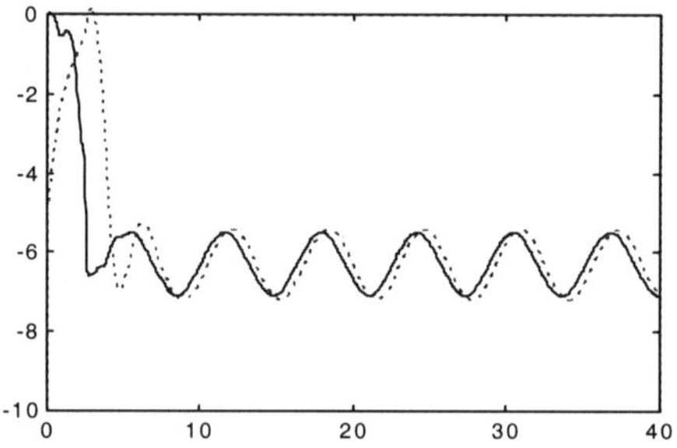

FIGURE 8.3. Identification results for θ_2.

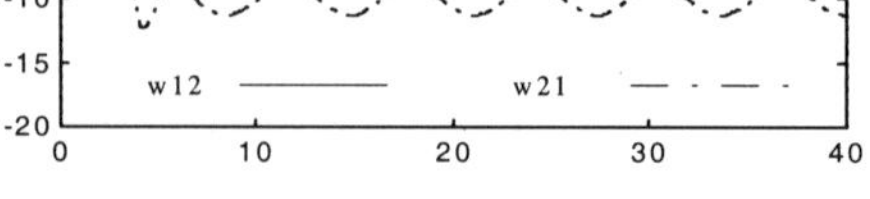

FIGURE 8.4. Time evaluation of W_t.

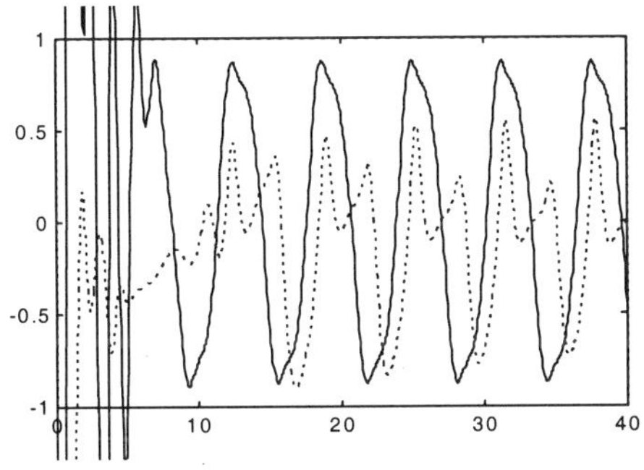

FIGURE 8.5. Identification results for $\dot{\theta}_1$.

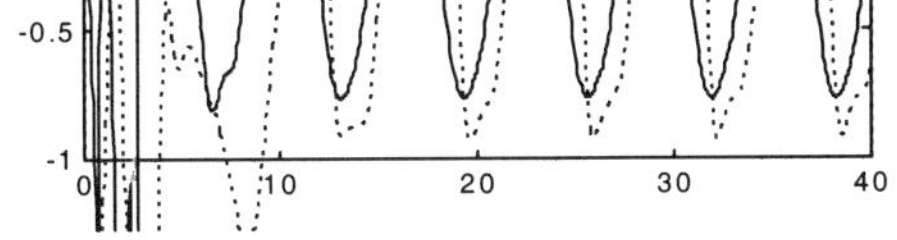

FIGURE 8.6. Identification results for $\dot{\theta}_2$.

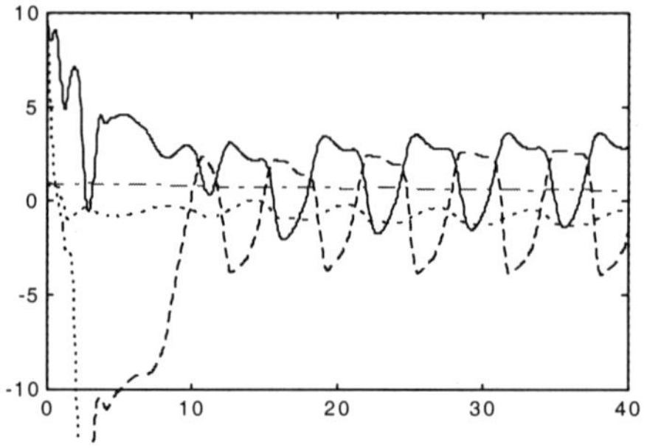

FIGURE 8.7. Time evalution for the weights $\overline{W}_t$.

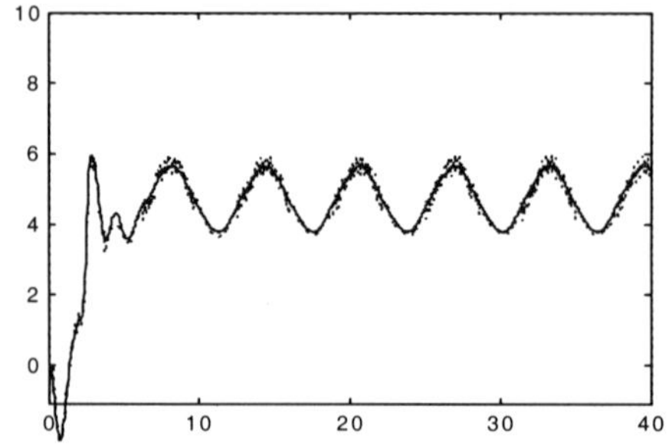

FIGURE 8.8. Sliding mode identification for θ_1.

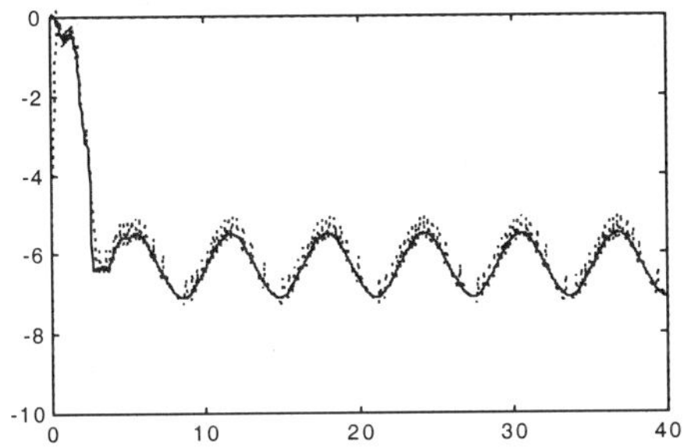

FIGURE 8.9. Sliding mode identification θ_2.

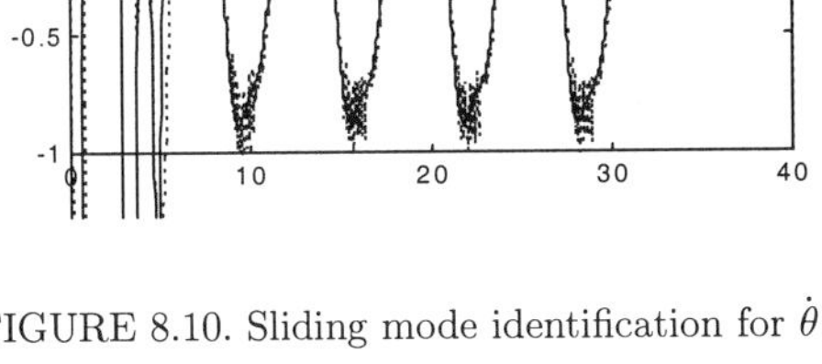

FIGURE 8.10. Sliding mode identification for $\dot{\theta}_1$.

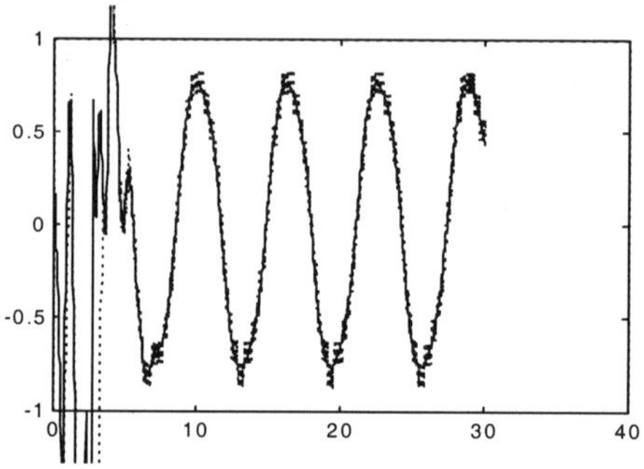

FIGURE 8.11. Sliding mode identification $\dot{\theta}_2$.

8.5.2 Neuro Control for Robot

The neural network for control is represented as

$$\begin{aligned} \dot{\widehat{\theta}}_1 &= -1.5\widehat{\theta}_1 + w_{11}\sigma\left(\widehat{\theta}_1\right) + w_{12}\sigma\left(\widehat{\theta}_2\right) + \tau_1 \\ \dot{\widehat{\theta}}_2 &= -1.5\widehat{\theta}_2 + w_{21}\sigma\left(\widehat{\theta}_1\right) + w_{22}\sigma\left(\widehat{\theta}_2\right) + \tau_2 \end{aligned} \tag{8.64}$$

The neuro control is same as in Chapter 5. If

$$t < 480$$

we use a PD-control as

$$u_t = \begin{bmatrix} 5 & 0 \\ 0 & 5 \end{bmatrix} (\theta - \theta^*) - \begin{bmatrix} 10 & 0 \\ 0 & 10 \end{bmatrix} \left(\dot{\theta} - \dot{\theta}^*\right)$$

to make the neural network (8.64) following the dynamic of the robot. After

$$t > 480$$

the controller is switched to neuro control (6.9) as

$$\tau = u_{1,t} + u_{2,t}$$

$$u_{1,t} = \varphi\left(\theta^*\right) - \begin{bmatrix} -1.5 & 0 \\ 0 & -1.5 \end{bmatrix} \theta^* - W_t\sigma(\widehat{\theta}_t)$$

We assume that the trajectories to be tracked are:

$$\theta_1^* = 3$$

and θ_2^* is square wave given by

$$\begin{array}{lll} \theta_2^* = -2 & \text{if} & 0 < t \leq 800 \\ \theta_2^* = 2 & \text{if} & 800 < t \leq 2000 \\ \theta_2^* = -2 & \text{if} & 2000 < t \leq 2800 \end{array}$$

So,

$$\varphi(\theta^*) = \dot{\theta}^* = 0$$

The control action $u_{2,t}$ is selected to compensate the unmodeled dynamics and has the following structure:

1. If the link velocity $\dot{\theta}$ is measurable, then the **exact compensation method** can be applied. As in (6.11), we put

$$u_{2,t} = A\left(\theta - \widehat{\theta}\right) - \left(\dot{\theta} - \dot{\widehat{\theta}}\right).$$

The results are shown in Figure 8.12 - Figure 8.14.

2. If $\dot{\theta}$ is not available, the **sliding mode** technique may be applied selecting $u_{2,t}$ as (6.16):

$$u_{2,t} = -10 \cdot sgn(\theta - \theta^*)$$

The results are shown in Figure 8.15-Figure 8.17.

3. **Local Optimal Control**. If we select

$$Q = \frac{1}{2}, \quad R = \frac{1}{20}, \quad \Lambda - 4.5$$

the solution of the following Riccati equation

$$A^T P_t + P_t A + P_t \Lambda P_t + Q = -\dot{P}_t$$

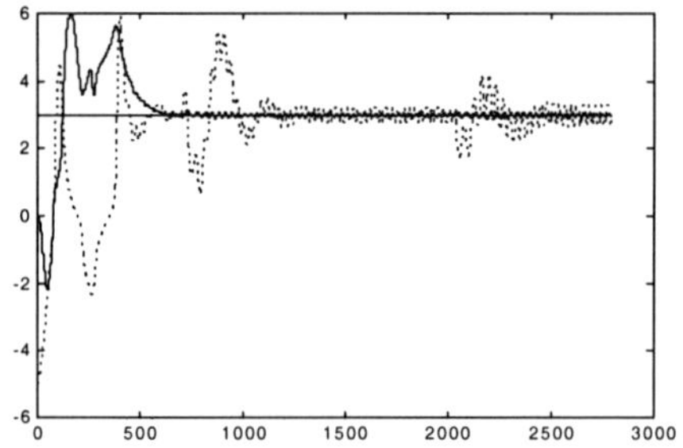

FIGURE 8.12. Control method 1 for θ_1.

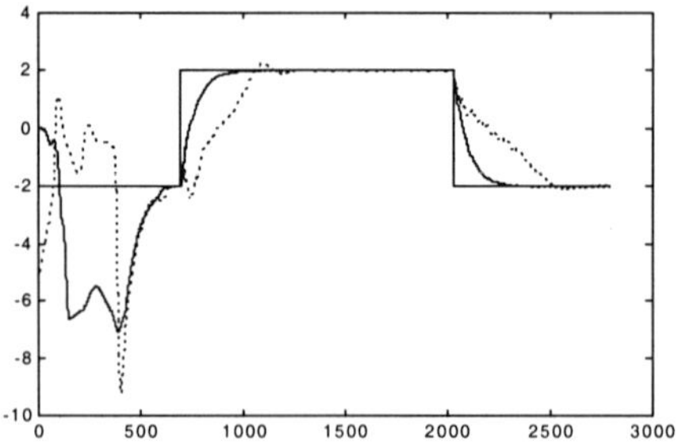

FIGURE 8.13. Control method for θ_2.

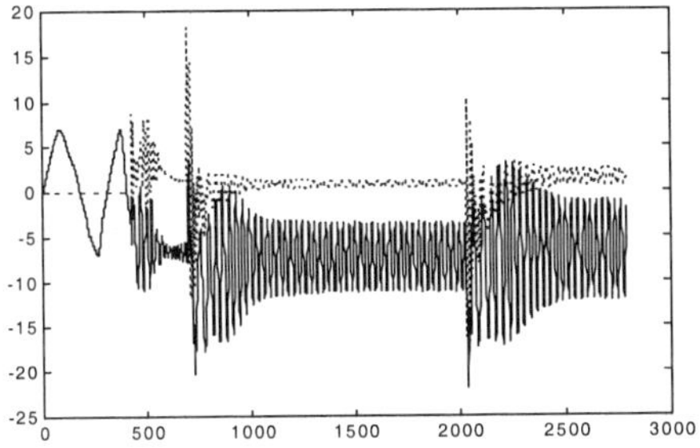

FIGURE 8.14. Control input for the method 1.

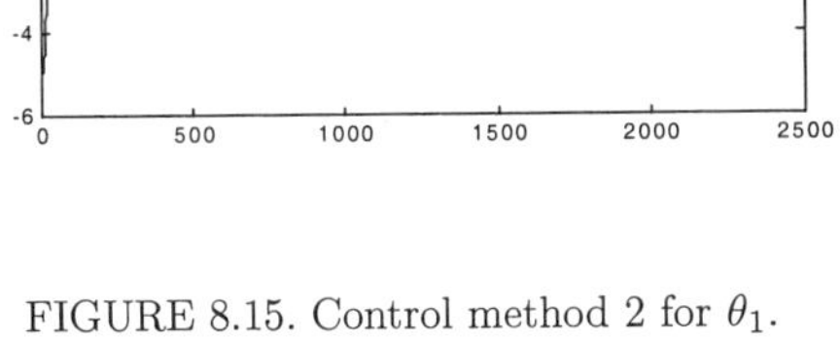

FIGURE 8.15. Control method 2 for θ_1.

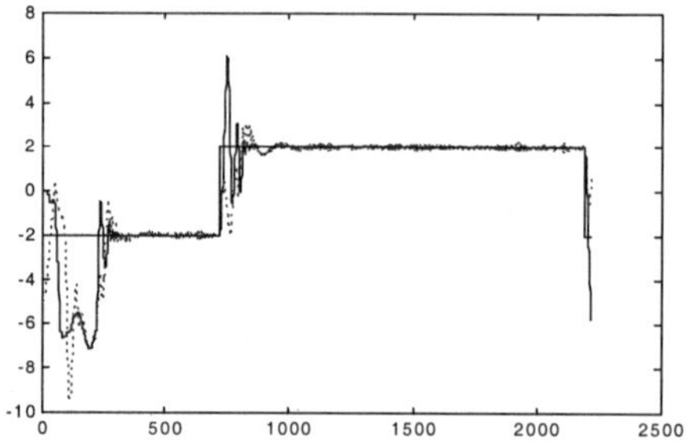

FIGURE 8.16. Control method 2 for θ_2.

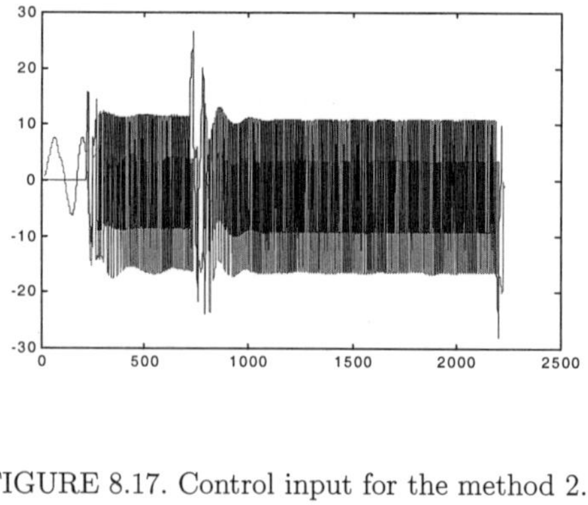

FIGURE 8.17. Control input for the method 2.

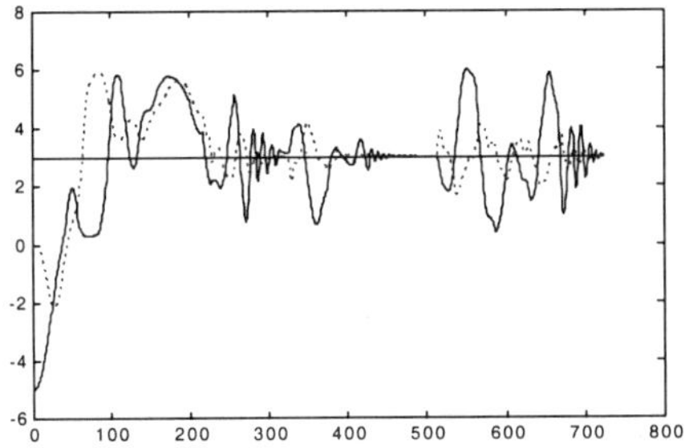

FIGURE 8.18. Control Method 3 for θ_1.

is

$$P = \begin{bmatrix} 0.33 & 0 \\ 0 & 0.33 \end{bmatrix}$$

In the case of no any restriction to τ, this control law turns out to be equal to the **linear squares optimal control** law (6.28):

$$u_{2,t} = -2R^{-1}P(\theta - \theta^*) = \begin{bmatrix} -20 & 0 \\ 0 & -20 \end{bmatrix} (\theta - \theta^*)$$

The results are shown in Figure 8.18-Figure 8.20

8.5.3 PD Control for robot

The following PD coefficients are chosen:

$$K_p = \begin{bmatrix} 31 & 0 \\ 0 & 45 \end{bmatrix}, \quad K_d = \begin{bmatrix} 60 & 0 \\ 0 & 80 \end{bmatrix}$$

The matrices P and Q are selected as

$$P = \begin{bmatrix} 5 & 0 & \frac{1}{2} & 0 \\ 0 & 5 & 0 & -\frac{1}{2} \\ -5 & 0 & 1 & 0 \\ 0 & -5 & 0 & 1 \end{bmatrix}, \quad Q = \begin{bmatrix} 45 & 0 & 0 & 0 \\ 0 & 45 & 0 & 0 \\ -45 & 0 & 5.5 & 0 \\ 0 & -45 & 0 & 5.5 \end{bmatrix}$$

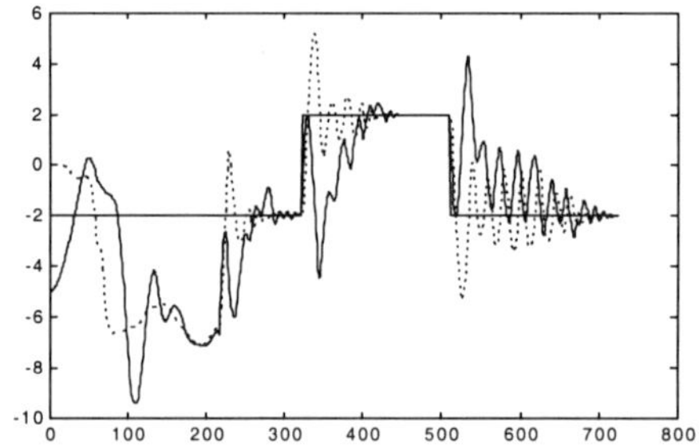

FIGURE 8.19. Control method 3 for θ_2.

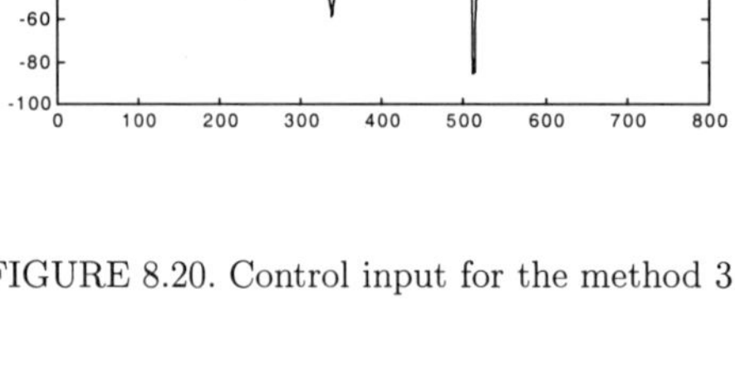

FIGURE 8.20. Control input for the method 3.

Let us calculate the constants in Theorem 5

$$\alpha_1 = \left\| \begin{bmatrix} 0 & 0 \\ 0 & K_d \end{bmatrix} \right\| = 80, \quad \alpha_2 = \|Q\| = 45$$

$$K_1 = \left\| P \begin{bmatrix} 0 & 0 \\ 0 & -2M(\overline{x}_1 + x_1^d)^{-1} \left(F - K_d + C(\overline{x}_1 + x_1^d, x_2^d) \right) \end{bmatrix} \right\|$$
$$= 42.71$$

$$K_2 = \left\| P \begin{bmatrix} 0 & 0 \\ -M(\overline{x}_1 + x_1^d)^{-1} K_p & -M(\overline{x}_1 + x_1^d)^{-1} \left(K_d + C(\overline{x}_1 + x_1^d, x_2^d) \right) \end{bmatrix} \right\|$$
$$= 84.3242$$

$$\beta_1 = \left\| \begin{bmatrix} 0 & 0 \\ 0 & -F + K_d - C(\overline{x}_1 + x_1^d, x_2^d) \end{bmatrix} \right\| = 80.0288$$

where $\|A\|$ denotes the absolute value of the real part of the maximum eigenvalue of the matrix A. Then (8.54) becomes:

$$\begin{gathered} f(\epsilon) = 3555.3d^2\epsilon^4 + 6748.4(1-d)d\epsilon^2 + \\ 6745.9(1-d)\epsilon + 3202.3(1-d)^2 - 3600(1-d)d \end{gathered} \tag{8.65}$$

With $d = 0.5$ the polynomial (8.65) is shown in Figure8.21. One can see that for

$$0 < \epsilon < 0.0558309561787$$

(8.65) is negative. So,

$$\overline{\epsilon} = 0.056, \Gamma = (0, 0.056)$$

The high-gain observer (8.12) is determined by $\epsilon = 0.003$.

Figure 8.22 shows a rapid convergence of the observer to the links velocities. The observer error is almost zero in this short interval.

Remark 8.9 *Rapid convergence of the observer values is essential for the PD-like controller, because they form part of the feedback. Accuracy of them is important*

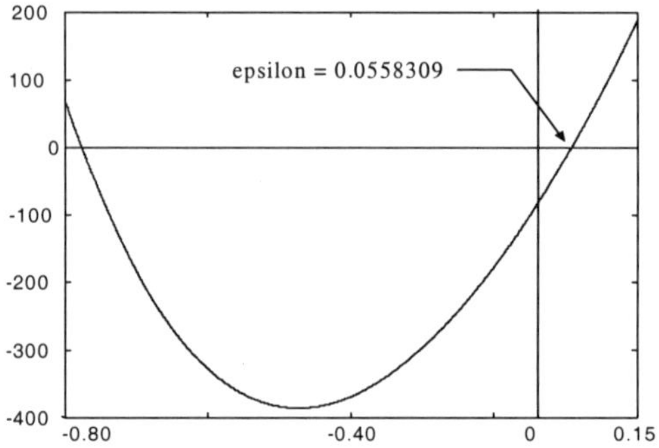

FIGURE 8.21. Polinomial of epsilon.

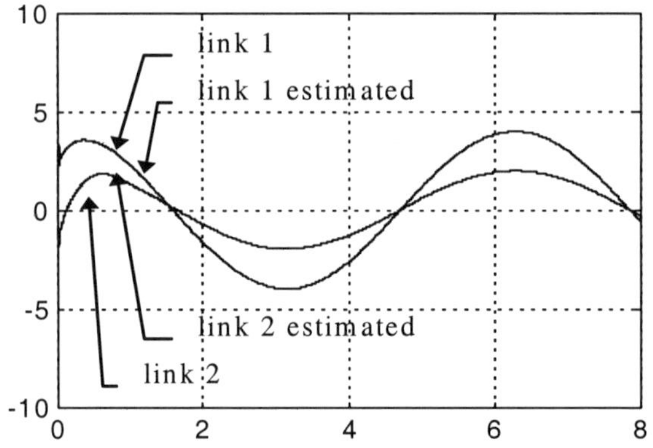

FIGURE 8.22. High-gain observer for links velocity

as well. This useful property of the high-gain observer permits us to use a simple controller, instead of a complicated one having to compensate the nonlinear dynamics of the robot plus the uncertainties of the observer response, or having to use a link position-only feedback controller. Here it is important to note, that the independence of the observer from robot dynamics makes it almost invariant to the perturbation. And both results, with and without the perturbation, are very similar.

Friction and gravity can be uniformly approximated by a radial basis function as in (8.25) with $N = 10$. The Gaussian function is

$$\phi_j(Vx) = \exp\left\{ -\frac{\sum_{i=1}^{n} (Vx_i - c_i)^2}{100} \right\}$$

where the spread σ_l was set to $\sqrt{50}$ and the center c_i is a random number between 0 and 1. The control law applied is

$$u = -K_p(x_1 - x_1^d) - K_d(x_2 - x_2^d) + \widehat{W}\Phi(\widehat{V}x)$$

starting with $W^* = 0.7$ and $V^* = 0.7$ as initial values. Even though some initial weights are needed for the controller to work, no special values are required nor a previous investigation on the robot dynamics for the implementation of this control. Figure 8.23 and Figure 8.24 give the comparison of the performance of PD controller neuro compensation. The continuous line is the exact position, the dashed line is general PD control without friction and gravity, the dashed-dotted line is neural networks compensation.

Let combine high-gain observer and neuro compensator. The PD control is

$$u = -K_p(x_1 - x_1^d) - K_d(\widehat{x}_2 - x_2^d) + \widehat{W}\Phi(\widehat{V}s)$$

The the tracking error are shown in Figure 8.25 and Figure 8.26. The continuous line is PD control with high-gain observer and neuro compensator, position, the dashed line is general PD control without friction and gravity, the dashed-dotted line is PD control with neural networks compensation.

We can see that the combination of high-gain observer and neuro compensator is a good way to improve the performance of the popular PD control.

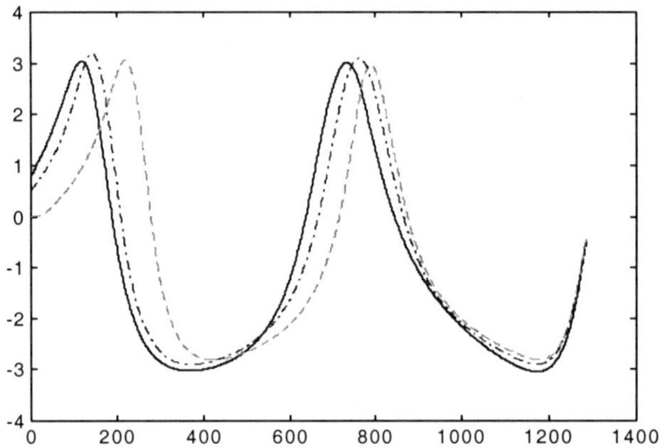

FIGURE 8.23. Positions of link 1.

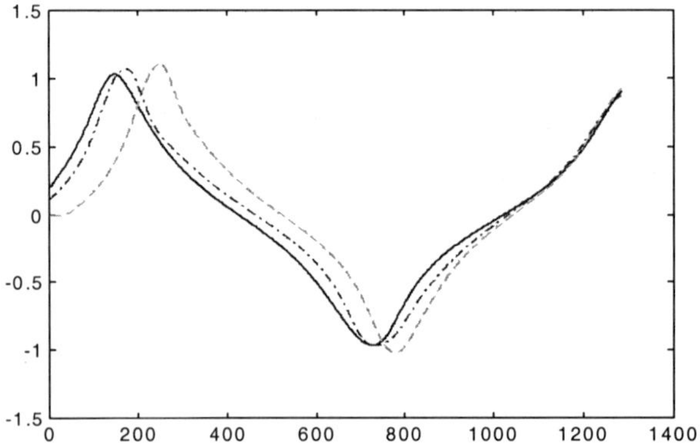

FIGURE 8.24. Positions of link 2.

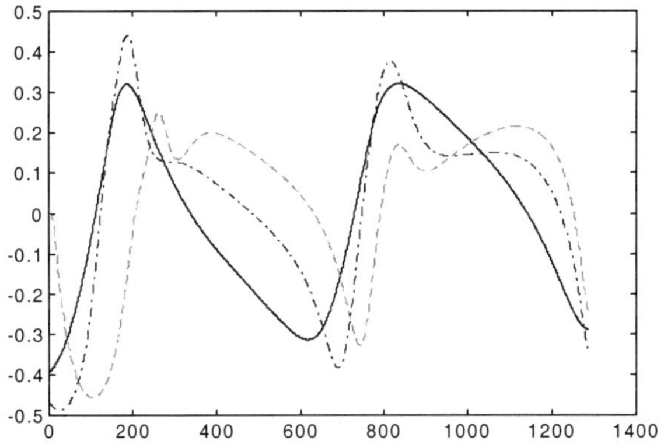

FIGURE 8.25. Tracking errors of link 1.

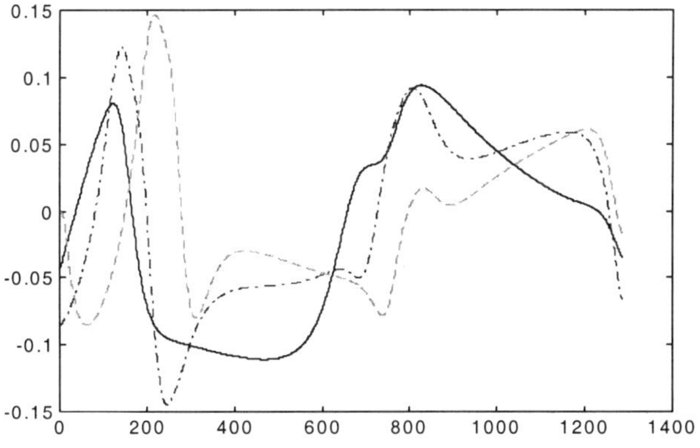

FIGURE 8.26. Tracking errors of link 2.

8.6 Conclusion

In this chapter a dynamic neural network was developed for the control of two-link robot manipulator. First, we use the parallel dynamic neural network to identify the dynamic of robot, then a direct linearization controller is applied based on this neuro identifier. Because of the modelling error, three types of compensators are presented and compared.

8.7 References

[1] H.Berghuis and H.Nijmeijer, A Passivity Approach to Controller-Observer Design for Robots, *IEEE Tran. on Robot. Automat.*, Vol. 9, 740-754, 1993.

[2] Brian Armstrong-Hélouvry, Pierre Dupont and Carlos Canudas de Wit,"A Survey of Models, Analysis Tools and Compensation Methods for the Control of Machines with Friction", *Automatica*, vol 30, no. 7, pp. 1083-1138, 1994.

[3] B. R. Barmish, M. Corless and G. Leitmann, "A new class of stabilizing controllers for uncertain dynamical systems". *SIAM J. Control and Optimization,* 21, 1983, pp. 246-255.

[4] M. M. Bridges, D. M. Dawson, Qu Z. and S. C. Martindale, "Robust control of rigid -link flexible-joint robots with redundant actuators," *IEEE Transactions Syst., Man, Cybern.*, vol. 24, pp. 961-970, 1994.

[5] C.Canudas de Wit and J.J.E.Slotine, Sliding Observers for Robot Manipulator, *Automatica*, Vol.27, No.5, 859-864, 1991.

[6] C.Canudas de Wit and N.Fixot, Adaptive Control of Robot Manipulators via Velocity Estimated Feedback, *IEEE Tran. on Automatic Control*, Vol. 37, 1234-1237, 1992.

[7] Carlos Canudas de Wit, Bruno Siciliano and Georges Bastin (Eds), *"Theory of Robot Control"*. Springer-Verlag, London, 1996.

[8] Y.H.Chen, "Adaptive robust model-following control and application to robot manipulators", *Transactions of ASME , Journal of Dynamic Systems, Measurement, and Control*, vol. 109, pp. 209-15, 1987.

[9] D.Dawson, Z.Qu and M.Bridge,"Hybrid adaptive control for the tracking of rigid-link flexible-joint robots", *in Modelling and Control of Compliant and rigid Motion Systems*, 1991 ASME *Winter Annual Meeting*, Atlanta GA, pp. 95-98, 1991.

[10] S.Gutman and G. Leitman, "Stailizing feedback control for dynamic systems with bounded uncertainty". *Proceedings of IEEE Conference on Decision and Control*, vol.1, pp. 836-842, 1976.

[11] S.Haykin, *Neural Networks- A comprehensive Foundation*, Macmillan College Publ. Co., New York, 1994.

[12] P.A.Ioannou and J.Sun, *Robust Adaptive Control*, Prentice-Hall, NJ:07458, 1996.

[13] S.Jagannathan and F.L. Lewis, Identification of nonlinear dynamical systems using multilayered neural networks, *Automatica*, Vol.32, no.12, 1707-1712, 1996.

[14] H.K.Khalil, Adaptive Output Feedback Control of Nonlinear Systems Represented by Input-Output Models, *IEEE Trans. Automat. Contr.*, Vol.41, No.2, 177-188, 1996.

[15] R.Kelly, Global Positioning on Robot Manipulators via PD control plus a Classs of Nonlinear Integral Actions, *IEEE Trans. Automat. Contr.*, vol.43, No.7, 934-938, 1998.

[16] R.Kelly, A Tuning Procedure for Stable PID Control of Robot Manipulators, *Robotica*, Vol.13, 141-148, 1995.

[17] Y.H.Kim and F.L.Lewis, Neural Network Output Feedback Control of Robot Manipulator, *IEEE Trans. Neural Networks*, Vol.15, 301-309, 1999.

[18] E.B.Kosmatopoulos, M.M.Polycarpou, M.A.Christodoulou and P.A.Ioannpu, High-Order Neural Network Structures for Identification of Dynamical Systems, *IEEE Trans. on Neural Networks*, Vol.6, No.2, 442-431, 1995.9.

[19] Frank L. Lewis, Aydin Yesildirek, and Kai Liu, Multilayer Neural-Net Robot Controller with Guaranteed Tracking Performance *IEEE Transactions on Neural Networks*, Vol.7, No.2, 1996.

[20] K. S. Narendra and A. M. Annaswamy, *"Stable Adaptive Systems"*, Prentice Hall, Englewood Cliffs, NJ, 1989.

[21] K.S.Narendra and K.Karthasarathy, Identification and Control of Dynamical Systems Using Neural Networks, *IEEE Trans. Neural Networks*, Vol.1, 4-27, 1990

[22] R. Ortega and M. W. Spong, "Adaptive motion control of rigid robots: A tutorial", *Automatica*, The Journal of IFAC, vol 25, pp 877-88, 1989.

[23] Zhihua Qu and Darren M. Dawson, *"Robust Tracking Control of Robot Manipulators"*, IEEE Press, New York, 1996.

[24] G.A.Rovithakis and M.A.Christodoulou, Adaptive Control of Unknown Plants Using Dynamical Neural Networks, *IEEE Trans. on Syst., Man and Cybern.*, Vol. 24, 400-412, 1994.

[25] M.M. Polycarpou and P.A. Ioannou, Neural networks as on-line approximators of nonlinear systems, *Proc. IEEE Conf. Decision and Control*, pp. 7-12, Tucson, Dec. 1992.

[26] P.Tomei, Adaptive PD Controller for Robot Manipulator, *IEEE Tran. on Automatic Control*, Vol. 36, 556-570, 1992.

[27] P. Tomei, "Tracking control of flexible joint robots with uncertain parameters and disturbances," *IEEE Transactions on Automatic Control*, vol. 39, pp. 1067-1072, 1994.

[28] M.Takegaki and S.Arimoto, A New Feedback Method for Dynamic control of Manipulator, *ASME J. Dynamic Syst. Measurement, and Contr.*, Vol.103, 119-125, 1981.

[29] S.N.Singh,"Adaptive model-following control of nonlinear robotic systems", *IEEE Transactions on Automatic Control*, vol. 30, pp. 1099-1100, 1985.

[30] M.W. Spong and M. Vidyasagar, *"Robot Dynamics and Control"*. John Wiley & Sons, 1989.

[31] J. J. Slotine and W. Li, "On the adaptive control of robot manipulators", *International Journal of Robotics Research*, vol. 6, pp. 49-59, 1987.

[32] V.Santibanez and R.Kelly, Global Asymptotic Stability of the PD Control with Computed Feedforward in Closed Loop with Robot Manipulators, *Proc.14th IFAC World Congress*, 197-203, Beijing, 1999

[33] Ali Saberi, Hassan Khalil, Quadratic-Type Lyapunov Functions for Singularly Perturbed Systems, *IEEE Transactions on Automatic Control*, Vol. AC-29, No. 6, June 1984.

[34] V.I.Utkin, "Variable structure systems with sliding modes: A survey", *IEEE Transactions on Automatic Control*, vol. 22, pp. 212-22, 1977.

[35] Wen Yu and Alexander S.Poznyak, Indirect Adaptive Control via Parallel Dynamic Neural Networks, *IEE Proceedings - Control Theory and Applications*, Vol.146, No.1, 25-30, 1999.

[36] Wen-Hong Zhu, Zeungnam Bien and Joris De Schutter," Adaptive Motion/Force Control of Multiple Manipulators with Joint Flexibility Based on Virtual Decomposition,", *IEEE Transactions on Automatic Control*, vol. 43, pp. 46-60, 1998.

9

Identification of Chemical Processes

A dynamic mathematical model of an ozonization reactor is derived using material balancing. Some concentrations are not measurable. They represent the unobservable states of the considered system. A dynamic neural network is used for states estimation. Some theoretical results concerning the bound of the observation error are presented. Based on the neuro-observer outputs, the continuous version of the least squares algorithm and a projection procedure are used to estimate the unknown chemical reaction constants. Several simulation results have been carried out to illustrate the feasibility and efficiency of the estimation approach.

9.1 Nomenclature

c_t^i $(mole/l)$ is the ith organic compound concentration at time $t \geq 0$, $i = 1, ..., N$ where N is the number of different organic compounds dissolved in the liquid phase of the given ozonation reactor;

c_t^{gas} $(mole/l)$ is the gas (moles) which do not react with organic compounds dissolved in the solvent and can be *directly measured* in the outlet of the ozonation reactor. Since this process is smooth enough the derivative $\frac{d}{dt}c_t^{gas}$ is also assumed to be available (or estimated from c_t^{gas});

w^{gas} (l/s) is the gas consumption assumed to be constant;

v^{gas} (l) is the volume of the gas phase which is also assumed to be constant;

Q_t $(mole)$ is the dissolved ozone;

v^{liq} (l) is the volume of the liquid phase assumed to be constant too;

$Q_{\max} = \alpha c_t^{gas} v^{liq}(mole)$ is the maximum of ozone being in the saturated liquid phase under the given conditions, α is the parameter deduced from the Henry constant;

K_{sat} (s^{-1}) is the ozone volumetric mass transfer coefficient;

k_i $(ls^{-1}mol^{-1})$ is the rate constant of the ozonation reaction of the ith organic compound.

9.2 Introduction

Ozone-liquid systems are extensively used in different industrial environmental processes such as wastewater, river and drinking water treatment, etc. The main aim of the ozonation treatment ("purification") is the quick and effective elimination of hydrocarbon contaminants (paraffins, olefins and aromatic compounds) from the given liquid mixture (for example, water) [1], [9], [20]. Such processes are usually carried out in ozonation reactors under specific temperature and pressure [2]. In general, a reactor is one of the major components in a chemical processing system [13], [14]. It is used to convert reactants into products.

The ozonation reactor, considered here, represents a semi-batch reactor where the ozone feed enters the bottom as shown in FIGURE 10.1.

Several parallel ozonation reactions take place in the reactor [6]:

$$O_3 + A_i \xrightarrow{k_i} B_i \ \left(i = \overline{1, N}\right)$$

where O_3 is ozone, A_i is one of the organic compounds and B_i is the corresponding ozonation product. The monotonic decreasing elimination curves for different organic compounds (c_t^i is the current concentration of the compound A_i) are shown in FIGURE 10.2.

The ozonation process can be stopped at time τ if the "contaminant level" for all contaminants does not exceed a given value, namely d, that is,

$$\tau := \max_{i=\overline{1,N}} \left\{ t_i : c_{t_i}^i = d \right\} \tag{9.1}$$

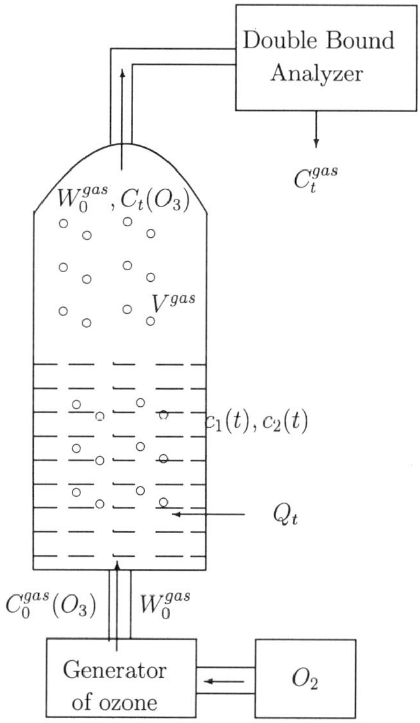

FIGURE 9.1. Schematic diagram of the ozonization reactor.

The mathematical model of these processes, developed in [19], is actively applied to ozonation reactor design [20], the efficiency optimization and the prediction of their actual performance [22], operation and maintenance, ensuring safety, and development of control strategies [3], [13]. Information on flow phenomena, rates at which the reactions proceed and estimations of the current concentrations of each compound is needed for successful realization of this treatment. Indeed, the current concentration estimation of the compounds provides the possibility to estimate τ (9.1) that can be considered as the "residence time" of the reactor and, hence, its volume $v = v^{liq} + v^{gas}$ can be calculated as $v \simeq \tau w^{gas}$ that is very important for the preliminary reactor design. The estimation of the rate constants k_i is needed for the

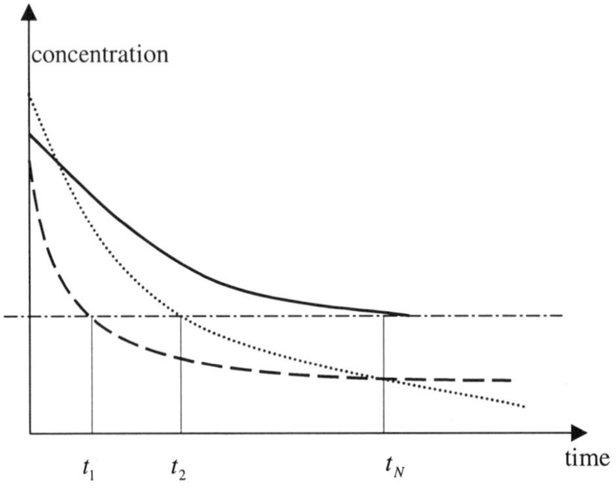

FIGURE 9.2. Concentration behaviour and ozonation times for different organic compounds.

selection of the corresponding temperature regime.

The design and control of ozonization reactors have always been the challenging tasks mostly because of the inadequacy of on-line sensors with fast sampling rate and small time delay (since ozone is the most quick oxidant), and the complex nonlinear strongly interactive behavior of ozonation reactions. There exists an extensive literature concerning the understanding of the qualitative and/or quantitative relations between easily-available on-line measurements. For these reasons, it can be assumed that control techniques will be exploited to a far greater extent in order to avoid these limitations (unavailable or very expensive sensors, complex models, etc.) in monitoring and control of chemical reactions. The nonlinear model of the ozonation process is developed using mass balance principle and consists of a set of nonlinear differential equations [18]. The only available measurement concerns the concentration of ozone in the gaseous phase of the reactor. To overcome the present limitations of sensors technology (for example, sensors for concentrations measurement are not available or are very expensive since special chromatography devices are required, etc.), *dynamic neural networks* are used to estimate the unmeasurable (inaccessible) states which are the unmeasured compound concentrations [7], [24], [11] and [12].

There are two general concepts of *recurrent structure training*. Fixed point learning is aimed at making the neural network reach the prescribed equilibria and performs steady-state matching. Trajectory learning trains the network to follow the desired trajectory in time. In this chapter we follow the second approach [14] since in the equilibria state (stationary regime), when the compound concentrations are equal (or close) to zero, it is impossible to estimate the reaction rate constants: the only nonstationary (transition) part of the process contains the sufficient information to identify them. This is the *main specific feature of the quick ozonation processes* under consideration.

Some authors have already discussed the application of neural networks techniques to construct state observers for nonlinear systems. In [4] a nonlinear observer based on the ideas of [5] is combined with a feedforward neural network, which is used to solve a matrix equation. [8] uses a nonlinear observer to estimate the nonlinearities of an input signal. As far as we know, the first observer for nonlinear systems using dynamic neural networks is presented in [10]. The stability of this observer with on-line updating of neural network weights is analyzed, but several restrictive assumptions are used: the nonlinear plant must contain a known linear part and a strictly positive real (*SPR*) condition must be fulfilled to proof the stability of the error. In [18] a robust neuro-observer with time delay term and adjusted weights in the hidden layer is suggested.

In this chapter the differential neuro observer (DNO) is considered to carry out the current estimates of the compound concentrations without any *a priori* knowledge on the corresponding rate constants. Based on the neuro-observer outputs, the continuous version of the least squares (LS) algorithm with a projection procedure is used to estimate the unknown chemical reaction rates. The theoretical analysis of this DNO is carried out using Lyapunov-like technique.

The *remainder* of this chapter is organized as follows: The model of the considered ozonization reactor is presented in the next section. Section 3 deals with observability condition for the partial (but more practice) case of $N = 3$ compounds mixture. The neuro-observer with the corresponding learning law for the weight matrix is described in section 4. The estimation of the reaction rate constants is discussed in section 5. Two numerical simulations are given in section 6. Section 7 concludes this study.

9.3 Process Modeling and Problem Formulation

9.3.1 Reactor Model and Measurable Variables

Ozone, a strong oxidant, is more and more often used, to treat and produce high quality drinking water in conjunction with chlorine. Ozone is a cost effective treatment for many types of industrial waste waters. Ozone is effective at reducing Chemical Oxygen Demand (COD), as well as making many compounds more amenable to biological treatment. The model of the considered semi-batch ozonation reactor is described in what follows.

Ozone Mass Balance

The mass balance consideration (with respect to ozone) leads to the following model [19] given in the integral form:

$$\int_{\tau=0}^{t} w^{gas} c_0^{gas} d\tau = \int_{\tau=0}^{t} w^{gas} c_\tau^{gas} d\tau + c_t^{gas} v^{gas} + Q_t + v^{liq} \sum_{i=1}^{N} \left(c_0^i - c_t^i\right) \tag{9.2}$$

or, in the equivalent differential form,

$$\frac{d}{dt} c_t^{gas} = \frac{1}{v^{gas}} \left\{ w^{gas} \left[c_0^{gas} - c_t^{gas}\right] - \frac{d}{dt} \left[Q_t + v^{liq} \sum_{i=1}^{N} \left(c_0^i - c_t^i\right) \right] \right\} \tag{9.3}$$

Ozone Dissolution Process

The differential equation associated with the ozone dissolution process is as follows [2]:

$$\frac{d}{dt} \left[Q_t + v^{liq} \sum_{i=1}^{N} \left(c_0^i - c_t^i\right) \right] = K_{sat} [Q_{\max} - Q_t] \tag{9.4}$$

Measurable Variables

Substitution of (9.4) into (9.3) leads to

$$Q_t = Q_{\max} + K_{sat}^{-1} v^{gas} \frac{d}{dt} c_t^{gas} - K_{sat}^{-1} w^{gas} \left[c_0^{gas} - c_t^{gas} \right] \tag{9.5}$$

Applying the "Euler-back" approximation

$$\frac{d}{dt} c_t^{gas} \simeq h^{-1} \left(c_t^{gas} - c_{t-h}^{gas} \right), h > 0 \tag{9.6}$$

the process Q_t can be estimated as

$$\begin{aligned} Q_t &= \hat{Q}_t + v^{liq} \xi_t \\ \hat{Q}_t &:= Q_{\max} + K_{sat}^{-1} v^{gas} h^{-1} \left(c_t^{gas} - c_{t-h}^{gas} \right) - K_{sat}^{-1} w^{gas} \left[c_0^{gas} - c_t^{gas} \right] \end{aligned} \tag{9.7}$$

where ξ_t is the unmeasurable process related to the approximation (9.6) and given by

$$\xi_t := \left(Q_t - \hat{Q}_t \right) / v^{liq} = K_{sat}^{-1} v^{gas} \left[\frac{d}{dt} c_t^{gas} - h^{-1} \left(c_t^{gas} - c_{t-h}^{gas} \right) \right] / v^{liq} \tag{9.8}$$

The integration of (9.4) directly leads to the following expression:

$$S_t := \sum_{i=1}^{N} c_t^i = \sum_{i=1}^{N} c_0^i + \frac{1}{v^{liq}} \left(Q_t - Q_0 \right) - \frac{1}{v^{liq}} \int_{s=0}^{t} K_{sat} [Q_{\max} - Q_s] ds \tag{9.9}$$

that, in view of (9.7), can be written as follows

$$S_t = y_t + \xi_t \tag{9.10}$$

where the measurable variable y_t is given by

$$\begin{aligned} y_t &:= \sum_{i=1}^{N} c_0^i + \left(\hat{Q}_t - Q_0 \right) / v^{liq} - \left[c_0^{gas} - c_t^{gas} \right] v^{gas} / v^{liq} \\ &\quad - w^{gas} / v^{liq} \int_{s=0}^{t} \left[c_0^{gas} - c_s^{gas} \right] ds \end{aligned} \tag{9.11}$$

So, the measurable processes, related to the considered ozonation reactor and constructed based on the measurements of c_t^{gas}, are $\hat{Q}_t$ and y_t. They satisfy

$$\begin{aligned} Q_t &= \hat{Q}_t + v^{liq} \xi_t \\ S_t &= y_t + \xi_t \end{aligned} \tag{9.12}$$

9.3.2 Organic Compounds Reactions with Ozone

The differential equations describing the bimolecular chemical reactions for each organic compound are as follows [22]:

$$\frac{d}{dt}c_t^i = -k_i c_t^i \left(\frac{Q_t}{v^{liq}}\right)^n \quad (i = 1, ..., N) \tag{9.13}$$

where n is the stoichiometric parameter. Below the only case $n = 1$ will be considered.

9.3.3 Problem Setting

The problem which we are dealing with can be formulated as follows: *based on the available data* $\{c_t^{gas}\}$ *(and, hence, on (9.12)) construct the estimates* $\widehat{c}_t^i$ *of the state vector* c_t^i *as well as the estimates* $\hat{k}_{i,t}$ *of unknown parameters* k_i $(1, ..., N)$ and derive their accuracy bounds.

Since, as already mentioned, reliable sensors for concentrations measurement are not available or are very expensive, an efficient estimation procedure (based only on $\{c_t^{gas}\}$) can contribute significantly to the improvement of the reactor monitoring and control. The model of this process given by

$$\begin{cases} \frac{d}{dt}c_t^i = -k_i c_t^i \left(\hat{Q}_t/v^{liq} + \xi_t\right) \quad (i = 1, ..., N) \\ y_t = \sum_{i=1}^{N} c_t^i - \xi_t \end{cases} \tag{9.14}$$

represents the basis for on-line estimation of the current compound concentrations c_t^i and the unknown rate constants k_i. Here $\hat{Q}_t$ and ξ_t are the measurable input signal and an unmodeled bounded unmeasured dynamics.

9.4 Observability Condition

Consider now the same problem assuming that c_t^{gas} as well as $\frac{d}{dt}c_t^{gas}$ are available, that is, put $\xi_t = 0$. It implies that Q_t is available too and the considered process can be abstracted as follows

$$\begin{cases} \frac{d}{dt}c_t^i = f_t^i\left(c_t^i\right) := -k_i c_t^i Q_t / v^{liq} \quad (i = 1, ..., N) \\ y_t = \sum_{i=1}^{N} c_t^i \end{cases} \tag{9.15}$$

where c_t^i and y_t are the states and output (now measurable) of the corresponding dynamic system whose model is given by (9.15).

Consider (for the simplicity) the partial case $N = 3$ which covers a lot of practical situations. The calculation of the Lee's derivatives $\dot{y}_t$ and $\ddot{y}_t$ along the trajectories of this system implies the relation

$$\begin{pmatrix} y_t \\ \dot{y}_t \\ \ddot{y}_t \end{pmatrix} = \mathcal{O}_t \begin{pmatrix} c_t^1 \\ c_t^2 \\ c_t^3 \end{pmatrix}$$

where $\mathcal{O}_t$ is the observability matrix given by

$$\mathcal{O}_t = \begin{pmatrix} 1 & 1 & 1 \\ -\tilde{k}_1 Q_t & -\tilde{k}_2 Q_t & -\tilde{k}_3 Q_t \\ \tilde{k}_1\left(\tilde{k}_1 Q_t^2 - \dot{Q}_t\right) & \tilde{k}_2\left(\tilde{k}_2 Q_t^2 - \dot{Q}_t\right) & \tilde{k}_1\left(\tilde{k}_2 Q_t^2 - \dot{Q}_t\right) \end{pmatrix} \tag{9.16}$$
$$\tilde{k}_i := k_i / v^{liq} \quad (i = 1, ..., N = 3)$$

The states c_t^i of the system (9.15) are globally observable if and only if

$$\begin{aligned} \det \mathcal{O}_t = Q_t \Big[& \tilde{k}_1\left(\tilde{k}_1 Q_t^2 - \dot{Q}_t\right)\left(\tilde{k}_2 - \tilde{k}_3\right) + \tilde{k}_2\left(\tilde{k}_2 Q_t^2 - \dot{Q}_t\right)\left(\tilde{k}_3 - \tilde{k}_1\right) \\ & + \tilde{k}_1\left(\tilde{k}_2 Q_t^2 - \dot{Q}_t\right)\left(\tilde{k}_1 - \tilde{k}_2\right)\Big] \\ = Q_t^3 & \left[\tilde{k}_1^2\left(\tilde{k}_2 - \tilde{k}_3\right) + \tilde{k}_2^2\left(\tilde{k}_3 - \tilde{k}_1\right) + \tilde{k}_3^2\left(\tilde{k}_1 - \tilde{k}_2\right)\right] \neq 0 \end{aligned}$$

That is, the process y_t contains the sufficient information to reinstall the states c_t^i if Q_t is not equal to zero and all reaction rates are different:

$$(Q_t \neq 0) \wedge (k_2 \neq k_3) \wedge (k_1 \neq k_2) \wedge (k_1 \neq k_3) \tag{9.17}$$

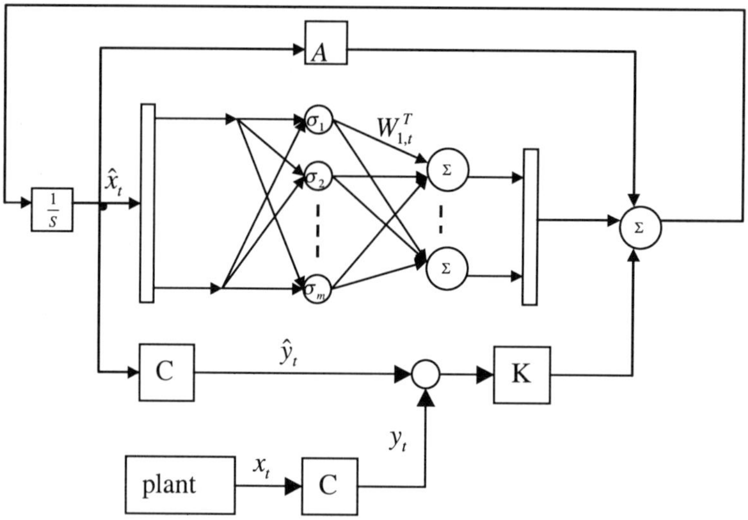

FIGURE 9.3. General structure of Dynamic Neuro Observer without hidden layers.

9.5 Neuro Observer

9.5.1 Neuro Observer Structure

According to [16][12][13], consider the dynamic neuro observer given by

$$\begin{cases} \frac{d}{dt}\widehat{x}_t = A\widehat{x}_t + W_t\sigma(\widehat{x}_t) + K\left[y_t - \widehat{y}_t\right] \\ \widehat{y}_t = C^T\widehat{x}_t, \qquad C^T = (1, ..., 1) \in \mathcal{R}^N \end{cases} \tag{9.18}$$

where $\widehat{x}_t \in \mathcal{R}^N$ is the state of the observer interpreted as the current estimate of the state vector $c_t = \left(c_t^1, ..., c_t^N\right)^{\intercal}$, $A \in \mathcal{R}^{n \times n}$ is a Hurwitz matrix to be selected, $\sigma : \mathcal{R}^N \to \mathcal{R}^k$ a smooth vector field usually represented by the sigmoid of the form

$$\sigma_i(x) = \frac{a_i}{1 + e^{-b_i^{\intercal} x}} - c_i \tag{9.19}$$

$W_t \in \mathcal{R}^{N \times k}$ is the weight matrix to be adjusted by a learning procedure, y_t is given by (9.11) and K is the observer gain matrix to be selected. The corresponding structure of this DNN is shown at FIGURE 10.3.

9.5.2 Basic Assumptions

A9.1: The sigmoid vector functions $\sigma(\cdot)$, commonly used in neural networks as the activation function, satisfy the Lipshitz condition $\left(\forall x^{'}, x^{''} \in \mathcal{R}^n\right)$

$$\widetilde{\sigma} := \sigma\left(x^{'}\right) - \sigma\left(x^{''}\right)$$
$$\widetilde{\sigma}^T \Lambda_{\sigma} \widetilde{\sigma} \leq \left(x^{'} - x^{''}\right)^{\mathsf{T}} D_{\sigma} \left(x^{'} - x^{''}\right)$$

where $\Lambda_{\sigma} = \Lambda_{\sigma}^{\mathsf{T}} > 0$, $D_{\sigma} = D_{\sigma}^{\mathsf{T}} > 0$ are known normalizing matrices.

A9.2: There exist strictly positive defined matrices $\bar{Q}$, Λ_1, Λ_{φ} and a positive $\delta > 0$ such that the matrix Riccati equation

$$Ric := P\mathcal{A} + \mathcal{A}^{\mathsf{T}} P + P\mathcal{R}P + \mathcal{Q} = 0 \tag{9.20}$$

with

$$\mathcal{A} := A - KC^T \text{ - a stable matrix}$$
$$\mathcal{R} := \Lambda_1^{-1} + \Lambda_{\varphi}^{-1} + W^* \Lambda_{\sigma}^{-1} W^{*\mathsf{T}}$$
$$W^* = W_0 \text{ - an initial weitght matrix}$$
$$\mathcal{Q} := D_{\sigma} + \bar{Q} + \delta I$$

has a positive definite solution P.

9.5.3 Learning Law

Let the weight matrix W_t be adjusted as follows:

$$\begin{aligned} \dot{W}_t = DP\left(N_{\delta}^{-1}\right)^{\mathsf{T}} [2\left(C^{+}\right)^{\mathsf{T}} e_t - \left[\left(C^{+}\right)^{\mathsf{T}} \Lambda_{\xi}^{-1} C^{+} + \delta I\right]^{\mathsf{T}} \\ N_{\delta}^{-1} P\left(W_t - W^*\right) \sigma\left(\widehat{x}_t\right)] \sigma^{\mathsf{T}}\left(\widehat{x}_t\right) \end{aligned} \tag{9.21}$$

where e_t is the observable (measurable) output error given by

$$e_t := y_t - C^T \widehat{x}_t$$
$$N_{\delta} := CC^{+} + \delta I, \quad \delta > 0$$
$$C^{+} = C^{\mathsf{T}} / \left\|C\right\|^2 = C^{\mathsf{T}} / N$$

9.5.4 Upper Bound for Estimation Error

Theorem 9.1 *If, under assumptions* ***A9.1-A9.2****, the updating law is given by (9.21), then the observation error satisfies the following performance:*

$$\begin{array}{c}\overline{\lim_{T\to\infty}}\frac{1}{T}\int\limits_0^T\Delta_t^T\bar{Q}\Delta_t dt\leq\bar{\eta}:=\overline{\lim_{T\to\infty}}\frac{1}{T}\int\limits_0^T\eta_t dt\\ \eta_t:=\xi_t^{\intercal}\left(K^{\intercal}\Lambda_1^{-1}K+\Lambda_\xi\right)\xi_t+\varphi_t^{\intercal}\Lambda_\varphi\varphi_t\\ \varphi_t:=\left[W^*\sigma\left(c_t\right)+Ac_t-f_t\left(c_t\right)\right]\end{array}\tag{9.22}$$

Below we will especially repeat the proof of the theorem to remind all steps of the suggested approach.

Proof. From (9.12) and (9.18), it follows

$$\begin{array}{c}\dot{\Delta}_t=A\widehat{x}_t+W_t\sigma\left(\widehat{x}_t\right)+K\left[y_t-C^T\widehat{x}_t\right]-f_t\left(c_t\right)\\ =\left(A-KC^T\right)\Delta_t-K\xi_t+\tilde{W}_t\sigma\left(\widehat{x}_t\right)\\ +W^*\left[\sigma\left(\widehat{x}_t\right)-\sigma\left(c_t\right)\right]+\left[W^*\sigma\left(c_t\right)+Ac_t-f_t\left(c_t\right)\right]\end{array}\tag{9.23}$$

where $\tilde{W}_t:=W_t-W^*$. Consider the Lyapunov function given by

$$V_t:=V\left(\Delta_t,\tilde{W}_t\right):=\Delta_t^{\intercal}P\Delta_t+\frac{1}{2}tr\left\{\tilde{W}_t^{\intercal}D^{-1}\tilde{W}_t\right\},\ D=D^{\intercal}>0\tag{9.24}$$

whose derivative, calculated over the trajectories of (9.23), satisfies

$$\dot{V}_t=\dot{V}\left(\Delta_t,\tilde{W}_t\right)=2\Delta_t^{\intercal}P\dot{\Delta}_t+tr\left\{\left(\frac{d}{dt}\tilde{W}_t^{\intercal}\right)D^{-1}\tilde{W}_t\right\}\tag{9.25}$$

The substitution of (9.23) into (9.25) implies:

1)

$$2\Delta_t^{\intercal}P\left(A-KC^T\right)\Delta_t=\Delta_t^{\intercal}\left[P\left(A-KC^T\right)+\left(A-KC^T\right)^{\intercal}P\right]\Delta_t\tag{9.26}$$

2)

$$-2\Delta_t^{\intercal}PK\xi_t\leq\Delta_t^{\intercal}P\Lambda_1^{-1}P\Delta_t+\xi_t^{\intercal}K^{\intercal}\Lambda_1^{-1}K\xi_t\tag{9.27}$$

(here the inequality

$$X^{\intercal}Y+Y^{\intercal}X\leq X^{\intercal}\Lambda^{-1}X+Y^{\intercal}\Lambda Y\tag{9.28}$$

which is valid for any $X, Y \in \mathcal{R}^{n\times m}$ and any $\Lambda = \Lambda^{\intercal} \in \mathcal{R}^{n\times n}$, is applied).

3) by (9.28) it follows

$$\begin{gathered}2\Delta_t^T P\tilde{W}_t\sigma(\widehat{x}_t) = 2\left(-e_t^{\intercal}C^{+}N_{\delta}^{-1} - \xi_t^{\intercal}C^{+}N_{\delta}^{-1} + \delta\Delta_t^{\intercal}N_{\delta}^{-1}\right)P\tilde{W}_t\sigma(\widehat{x}_t) \\ = -tr\left\{2\sigma(\widehat{x}_t)e_t^{\intercal}C^{+}N_{\delta}^{-1}P\tilde{W}_t\right\} - 2\xi_t^{\intercal}C^{+}N_{\delta}^{-1}P\tilde{W}_t\sigma(\widehat{x}_t) + 2\delta\Delta_t^{\intercal}N_{\delta}^{-1}P\tilde{W}_t\sigma(\widehat{x}_t) \\ \leq -tr\left\{2\sigma(\widehat{x}_t)e_t^{\intercal}C^{+}N_{\delta}^{-1}P\tilde{W}_t\right\} + \sigma^{\intercal}(\widehat{x}_t)\tilde{W}_t^{\intercal}P\left(N_{\delta}^{-1}\right)^{\intercal}(C^{+})^{\intercal}\Lambda_{\xi}^{-1}C^{+}N_{\delta}^{-1}P\tilde{W}_t\sigma(\widehat{x}_t) \\ +\xi_t^{\intercal}\Lambda_{\xi}\xi_t + \delta\Delta_t^{\intercal}\Delta_t + \delta\sigma^{\intercal}(\widehat{x}_t)\tilde{W}_t^{\intercal}P\left(N_{\delta}^{-1}\right)^{\intercal}N_{\delta}^{-1}P\tilde{W}_t\sigma(\widehat{x}_t) \\ = tr\left\{\sigma(\widehat{x}_t)\left[-2e_t^{\intercal}C^{+} + \sigma^{\intercal}(\widehat{x}_t)\tilde{W}_t^{\intercal}P\left(N_{\delta}^{-1}\right)^{\intercal}\left[(C^{+})^{\intercal}\Lambda_{\xi}^{-1}C^{+} + \delta I\right]\right]N_{\delta}^{-1}P\tilde{W}_t\right\} \\ +\xi_t^{\intercal}\Lambda_{\xi}\xi_t + \delta\Delta_t^{\intercal}\Delta_t\end{gathered}$$

since

$$\begin{gathered}e_t := y_t - C^T\widehat{x}_t = -C^{\intercal}\Delta_t - \xi_t \\ \Delta_t^{\intercal}C = -e_t^{\intercal} - \xi_t^{\intercal}\end{gathered}$$

and

$$\begin{gathered}\Delta_t^{\intercal} = \Delta_t^{\intercal}N_{\delta}N_{\delta}^{-1} = \Delta_t^{\intercal}\left(CC^{+} + \delta I\right)N_{\delta}^{-1} \\ = -e_t^{\intercal}C^{+}N_{\delta}^{-1} - \xi_t^{\intercal}C^{+}N^{-1} + \delta\Delta_t^{\intercal}N_{\delta}^{-1} \\ N_{\delta} := CC^{+} + \delta I,\ \delta > 0\end{gathered}$$

(C^{+} represents the pseudo-inverse of C).

4) by **A9.1**, we derive

$$\begin{gathered}2\Delta_t^{\intercal}PW^{*}\left[\sigma(\widehat{x}_t) - \sigma(c_t)\right] \leq \Delta_t^{\intercal}PW^{*}\Lambda_{\sigma}^{-1}W^{*\intercal}P\Delta_t \\ + \left[\sigma(\widehat{x}_t) - \sigma(c_t)\right]^{\intercal}\Lambda_{\sigma}\left[\sigma(\widehat{x}_t) - \sigma(c_t)\right] \\ \leq \Delta_t^{\intercal}\left(PW^{*}\Lambda_{\sigma}^{-1}W^{*\intercal}P + D_{\sigma}\right)\Delta_t\end{gathered}$$

5) for the term $\varphi_t := \left[W^{*}\sigma(c_t) + Ac_t - f_t(c_t)\right]$, tending to zero in view of (9.13), it follows:

$$2\Delta_t^{\intercal}P\varphi_t \leq \Delta_t^{\intercal}P\Lambda_{\varphi}^{-1}P\Delta_t + \varphi_t^{\intercal}\Lambda_{\varphi}\varphi_t$$

Adding and substituting the term $\Delta_t^{\intercal}\bar{Q}\Delta_t$ ($\bar{Q} = \bar{Q}^{\intercal} > 0$) from the right-hand side of (9.25), we obtain:

$$\dot{V}_t \leq \Delta_t^{\intercal}Ric\Delta_t + tr\left[L_t + \tfrac{d}{dt}\left(\widetilde{W}_t\right)^{\intercal}D^{-1}\right] - \Delta_t^{\intercal}\bar{Q}\Delta_t + \eta_t$$

where

$$L_t := \sigma(\widehat{x}_t)\left[-2e_t^{\intercal}C^{+} + \sigma^{\intercal}(\widehat{x}_t)\tilde{W}_t^{\intercal}P\left(N_{\delta}^{-1}\right)^{\intercal}\left[(C^{+})^{\intercal}\Lambda_{\xi}^{-1}C^{+} + \delta I\right]\right]N_{\delta}^{-1}P$$
$$\eta_t := \xi_t^{\intercal}\left(K^{\intercal}\Lambda_1^{-1}K + \Lambda_{\xi}\right)\xi_t + \varphi_t^{\intercal}\Lambda_{\varphi}\varphi_t$$

By **A9.2** and by the updating law (9.21) it follows that $Ric = 0$ and $DL_t^{\intercal} = -\frac{d}{dt}\widetilde{W}_t$, that implies

$$\dot{V}_t \leq -\Delta_t^{\intercal}\bar{Q}\Delta_t + \eta_t \tag{9.29}$$

Integrating (9.29) over the interval $[0, T]$, and dividing both side by T, we finally obtain

$$T^{-1}(V_T - V_0) \leq -T^{-1}\int_{s=0}^{T}\Delta_s^{\intercal}\bar{Q}\Delta_s ds + T^{-1}\int_{s=0}^{T}\eta_s ds$$

that leads to the following estimate:

$$\begin{aligned} T^{-1}\int_{s=0}^{T}\Delta_s^{\intercal}\bar{Q}\Delta_s ds &\leq T^{-1}\int_{s=0}^{T}\eta_s ds - T^{-1}(V_T - V_0) \\ &\leq T^{-1}\int_{s=0}^{T}\eta_s ds + T^{-1}V_0 \end{aligned}$$

Theorem is proven. ■

A remark is in order at this point.

Remark 9.1 *The upper bound $\overline{\eta}$ for the estimation error can be done less than any $\varepsilon > 0$ since $\varphi_t \to 0$ by (9.13) and ξ_t can be done small enough by the corresponding selection of the time-interval h in the approximation (9.6).*

The main concern of the next section is the estimation of the reaction rate constants.

9.6 Estimation of the Reaction Rate Constants

From the previous sections, it follows that the neuro-observer (9.18) can be used to approximate the chemical kinetics system (9.15). Observe that $\widehat{x}_t = \widehat{c}_t$ as well as

$d\widehat{x}_t/dt$ are available. In these conditions, we are ready to define the LS-estimates $\hat{k}_i(t)$ as the following optimization problem:

$$\left(\hat{k}_{1,t},...,\hat{k}_{N,t}\right) := \min_{k_l} \int_{s=0}^{t} \sum_{i=1}^{N} \left(\frac{d\widehat{x}_s^i}{ds} + k_i \widehat{x}_s^i \hat{Q}_s v_{liq}^{-1}\right)^2 ds \tag{9.30}$$

whose solution satisfies the following differential equation:

$$\begin{gathered} \frac{d}{dt}\tilde{K}_t = \left(\frac{d}{dt}\widehat{x}_t - \widehat{K}_t\widehat{x}\hat{Q}_t/v^{liq}\right)\widehat{x}_t^T\Gamma_t\hat{Q}_t/v^{liq} \\ \dot{\Gamma}_t = -\Gamma_t\widehat{x}_t\widehat{x}_t^T\Gamma_t\left(\hat{Q}_t/v^{liq}\right)^2, \quad \Gamma_0 = \eta^{-1}I \\ \tilde{K}_t := diag\left(\tilde{k}_{1,t},...,\tilde{k}_{N,t}\right) \\ \hat{K}_t = \left\| \begin{matrix} \hat{k}_{1,t} = \left[\tilde{k}_{1,t}\right]_+ & 0 & \cdot & 0 \\ 0 & \cdot & 0 & 0 \\ 0 & 0 & \cdot & 0 \\ 0 & \cdot & 0 & \hat{k}_{N,t}\left[\tilde{k}_{N,t}\right]_+ \end{matrix} \right\| \end{gathered} \tag{9.31}$$

where η is a small enough positive constant and the function $[\cdot]_+$ is defined as follows

$$[z]_+ := \begin{cases} z & \text{if } z \geq 0 \\ 0 & \text{if } z < 0 \end{cases} \tag{9.32}$$

From an initial condition $\widehat{K}_0$, we calculate the diagonal elements of $\widehat{K}_t$ on-line using $\hat{Q}_t$ and $\widehat{x}_t$ generated by (9.7) and (9.18), correspondingly.

Remark 9.2 *This algorithm generates the continuous-time Least Square estimates with a special projection procedure (9.32).*

The next section presents the simulation results.

9.7 Simulation Results

The simulation experiments described in this section are intended to illustrate the main results presented previously. The estimation algorithms described and analyzed in this paper are easy to code since they have few adjustable (design) parameters: the matrices A, $\Lambda_1, \Lambda_\varphi, W^*, \Lambda_\sigma, K, D$ and the scalars δ and η.

9.7.1 *Experiment 1 (standard reaction rates)*

This experiment has been carried out using the following values of the parameters associated with the system (9.3), (9.4) and (9.13):

$c_0^1 = 10^{-5}$	$c_0^2 = 10^{-5}$	$Q_0 = 10^{-8}$	$c_0^{gas} = 10^{-6}$
$k_1 = 10^5$	$k_2 = 10^4$	$K_{sat} = 0.2$	$Q_{\max} = 1.68 \cdot 10^{-8}$
$v^{liq} = 8 \times 10^{-3}$	$v^{gas} = 6 \times 10^{-3}$	$w^{gas} = 1.94 \cdot 10^{-3}$	$h = 1$

- First, we check the neuro-observer ability to estimate the states of the ozonization reactions without a priory knowledge of the rate constants k_1 and k_2. Apply DNN (9.18) with $N = 2$. The design parameters have been selected as follows:

$$A = diag[-1.5, -1.5], \quad C = [1,1], \quad \delta = 0.1, \quad N_\delta = \frac{1}{2}\begin{bmatrix} 1.1 & 1 \\ 1 & 1.1 \end{bmatrix},$$

$$\sigma_i(x) = 1/\left(1 + e^{-2x_i}\right) - \frac{1}{2}, \quad \widehat{x}_0 = \left[10^{-5}, 10^{-5}\right], \quad K = [3,3]^T,$$

$$W^* \Lambda_\sigma^{-1} W^{*\intercal} = \begin{bmatrix} 3 & 0.1 \\ 0.1 & 3 \end{bmatrix}, \quad \Lambda_1^{-1} = \Lambda_\varphi^{-1} = 2I_2,$$

$$\bar{Q} = I_2, \quad D_\sigma = \begin{bmatrix} 1 & 0.1 \\ 0.1 & 1 \end{bmatrix}.$$

Observe that

$$\mathcal{A} = A - KC = \begin{bmatrix} -4.5 & 3 \\ -3 & -4.5 \end{bmatrix}$$

is stable with the eigenvalues equal to (-1.5) and (-7.5). The corresponding solution of the Riccati equation (9.20) is

$$P = \begin{bmatrix} 0.3614 & 0.01 \\ 0.01 & 0.3614 \end{bmatrix}.$$

The weight matrix W_t is updated according to (9.21) with

$$W^* = W_0 = \begin{bmatrix} 7.25 \times 10^{-6} & -1.09 \times 10^{-6} \\ -1.75 \times 10^{-6} & 7.91 \times 10^{-6} \end{bmatrix}$$

and $D = 5I$. The observer results are shown in FIGURE 10.6. The continuous lines correspond to the estimate $\hat{x}_t^1$ and $\hat{x}_t^1$ of the states c_t^1 and c_t^2. One can see that the neuro-observer is able to generate good estimates of the non measurable states of the considered chemical reactions. In view of this ability, the neuro-observer can be helpfully used for the estimation of the rate constants.

- Second, we test the least squares procedure (9.31) to estimate the reaction rate constants using the observations $\hat{x}_t^1$ and $\hat{x}_t^1$ and $\hat{Q}_t$ to estimate the reaction rate constants k_1 and k_2. The design parameters are selected as follows:

$$\eta^{-1} = 10^{16}, \quad \widehat{K}_0 = \begin{bmatrix} 0.1 & 0 \\ 0 & 0.1 \end{bmatrix}.$$

The estimation results are depicted in FIGURE 10.7. The estimations $\hat{k}_{1,t}$ and $\hat{k}_{2,t}$ converge to their real values k_1 and k_2 after approximately 100 iterations.

9.7.2 *Experiment 2 (more quick reaction)*

Now consider a more quick reaction with the rates

$$k_1 = 10^6, \quad k_2 = 10^5$$

(the other parameters remain unchanged). We apply the same neuro-observer as well as the same LS- method (both with the same parameters as in the previous experiment). The simulation results are shown in FIGURE 10.8. One can see that the unknown constants k_1 and k_2 are well estimated.

9.8 Conclusions

A neuro-observer and a continuous least squares algorithm with a projection procedure have been presented to estimate, respectively, the states associated with the

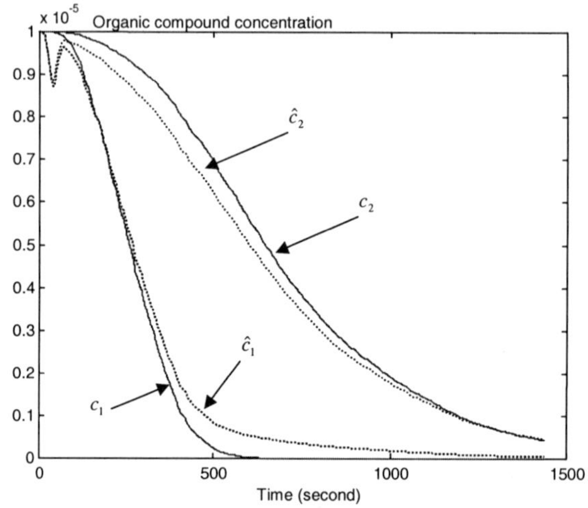

FIGURE 9.4. Current concentration estimates obtained by Dynamic Neuro Observer.

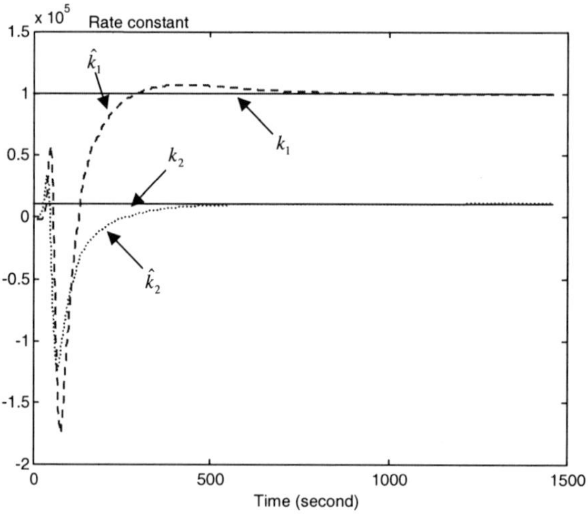

FIGURE 9.5. Estimates of k_1 and k_2.

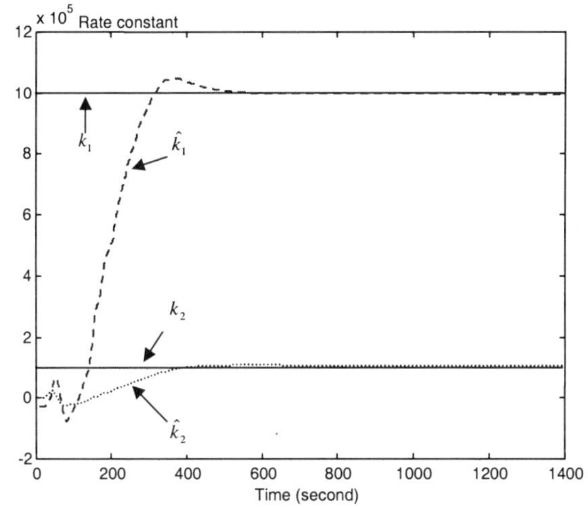

FIGURE 9.6. Estimates of k_1 and k_2.

current compound concentrations, and their reaction rate constants. The analysis of the estimation error for the suggested dynamic neuro-observer is shown to be carried out successfully based on Lyapunov-like analysis that provides also a regular procedure for the design of the learning procedure for the given DNN.

The algorithms presented in this paper seems to be useful

i) when the use of the current concentration sensors are not available or are very expensive;

ii) for the implementation of state-feedback controllers;

iii) for the simultaneous estimation of the reaction rate constants.

On the authors opinion, the designed estimation algorithm, analyzed in this paper, can find potential applications in many industries such as chemical, mineral, etc.

9.9 References

[1] Bailey, P. S., Reactivity of ozone with various organic functional groups important to water purification. *Proceedings of the 1-st Int. Symposium. on Ozone for Water and Wastewater Treatment*, Stamford, CT: Int. Ozone Assoc., Pan

American Group,101-117, 1975.

[2] Bin A. K. and M. Roustan, Mass Transfer in Ozone Reactors. *Proceedings of the International Specialized Symposium of IOA/ EA3G* "Fundamental and Engineering Concepts for Ozone Reactor Design" (March 1-3, Toulouse), France, 99-131, 2000.

[3] Bonvin, D., R. G. Rinker and D. A. Mellichamp, On controlling an autothermal fixed-bed reactor at an unstable state - 1. Chemical Engineering Science, **38**, 233-244, 1983.

[4] de Leon, J., E. N. Sanchez and A. Chataigner, Mechanical system tracking using neural networks and state estimation simultaneously. *Proc.33rd IEEE CDC*, 405-410, 1994.

[5] Gauthier, J. P., H. Hammouri and S. Othman, A simple observer for nonlinear systems: applications to bioreactors. *IEEE Trans. on Automatic Control*, **37**, 875-880, 1992.

[6] Hoigné J. and H. Barder, Rate Constants of Reactions of Ozone with Organic and Inorganic Compounds in Water -I., *Water Res.*, **17**, 185 - 192, 1981.

[7] Hunt, K. J., D. Sbarbaro, R. Zbinkowski and P. J. Gawthrop, Neural networks for control systems - A survey. *Automatica*, **28**, 1083-1112, 1992.

[8] Keerthipala, W. L., H. C. Miao and B. R. Duggal, An efficient observer model for field oriented induction motor control. *Proc. IEEE Trans. SMC*, 165-170, 1995.

[9] Kerwin L. Rakness, Glenn F. Hunter and Larry D. DeMers, Drinking Water Ozone Process Control and Optimization. *Proceedings of the International Specialized Symposium of IOA/ EA3G* "Fundamental and Engineering Concepts for Ozone Reactor Design" (March 1-3, Toulouse), France,231-254, 2000.

[10] Kim, Y. H., F. L. Lewis and C. T. Abdallah, Nonlinear observer design using dynamic recurrent neural networks. *Proc. 35th CDC,* 1996.

[11] Lera, G., A state-space-based recurrent neural network for dynamic system identification. *J. of Systems Engineering*, **6**, 186-193, 1996.

[12] Levin, A. U. and K. S. Narendra, Control of nonlinear dynamical systems using neural networks - Part II: Observability, identification, and control. *IEEE Trans. on Neural Networks*, **7**, 30-42, 1996.

[13] Najim, K., *Control of Liquid-liquid Extraction Columns.* Gordon and Breach, London, 1988.

[14] Najim, K., *Process Modeling and Control in Chemical Engineering.* Marcel Dekker, New York, 1989.

[15] Poznyak, A. S., Learning for dynamic neural networks. 10th Yale Workshop on Adaptive and Learning System, 38-47, 1998.

[16] Poznyak, A. S., W. Yu , E. N. Sanchez and J. P. Perez, Stability analysis of dynamic neural control. *Expert System with Applications*, **14**, 227-236, 1998.

[17] Poznyak A., Wen Yu, Edgar N. Sanchez, and Jose P. Perez, Nonlinear Adaptive Trajectory Tracking Using Dynamic Neural Networks. Identification via dynamic neural control. *IEEE Transactions on Neural Networks*, **10** (6), 1402-1411, 1999.

[18] Poznyak A. and Wen Yu, Robust Asymptotic Neuro Observer with Time Delay Term. *Int.J. of Robust and Nonlinear Control.* **10**, 535-559, 2000.

[19] Poznyak, T. I., D. M. Lisitsyn and F. S. D'yachkovskii, Selective detector for unsaturated compounds for gas chromatography. *J. Anal. Chem.*, **34**, 2028-2034, 1979.

[20] Poznyak T and J. L. Vivero Escoto, Simulation and Optimization of Phenol and Chlorophenols Elimination from Wastewater. *Proceedings of the IOA/PAG Conference*, Vancouver, Canada, (18-21 October), 615 - 628, 1998.

[21] Poznyak T. I. and J. L. Vivero Escoto, Modeling and Optimization of Ozone Mass Transfer in Semibatch Reactor. *Proceedings of the International Specialized Symposium of IOA/ EA3G* "Fundamental and Engineering Concepts for Ozone Reactor Design" (March 1-3, Toulouse), France, 133-136, 2000.

[22] Poznyak T. and A. Manzo Robledo, Kinetic Study of the Unsaturated Hydrocarbon Pollutants Elimination by Ozonation Method: Simulation and Optimization. Proceedings of the IOA/ PAG Conference, Vancouver, Canada, (18-21 October), 301 - 311, 1998.

[23] Rovithakis, A. and M. A. Christodoulou, Adaptive control of unknown plants using dynamical neural networks. *IEEE Trans. on Syst., Man and Cybern.*, **24**, 400-412, 1994.

[24] Sjöberg, J., Q. Zhang, L. Ljung, A. Benveniste and B. Delyon, Non-linear black-box modelling in system identification: a unified overview. *Automatica*, **31**, 1691-1724, 1995.

10

Neuro-Control for Distillation Column

The control problem of a multicomponent nonideal distillation column is proposed. It is based on Dynamic Neural Network Approach discussed before. The holdup, liquid and vapor flow rates are assumed to be time-varying (nonideal). The technique proposed in this chapter is based on two central notions: a dynamic neural identifier and a neuro-controller for output trajectory tracking. The first one guarantees boundness of the state estimation error with a small enough tolerance level. The tracked trajectory is generated by a nonlinear reference model, and we derive a control law to minimize the trajectory tracking error. The controller structure which we propose is composed of two parts: the neuro-identifier and the local optimal controller. Numerical simulations, concerning a 5 components distillation column with 15 trays, illustrate the effectiveness of the suggested approach [21].

10.1 Introduction

The *distillation column* is probably the most popular and important plant, intensively studied in the chemical engineering field during the last three decades [5, 8]. The general objective of distillation is the separation of substances under different vapor pressure at any given temperature. The word *distillation* refers to physical separation of a mixture into two or more fractions that have different boiling points. Distillation column is implemented to separate the feed flow and to purify the final and intermediate products in many chemical processes such as oil fractionating, water and air purification, etc. [9].

Since this process is strongly nonlinear, has large system uncertainty, large input-output interaction and there exists a lack of measurement of some key variables, it is very difficult to obtain a suitable model for the controller design. On the other hand the mathematical models for these systems are almost always too complex to be handled analytically. So many attempts were made to introduce simplified models in order to construct "model-based" controllers [8]. Most of these controllers use the *ideal binary distillation* column model (see, for example, [7, 19]). They approximate the *multicomponent* feeds (practically, most of real columns handle multicomponent feeds) as binary or pseudo-binary mixtures. As a result, this leads only to a crude approximation because of several restrictive assumptions which are valid only in some special cases, and in practical situation the column usually appears as *non-ideal (*holdup, liquid flow rate and vapor rate are time-varying). So the realistic mathematical model of *multicomponent nonideal* distillation column seems to be very important to design an advanced controller.

There exist few publications dealing with modelling of *multicomponent distillation* columns. In [8] several special algebraic equations related to physical and chemical properties of a process are used to calculate the vapor and liquid flow rates. In [6] the author uses an additional assumption that the holdups are independent of the vapor and liquid flow rates, so flow rates can be directly calculated from a system of algebraic equations. In [10] they assume that at each plate the molal holdups stay to be constant and that leads to an ideal simplified model. This is a very simplified method and can not cover many realistic processes. An analytical simulation of a multiple effect distillation plant is presented in [9] which is based on some alternative assumptions.

In this chapter we derive a dynamic mathematical model for multicomponent nonideal distillation column. We only assume that the liquid on the tray is perfectly mixed, the tray vapor holdups are negligible and the vapor and liquid are in thermal equilibrium. These assumptions are often used by everyone who studies these process and , virtually, seems not to be very restrictive. So, the model derived here is suitable for a large class of distillation columns with multicomponent mixtures of different physical nature.

The scheme and one-plate diagram of the multicomponent distillation column studied is shown in Figure 10.1.

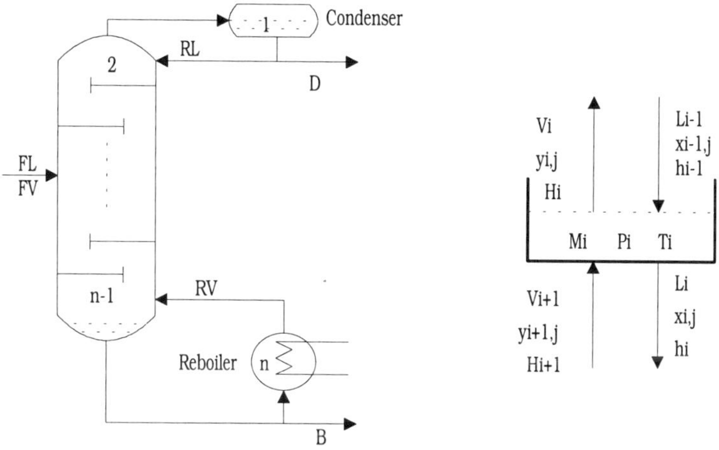

FIGURE 10.1. The scheme and one-plate diagram of a multicomponent distillation column.

The following nomenclature is used here and throughout this paper:

M_i	holdup of ith tray(mol),	F_V	feed vapor flow rate (mol/s),
L_i	liquid flow rate (mol/s),	F_L	feed liquid flow rate (mol/s),
V_i	vapor rate (mol/s),	R_L	reflux flow (mol/s),
h_i	molar liquid enthalpy (cal/mol),	R_V	boilup in Reboiler (mol/s),
H_i	molar vapor enthalpy (Btu/mol),	M_i	molar tray holdups (mol),
f	the tray of feed input,	T_i	temperture of i-th tray (°F)
B	bottom flow (mol/s),	P_i	pressure of i-th tray (psia)
D	distillate flow (mol/s),	$x_{i,j}$	liquid composition of the j-th component on i-th tray,
$y_{i,j}$	vapor composition.		

The objectives of the distillation columns *control* are twofold:

1. *Product Quality Control*: to maintain the product compositions at desired values despite the disturbances (the main part is feed flow). This needs a disturbance rejection controller.

2. *Optimal Operation Control*: to force the column to track an optimal set-point; that needs a trajectory tracking controller to provide the operations of a dis-

tillation column close to the economic optimum (energy saving and higher product yield).

These lead to two kinds of control: *optimal control* and *feedback control*. The former is based on an optimization technique to find optimal trajectories [4]. The later until now is mainly based on classical industrial controllers for distillation columns, such as PID controllers [1]. Even for ideal binary distillation columns the plant shows an "ill-conditioned" property (sensitive to changes in external flow but insensitive to changes in internal flows), so it is very difficult to design a model-based controller. Recently many researchers manage to use modern elegant theories to control distillation columns. In [18] they use the H_∞/structured singular value framework to construct a robust controller. In [7] the sensor selection and inferential controller design problems are solved by using structured singular value analysis theory. In [19] a Lyapunov-based controller and a high-gain observer were developed for a binary distillation column. [3] suggests a generic model control tool to design a controller for a crude tower (the properties are similar to multicomponent distillation columns). An application of nonlinear feedback theory to a binary distillation column is presented in [2]. The neural networks are used in [17] to model the ideal binary distillation column. The authors also use the input/output linearization method to make the nonlinear neural network model a linear system, based on this linear system the internal model control is implemented.

For multicomponent distillation columns, the mathematical model seems to be much more complicated than the binary one and, in fact, designing a model-based controller is practically impossible. To the best of our knowledge only a few advanced controllers were applied to multicomponent distillation columns (see, for example, [3] and its references).

Here we consider the model of a nonideal multicomponent distillation column which is unknown completely, we assume that only the basic structure characteristics are known (a number of multicomponent and plates, etc.). The main point of this study consists in applying an adaptive learning of dynamic neural networks to minimize the error between a real dynamic and a neural identifier, and a *local optimal controller* [11], which is based on the neural network identifier, is implemented.

The controller uses a solution of a corresponding algebraic Riccati equation.

The chapter is *organized* as follows: first, the mathematical model for a multicomponent distillation column is developed. Then, a local optimal controller based on a neural network identifier is presented. Next, the application of this technique to the distillation column is illustrated by showing the disturbance rejection and output trajectory tracking performances. Finally, the relevant conclusions are established.

10.2 Modeling of A Multicomponent Distillation Column

To compute the composition of the top and bottom products (at upper and down levels) which may be expected by use of a given distillation column operated at a given set of conditions, it is necessary to obtain a solution to the following fundamental equations:

1. Component-material balance.
2. Total-material balance.
3. Energy balance.
4. Vapor-liquid equilibrium relationships.

The following *mass* and *energy balances* are obtained by applying the basic conservation equations to each tray.

(1) A component-material balance may be written for each component *(Composition)*

$$\frac{d(M_i x_{i,j})}{dt} = L_{i-1}x_{i-1,j} + V_{i+1}y_{i+1,j} - L_i x_{i,j} - V_i y_{i,j}; \tag{10.1}$$

(2) A total material balance is also frequently useful and is just the sum of the component balances *(Holdup)*

$$\frac{dM_i}{dt} = L_{i-1} + V_{i+1} - L_i - V_i; \tag{10.2}$$

(3) The energy balance *(Enthalpy)*

$$\frac{d(M_i h_i)}{dt} = L_{i-1}h_{i-1} + V_{i+1}H_{i+1} - L_i h_i - V_i H_i \tag{10.3}$$

where $i = 1 \cdots n$, $j = 1 \cdots m$. Here n is the number of the tray, m is the number of total components. Following the standard assumptions [8], the kinetic and potential energy terms are neglected.

In *Feed plate*: ($i = f$) we have

$$\begin{aligned} \tfrac{d}{dt}(M_f x_{f,j}) &= L_{f-1}x_{f-1,j} + V_{f+1}y_{f+1,j} - L_f x_{f,j} - V_f y_{f,j} + xf_j \cdot F_L + yf_j \cdot F_V \\ \tfrac{d}{dt}(M_f h_f) &= L_{f-1}h_{f-1} + V_{f+1}H_{f+1} - L_f h_f - V_f H_f + hf \cdot F_L + Hf \cdot F_V \\ \tfrac{d}{dt}M_f &= L_{f-1} + V_{f+1} - L_f - V_f + F_L + F_V \end{aligned}$$

where Hf and hf are feed molar vapor enthalpy and feed molar liquid enthalpy. xf_j and yf_j are liquid and vapor compositions in the feed flows.

In *Condenser* $(i = 1)$ the following equation holds:

$$M_D \frac{dx_{1,j}}{dt} = V_2 y_{2,j} - R_L x_{1,j} - D x_{1,j}, \quad R_L = L_1.$$

In *Reboiler* $(i = n)$

$$M_B \frac{dx_{n,j}}{dt} = L_{n-1}x_{n-1,j} - R_V y_{n,j} - B x_{n,j}, \quad R_V = V_n$$

where

$$M_D = M_1, M_B = M_n$$

are top and bottom holdups. D and B are used for level control, R_V and R_L are used for quality control [15]. So, the total mass in Reboiler and Condenser are constants.

In order to obtain the solutions of holdups and flow rates, we need the following assumptions.

A9.1: The dynamic changes in internal energy on the trays are negligible:

$$h_i = T_i \sum_{j=1}^{N_c} x_{i,j} C_{3j} \qquad H_i = \sum_{j=1}^{N_c} y_{i,j}\left(C_{1j} + C_{2j}T_i\right), i = 1 \cdots n \tag{10.4}$$

where

$$\begin{aligned} C_{1j} &= B_j(C_{3j} - C_{2j}) + HVAP_jMW_j \\ C_{2j} &= HCAPV_jMW_j, \\ C_{3j} &= HCAPL_jMW_j, \\ B_j &= \left(\ln\left(VP_{2,j}\right) - A_j\right)/T_{2,j} \\ A_j &= T_{1,j}T_{2,j}\ln\left(\frac{VP_{1,j}}{VP_{2,j}}\right)/\left(T_{1,j} - T_{2,j}\right) \end{aligned}$$

$VP_{k,j}$ (psia) is vapor pressure at temperature $T_{k,j}$ (K), $k = 1, 2$. $HVAP_j$ (Btu/lb$_m$) is the heat of vaporization at normal boiling point and $HCVAP_j$ (Btu/lb$_m$) is the heat capacity of vapor. $HCAPL_j$ (Btu/lb$_m$) is the heat capacity of the liquid. MW_j (mol) is molecular weight of j-th component. T_i is the temperature of the i-th tray. A_i, B_i and C_i are constants related to the composition of the feed flows. So, the pressure $P_i (i = 2 \cdots n-1)$ is constant for each tray but varies linearly from top to bottom.

A9.2: The holdups are calculated from Francis weir formula [8]

$$M_i = CS_i \frac{\sum_{j=1}^{N_c} x_{i,j}\rho_j}{\sum_{j=1}^{N_c} x_{i,j}MW_j} = CS_i g(\overline{x}_i)$$

where CS_i is a known constant related with the physical size of the i-th tray,

$$\overline{x}_i = \left[x_{i,1} \cdots x_{i,n}\right], i = 2 \cdots n-1$$

So,

$$\frac{dM_i}{dt} = CS_i \frac{dg(\overline{x}_i)^T}{d\overline{x}_i}\frac{d\overline{x}_i}{dt} = L_{i-1} + V_{i+1} - L_i - V_i + L_f + V_f \tag{10.5}$$

A9.3: Vapor phase behaviors obey Raoult's law [8]

$$P_i = \sum_{j=1}^{N_c} x_{i,j}\exp(B_j + \frac{A_j}{T_i}) \tag{10.6}$$

$$y_{i,j} = \frac{x_{i,j}}{P_i}\exp(B_j + \frac{A_j}{T_i}) \tag{10.7}$$

where P_i is the total pressure of the i-th tray,

$$p_{i,j} := \exp(B_j + \frac{A_j}{T_i})$$

is the partial pressure of the j-th component in the i-th tray ($i = 2 \cdots n-1$).

The vapor compositions $y_{i,j}$ can be calculated from the vapor-liquid equilibrium (10.7). Temperatures T_i can be calculated also from (10.6) and (10.7) by the Newton method, because P_i is known and $x_{i,j}$ can be found from (10.1). Since the temperature may not change largely in each simulation step, we may use the previous temperature as the initial temperature of this time, that is

$$T_0 := T_{k-1}$$

Consequently, to use the Newton method, we need only 3 recursion steps to obtain the convergence of T_k.

The relations (10.2) and (10.1) lead to

$$M_i \frac{dx_{i,j}}{dt} = L_{i-1}\left(x_{i-1,j} - x_{i,j}\right) + V_{i+1}\left(y_{i+1,j} - x_{i,j}\right) - V_i\left(y_{i,j} - x_{i,j}\right) \tag{10.8}$$

where $i = 2 \cdots n-1$.

From (10.2) and (10.3) we can get

$$M_i \frac{dh_i^T}{d\overline{x}_i} \frac{d\overline{x}_i}{dt} = L_{i-1}\left(h_{i-1} - h_i\right) + V_{i+1}\left(H_{i+1} - h_i\right) - V_i\left(H_i - h_i\right)$$

for $i = 2 \cdots n-1$. Using (10.4) and (10.8), we obtain the first type algebraic equations:

$$\begin{aligned} L_{i-1} &\left[T_i \sum_{j=1}^{N_c} C_{3j}\left(x_{i-1,j} - x_{i,j}\right) - \left(h_{i-1} - h_i\right)\right] + \\ V_{i+1} &\left[T_i \sum_{j=1}^{N_c} C_{3j}\left(y_{i+1,j} - x_{i,j}\right) - \left(H_{i+1} - h_i\right)\right] - \\ V_i &\left[T_i \sum_{j=1}^{N_c} C_{3j}\left(y_{i,j} - x_{i,j}\right) - \left(H_i - h_i\right)\right] = 0. \end{aligned} \tag{10.9}$$

Based on (10.5) and on (10.8), we derive the second type algebraic equations:

$$L_{i-1}\left[\frac{C_i}{M_i}\sum_{j=1}^{N_c}\frac{dg(\overline{x}_i)}{dx_j}\left(x_{i-1,j}-x_{i,j}\right)-1\right]+V_{i+1}\left[\frac{C_i}{M_i}\sum_{j=1}^{N_c}\frac{dg(\overline{x}_i)}{dx_j}\left(y_{i+1,j}-x_{i,j}\right)-1\right]$$
$$-V_i\left[\frac{C_i}{M_i}\sum_{j=1}^{N_c}\frac{dg(\overline{x}_i)}{dx_j}\left(y_{i,j}-x_{i,j}\right)-1\right]+L_i=0 \tag{10.10}$$

for $i = 2\cdots n-1$, and

$$\frac{dg(\overline{x}_i)}{dx_j}=\frac{\rho_j}{\sum\limits_{j=1}^{N_c}x_{i,j}MW_j}-MW_j\frac{\sum\limits_{j=1}^{N_c}x_{i,j}\rho_j}{\left(\sum\limits_{j=1}^{N_c}x_{i,j}MW_j\right)^2}$$

Let us denote

$$\begin{aligned}
a_i &:= T_i\sum_{j=1}^{N_c}C_{3j}\left(x_{i-1,j}-x_{i,j}\right)-\left(h_{i-1}-h_i\right)\\
b_i &:= T_i\sum_{j=1}^{N_c}C_{3j}\left(y_{i+1,j}-x_{i,j}\right)-\left(H_{i+1}-h_i\right)\\
c_i &:= T_i\sum_{j=1}^{N_c}C_{3j}\left(y_{i,j}-x_{i,j}\right)-\left(H_i-h_i\right)\\
d_i &:= \frac{C_i}{M_i}\sum_{j=1}^{N_c}\frac{dg(\overline{x}_i)}{dx_j}\left(x_{i-1,j}-x_{i,j}\right)-1\\
e_i &:= \frac{C_i}{M_i}\sum_{j=1}^{N_c}\frac{dg(\overline{x}_i)}{dx_j}\left(y_{i+1,j}-x_{i,j}\right)-1\\
f_i &:= \frac{C_i}{M_i}\sum_{j=1}^{N_c}\frac{dg(\overline{x}_i)}{dx_j}\left(y_{i,j}-x_{i,j}\right)-1
\end{aligned}$$

So, the equations (10.9) and (10.10) can be rewritten as

$$\begin{aligned}
a_iL_{i-1}+b_iV_{i+1}-c_iV_i&=0\\
d_iL_{i-1}+e_iV_{i+1}-f_iV_i+L_i&=0
\end{aligned} \tag{10.11}$$

for $i = 2\cdots n-1$. Each tray have two variables (V_i, L_i) which are described by two algebraic equations. So, using vector presentation for V_i and L_i, the system (10.11) can be represented as

$$A\Phi = B$$

where

$$A_{2(n-2)\times 2(n-2)} =$$

$$\left[\begin{array}{ccc}
\begin{matrix} -c_2 & 0 & b_2 & 0 & 0 & 0 \\ -f_2 & 1 & e_2 & 0 & 0 & 0 \end{matrix} & 0 & 0 \\
\begin{matrix} 0 & a_3 & -c_3 & 0 & b_3 & 0 \\ 0 & d_3 & -f_3 & 1 & e_3 & 0 \end{matrix} & 0 & 0 \\
0 & \begin{matrix} 0 & a_f & -c_f & 0 & b_f & 0 \\ 0 & d_f & -f_f & 1 & e_f & 0 \end{matrix} & 0 \\
0 & 0 & 0 \\
0 & 0 & \begin{matrix} 0 & a_{n-2} & -c_{n-2} & 0 & b_{n-2} & 0 \\ 0 & d_{n-2} & -f_{n-2} & 1 & e_{n-2} & 0 \end{matrix} \\
0 & 0 & \begin{matrix} 0 & 0 & 0 & a_{n-1} & -c_{n-1} & 0 \\ 0 & 0 & 0 & d_{n-1} & -f_{n-1} & 1 \end{matrix}
\end{array}\right]$$

where

$$\Phi = [V_2, L_2, V_3, L_3, \cdots V_{n-2}, L_{n-2}, V_{n-1}, L_{n-1}]$$
$$B = [-a_2 R_L, d_2 R_L, 0 \cdots 0, L_f(hf - h_i), Lf, 0, \cdots 0, -b_{n-1} R_V, -e_{n-1} R_V]$$

As a result, when

$$\det(A) \neq 0$$

we have

$$\Phi = A^{-1} B \tag{10.12}$$

By substitution (10.12) and (10.7) into (10.1), we finally obtain a complete system of differential equations describing the given distillation process.

10.3 A Local Optimal Controller for Distillation Column

The distillation column model derived in Section II is complicated enough. As a fact, we have to deal with $m \times n$ differential equations and $2 \times m \times n$ algebraic equations to model $n-$trays and $m-$components column.

In the case when we wish to control the liquid $x_{n,k}$ and $x_{1,l}$ compositions of the most important components we are interested in (k and l are assumed to be fixed), we can introduce the following vectors:

$$\mathbf{x}_t = [x_{1,1} \cdots x_{1,m}, x_{2,1} \cdots x_{2,m}, x_{n,1} \cdots x_{n,m}],$$
$$\mathbf{y}_t = [x_{n,k}, x_{1,l}], \quad \mathbf{u}_t = [R_L, R_V]$$

where $x_{n,k}$ and $x_{1,l}$ are measurable compositions in the top and bottom tray.

Based on these definitions, the multicomponent nonideal distillation column model, derived above, can be represented in a standard form as follows:

$$\begin{aligned} \dot{\mathbf{x}}_t &= f(\mathbf{x}_t, \mathbf{u}_t, t), \\ \mathbf{y}_t &= C^T \mathbf{x}_t \end{aligned} \tag{10.13}$$

where

$\mathbf{x}_t \in \Re^p$ $(p := m \times n)$ is the system state vector at time $t \in R^+ := \{t : t \geq 0\}$;

$\mathbf{y}_t \in \Re^r$ $(r = 2)$ is the output vector that is assumed to be measurable at each time t;

$\mathbf{u}_t \in \Re^q$ $(q = 2)$ is a given control action;

$f(\cdot) : \Re^{p+q+1} \rightarrow \Re^p$ is a nonlinear function describing the dynamics of this system.

The matrix

$$C^T = \begin{bmatrix} 0 & \ldots & 1 & 0 & 0 & \ldots & 0 & 0 \\ 0 & \ldots & 0 & 0 & 0 & 1 & \ldots & 0 \end{bmatrix} \in \Re^{p \times r}$$

is the known matrix connecting the state and control vectors with the regular part of output.

Even in the case of a complete a priori information about all parameters of this model, we have not an analytical expression for the function $f(\cdot)$ because **some parameters of the differential equation** (10.1) **are calculated based on a recursive procedure.**

So, to implement any control method we need to identify this process and construct a model, which then can be used in an applied control algorithm.

The model of this nonlinear system is selected as following dynamic neural networks (and [16], [12] ,[13] and Chapter 2):

$$\dot{\widehat{\mathbf{y}}}_t = A\widehat{\mathbf{y}}_t + W_{1,t}\sigma(\widehat{\mathbf{y}}_t) + W_{2,t}\phi(\widehat{\mathbf{y}}_t)\mathbf{u}_t \tag{10.14}$$

where

$A \in \Re^{r\times r}$ is a known Hurtwitz (stable) matrix,

$W_{1,t} \in \Re^{r\times r}$ is the weight matrix for nonlinear state feedback,

$W_{2,t} \in \Re^{r\times r}$ is the input weight matrix,

$\widehat{\mathbf{y}}_t \in \Re^{r}$ is the state of the neural network.

The matrix function $\phi(\cdot)$ is assumed to be $\Re^{r\times r}$ diagonal:

$$\phi\left(\widehat{\mathbf{y}}_t\right) = diag(\sigma_1(\widehat{y}_1)\cdots\sigma_r(\widehat{y}_r))$$

The vector functions $\sigma(\cdot)$ are assumed to be $r-$dimensional with the elements increasing monotonically. The typical presentation of the elements $\sigma_i(\cdot)$ and $\phi_i(.)$ are as sigmoid functions, i.e.

$$\sigma_i(y_i) = \frac{a_i}{1+e^{-b_i y_i}} - c_i \tag{10.15}$$

The functions $\sigma_i(\cdot)$ and $\phi_i(.)$ satisfy the assumptions:

A9.4: The functions $\sigma\left(\mathbf{y}\right)$ and $\phi\left(\mathbf{y}\right)$ are Lipschitzian with constant L_ϕ and L_σ.

A9.5: $\sigma(\mathbf{y})$ satisfies:

$$\sigma^T\left(\mathbf{y}\right) Z\sigma\left(\mathbf{y}\right) \le \mathbf{y}^T\Lambda_\sigma\mathbf{y} + \lambda_{\sigma o} \quad \forall \mathbf{x}\in\Re^p$$

where

$$Z = Z^T > 0, \Lambda_\sigma = \Lambda_\sigma^T > 0, \lambda_{\sigma o} > 0$$

are known normalizing matrices.

The identification error is defined by

$$\Delta_t := \mathbf{y}_t - \widehat{\mathbf{y}}_t \tag{10.16}$$

Adding and subtracting the term $A_0\Delta_t$, we obtain:

$$\dot{\Delta}_t = A_0\Delta_t + h\left(\Delta_t, \mathbf{x}_t, \mathbf{u}_t, t\right) \tag{10.17}$$

with

$$h\left(\Delta_t, \mathbf{x}_t, \mathbf{u}_t, t\right) := C^T f(\mathbf{x}_t, \mathbf{u}_t, t) - W_{1,t}\sigma\left(\widehat{\mathbf{y}}_t\right) - W_{2,t}\phi\left(\widehat{\mathbf{y}}_t\right)\mathbf{u}_t - \left(A_0 - A\right)\Delta_t - AC^T\mathbf{x}_t$$

A_0 is any Hurtwitz matrix which we can select. Because $\sigma(\widehat{\mathbf{y}}_t)$ and $\phi(\widehat{\mathbf{y}}_t)$ are bounded, it is not difficult to check (see [19]) that $h(\cdot)$ satisfies the following "cone" conditions:

A9.6: There exist positive defined matrices H_h and H_Δ such that:

$$h^T(\Delta_t, \mathbf{x}_t, \mathbf{u}_t, t) H_h h(\Delta_t, \mathbf{x}_t, \mathbf{u}_t, t) \leq \varepsilon_0(\mathbf{x}_t, \mathbf{u}_t, t) + \varepsilon_1(\mathbf{x}_t, \mathbf{u}_t, t) \Delta_t^T H_\Delta \Delta_t \quad (10.18)$$

for $\varepsilon_0(.,.,.)$ and $\varepsilon_1(.,.,.)$ positive bounded functions, with ε^0 and ε^1 the respective bounds, i.e.

$$\sup \varepsilon_i(\mathbf{x}_t, \mathbf{u}_t, t) = \varepsilon^i < \infty,\ i = 0, 1, \forall \mathbf{x} \in \Re^p,\ \forall \mathbf{u} \in \Re^r,\ \forall t \geq 0.$$

All matrices H_h and H_Δ normalize the respective vector to one with dimensionless components.

Indeed, A6 holds because of the following inequality:

$$\|h(\cdot)\| \leq \left\|C^T f(\mathbf{x}_t, \mathbf{u}_t, t) - W_{2,t}\phi(\widehat{\mathbf{y}}_t)\mathbf{u}_t + AC^T\mathbf{x}_t\right\| + \|W_{1,t}\sigma(\widehat{\mathbf{y}}_t)\| + \|(A + A_0)\Delta_t\|$$

To adjust the weights (W_1, W_2) of this neural dynamic network, we use the following algorithm:

$$\dot{W}_{1,t} = -B\Delta_t\sigma(\widehat{\mathbf{y}}_t)^T, \quad \dot{W}_{2,t} = -B\Delta_t u_t^T\phi(\widehat{\mathbf{y}}_t)^T$$

where

$$B = diag\{b_1, ..., b_r\}$$

is a positive diagonal matrix. So, as it was demonstrated above, W_1 and W_2 stay to be bounded:

$$\|W_{1,t}\| \leq \overline{W}_1, \|W_{2,t}\| \leq \overline{W}_2, \forall t \geq 0.$$

Using sigmoid functions means that $\sigma(\widehat{\mathbf{y}}_t)$ and $\phi(\widehat{\mathbf{y}}_t)$ are bounded. If $h(\cdot)$ deviates from a linear function (for each fixed $\mathbf{u}_t$ and t) no more than a uniformly constant, then we obtain (10.18) which defines the class of bounded vector functions and the functions uniformly increasing not quicker than a linear one.

A9.7: There exists a strictly positive definite matrix Q_0 such that, for the given matrix A_0 the following algebraic matrix Riccati equation

$$A_0^T P + PA_0 + PRP + Q = 0 \tag{10.19}$$

with

$$R := H_h^{-1},\ Q := \varepsilon^1 H_\Delta + Q_0,$$

has a positive solution $P = P^T > 0$.

Such a solution exists if the matrix A_0 is stable, the pair $\left(A_0, R^{\frac{1}{2}}\right)$ is controllable, the pair $\left(Q, R^{\frac{1}{2}}\right)$ is observable and special matrix inequality holds (see Appendix A). These conditions are easily fulfilled by selecting A_0 as a diagonal matrix.

As it shown in Chapter 2, for the system (10.13) and the given neural network (10.14), the following property holds:

$$\lim_{t\to\infty} \sup \|\Delta(t)\| \leq \overline{\Delta}(P) = \sqrt{\frac{\varepsilon^0}{\lambda_{\max}(P)\,\lambda_{\min}(R_p)}} \tag{10.20}$$

where

$$R_p = P^{-\frac{1}{2}} Q_0 P^{-\frac{1}{2}}$$

and $\lambda_{min}(.)$ is the minimum eigenvalue of the corresponding matrix.

We will design a local optimal controller based on the neural network identifier (10.14). The control goal is to force the system states to track an optimal trajectory $\mathbf{y}_t^* \in \Re^r$ which is assumed to be smooth enough. If the trajectory has points of discontinuity in some fixed moments, we can use any approximating trajectory which is smooth, for example, the step function can be approximated by a sigmoid function. So, this trajectory can be regarded as a solution of a nonlinear reference model:

$$\dot{\mathbf{y}}_t^* = \varphi(\mathbf{y}_t^*, t) \tag{10.21}$$

where $\mathbf{y}_t^*(\cdot) \in \Re^r$ is the state of the desired trajectory. The function $\varphi(\cdot) : \Re^{r+1} \to \Re^r$ is a nonlinear function describing the dynamics of this system.

In other words, we would like to force the distillation column to follow a given reference dynamic given by (10.21). Defining the following semi-norms:

$$
\begin{aligned}
\|\mathbf{y}_t\|_Q^2 &= \lim_{t\to\infty} \sup \frac{1}{\tau} \int_0^{\tau} \mathbf{y}_t^T Q \mathbf{y}_t dt \\
\|\mathbf{u}_t\|_R^2 &= \lim_{t\to\infty} \sup \frac{1}{\tau} \int_0^{\tau} \mathbf{u}_t^T R \mathbf{u}_t dt
\end{aligned}
\tag{10.22}
$$

where

$$Q_C = Q_C^T > 0, R_C = R_C^T > 0$$

are the given weighting matrices, the output trajectory tracking can be formulated as the following optimization problem:

$$J_{\min} = \min_{u_t} J, \quad J = \|\mathbf{y}_t - \mathbf{y}_t^*\|_Q^2 + \|\mathbf{u}_t\|_R^2 \tag{10.23}$$

We can estimate from above the functional J (10.23) as:

$$J \le (1+\eta)\|\mathbf{y}_t - \widehat{\mathbf{y}}_t\|_Q^2 + \left(1+\eta^{-1}\right)\|\widehat{\mathbf{y}}_t - \mathbf{y}_t^*\|_Q^2 + \|\mathbf{u}_t\|_R^2 \tag{10.24}$$

The minimum of the term

$$\|\mathbf{y}_t - \widehat{\mathbf{y}}_t\|_Q^2 = \|\Delta_t\|_Q^2$$

has already been solved before dealing with the identification (observability) analysis. If we define

$$R := \left(1+\eta^{-1}\right) R_0$$

we can rewrite (10.24) as follows:

$$J \le (1+\eta)\|\mathbf{y}_t - \widehat{\mathbf{y}}_t\|_Q^2 + \left(1+\eta^{-1}\right) J^+$$

where

$$J^+ := \|\widehat{\mathbf{y}}_t - \mathbf{y}_t^*\|_Q^2 + \|\mathbf{u}_t\|_{R_0}^2$$

So, the control as in (6.8) is

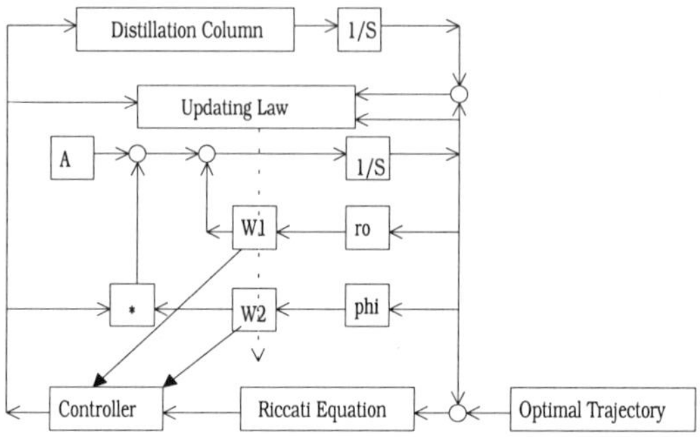

FIGURE 10.2. Identification and control scheme of a distillation column.

$$\mathbf{u}_t^* = -R_C^{-1}\phi\left(\widehat{\mathbf{x}}_t\right)\widehat{W}_{2,t}^T P_c\left(t\right)\Delta_t^* \tag{10.25}$$

will minimize J^+. The final structure of the neural network identifier and the tracking controller is shown in Figure 10.2. It is considered that the neural network weights are trained on line and the controller is based on the weights of that neural network, but not on the model of the distillation column directly.

10.4 Application to Multicomponent Nonideal Distillation Column

In this section we consider a multicomponent nonideal distillation column whose features are given in Table 1. Although we use the same chemical and physical properties of multicomponent distillation column as in [8], we do not calculate the flow rates by some special algebraic equations. So, the model presented here seems to be more general than in [8] and [6] etc.

n	15	number of trays
m	5	number of components
f	8	number of the feed tray
PD	19.7	pressure in top [psia]
PB	21.2	pressure in bottom [psia]
WHS, WLS, DS	$0.75, 48, 72$	weir height,weir length and column diameter [m]
MW_i	$30, 50, 90, 130, 300$	molecular weight [mol]
$HVAP_i$	$100, 90, 70, 80, 80$	heat of vaporization at normal boiling point [Btu/mol]
$HCAPV_i$	$0.2, 0.4, 0.3, 0.3, 0.3$	heat capacity of vapor [Btu/mol]
$HCAPL_i$	$0.6, 0.6, 0.5, 0.4, 0.4$	heat capacity of liquid [Btu/mol]
$VP_{1,i}$	$14.7, 14.7, 14.7, 14.7, 14.7$	vapor pressure at temperature T_1 [psia]
$VP_{2,i}$	$50, 500, 150, 150, 150$	vapor pressure at temperature T_2 [psia]
$T_{1,i}$	$470, 550, 610, 670, 820$	temperature T_1 [°F]
$T_{2,i}$	$500, 660, 660, 760, 880$	temperature T_2 [°F]

Table 1: The features of a multicomponent nonideal distillation column.

This multicomponent nonideal distillation column model includes

- 75 differential equations (correspond to (10.1)),
- 150 algebraic equations (correspond to (10.9) and (10.10)),
- 14 Newton recursion equations to calculate T_i.

It is impossible to design an advanced controller based on this model. According to our approach presented above, first we will use a neural network to identify this model. In (10.13) we select

$$n = 75, m = 2.$$

That means we only care about two important outputs: in Reboiler we select $x_{15,5}$ as an output, in the condenser we select $x_{1,1}$ as another output. The control inputs

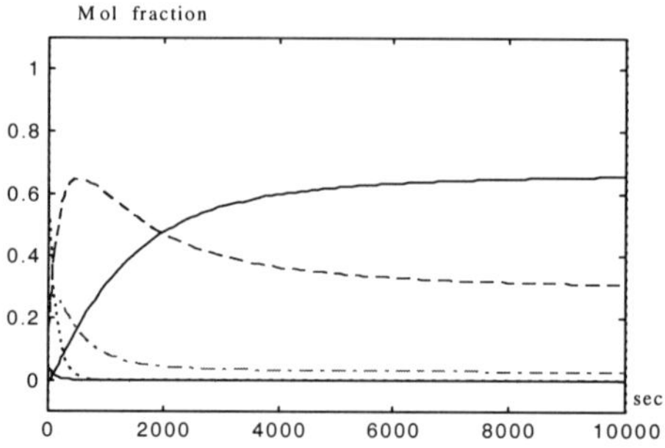

FIGURE 10.3. Compositions in the top tray.

are

$$u_1(t) = R_L, u_2(t) = R_V.$$

If the distillation column operates at the steady state,

$$R_L = 4.0(mol/s), \quad R_V = 3.5(mol/s), \quad L_f = V_f = 2.0(mol/s),$$

the restriction for the ideal case is

$$R_L + L_f \geq R_V \geq R_L - V_f$$

that means D and B are positive. The main dynamic characteristics of this distillation column can be seen from its open loop responses (the products in the end points $x_{15,j}$ and $x_{1,j}$), which are shown in Figure 10.3 and Figure 10.4.

First, we use a neural network to estimate the desired output $x_{15,5}$ and $x_{1,1}$. The sigmoid function is chosen as:

$$\sigma(x) = \frac{2}{1+e^{-2x}} - 0.5.$$

The structure of the neural network is as (10.14). We select

$$Q_C = Q_0 = H_{n0} = H_{nf} = H_{\Delta 0} = H_{\Delta f} = I, \varepsilon^{h1} = \varepsilon^{f1} = 3,$$

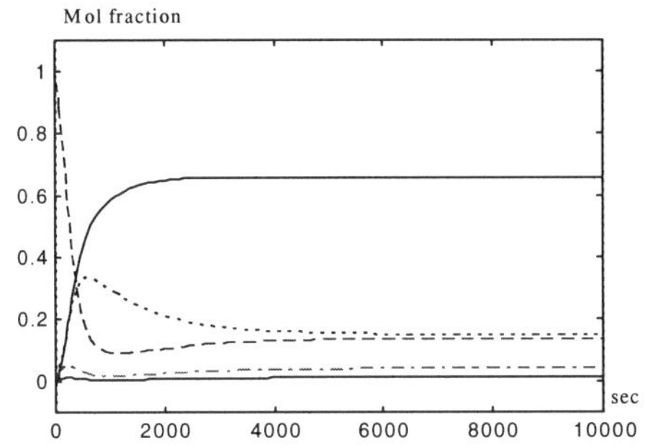

FIGURE 10.4. Compositions in the bottom tray.

and obtain the solution P of the corresponding Riccati Equation (10.19)

$$P = \begin{bmatrix} 0.84 & 0 \\ 0 & 0.84 \end{bmatrix}.$$

Choose

$$W_2 = I, \quad \overline{\eta} = 0.1$$

and

$$R_C = \begin{bmatrix} 0.05 & 0 \\ 0 & 0.05 \end{bmatrix}.$$

To adapt on-line the dynamic neural network weights we use the same learning algorithm as in Chapter 6 (6.45). The identification results are shown in Figure 10.5 and Figure 10.6. The solid lines correspond to distillation column responses $x_{15,5}$, $x_{1,1}(t)$, and the dashed line to neural network ones $\widehat{x}_1(t)$, $\widehat{x}_2(t)$. It can be seen that the neural network output time evolution follows the ones of the multicomponent distillation column.

At time $t = 2800s$, R_L is changed from 5.0(mol/s) to 3.5(mol/s).

At time $t = 5100s$, R_V is changed from 4.5(mol/s) to 2.5(mol/s).

At time $t = 7800s$, L_f and V_f are changed from 1.5(mol/s) to 2.0(mol/s).

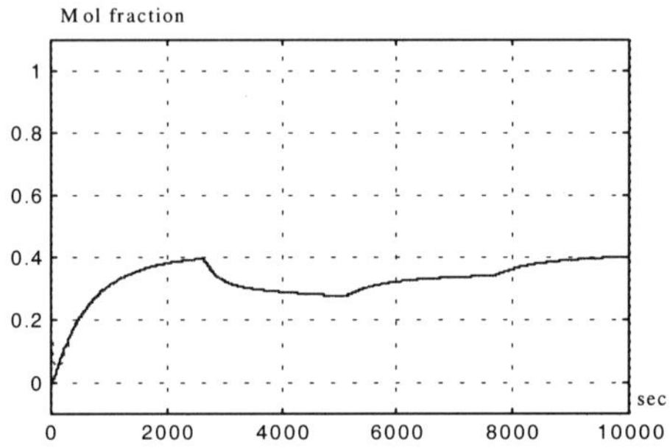

FIGURE 10.5. Identification results for $x_{1,1}$.

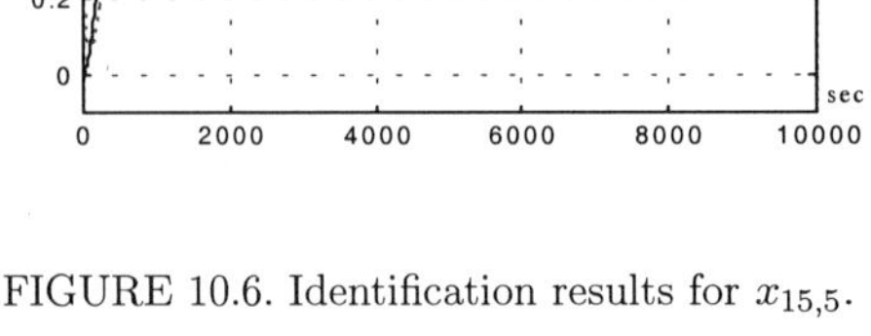

FIGURE 10.6. Identification results for $x_{15,5}$.

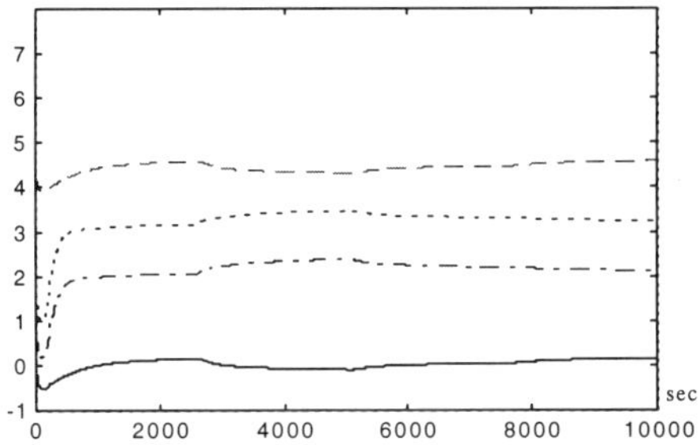

FIGURE 10.7. Time evolution of $W_{1,t}$.

The time evolution for $W_{1,t}$ in (10.14) is shown in Figure 10.7. One can see that the weights of the identifier (neural network) are changed when the operation conditions of the distillation column are changed, so our controller should be adaptive to be able to cope with this variations.

Based on the neural network identifier, we use the local optimal controller (10.25) (see Chapter 6). This controller has two objects: trajectory tracking and disturbance rejection. So, we generate the trajectory as

$$x_{1,1}^* = 0.3, x_{15,5}^* = 0.8.$$

At time $t = 8000s$, it is changed to

$$x_{1,1}^* = 0.3, x_{15,5}^* = 0.9.$$

At time $t = 20000s$, it is changed again to

$$x_{1,1}^* = 0.2, x_{15,5}^* = 0.8.$$

The perturbations on the feed flow are as follows: at time $t = 32000s$, they are changed from

$$L_f = V_f = 4.0(mol/s)$$

to

$$L_f = V_f = 6.0(mol/s).$$

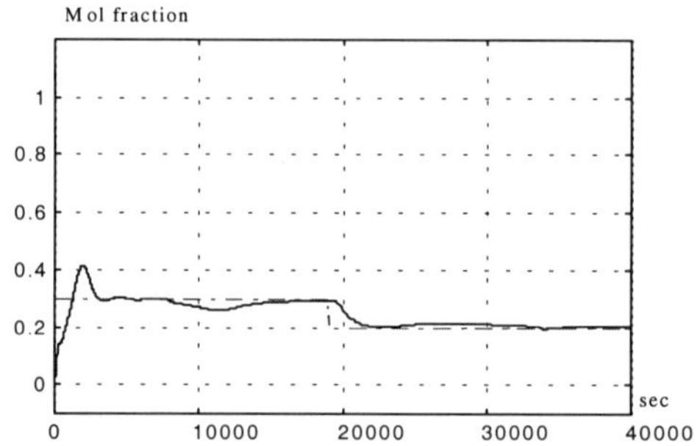

FIGURE 10.8. Top composition ($x_{1,1}$).

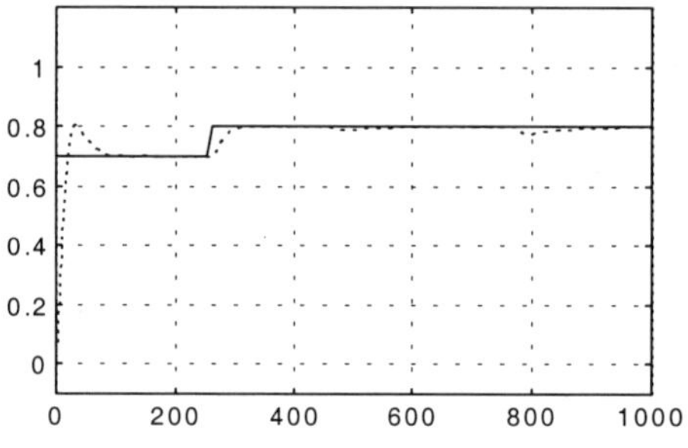

FIGURE 10.9. Bottom composition ($x_{15,5}$).

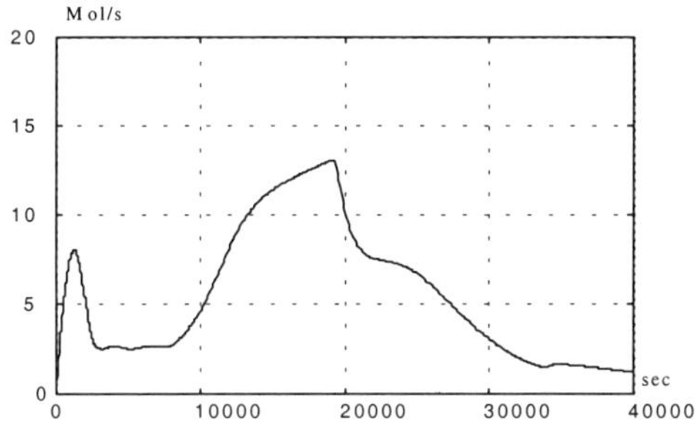

FIGURE 10.10. Reflux rate R_L.

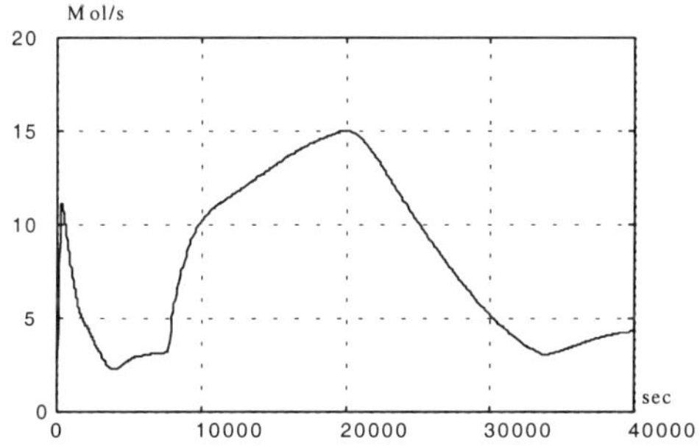

FIGURE 10.11. Vapor rate R_V.

The corresponding control results are shown in Figure 10.8 and Figure 10.9. The control inputs (R_L and R_V) are shown in Figure 10.10 and Figure 10.11 .

We can see form the illustration given above that the controllers are effective for both trajectory tracking and disturbance rejection.

10.5 Conclusion

By means of a Lyapunov-like analysis, discussed in details in Chapter 2, we determine stability conditions for the identification error. Then we analyze the trajectory tracking error when the adaptive controller is utilized. For the identification analysis, an algebraic Riccati equation has been used for the tracking error another one.

We also have derived a control law to guarantee a bound on the trajectory error. To establish this bound, we utilize a Lyapunov-like analysis (see Chapter 5). The final structure which we have proposed is composed by two parts: *the neuro-identifier* and *the tracking controller.*

The applicability of the proposed scheme is illustrated by a distillation column via simulations. The results show the good performances of the proposed scheme.

10.6 REFERENCES

[1] B.N.Bequette, Nonlinear Control of Chemical Process: A Review, *Ind. Eng. Chem. Res.*, Vol.30, 1391-1411, 1991.

[2] R.Castro, Jaime Alvarez and Joaquin Alvarez, Nonlinear Disturbance Decoupling Control of a Binary Distillation Column, *Automatica*, Vol.26, 567-572, 1990.

[3] C-B.Chung and J.B.Riggs, Dynamic Simulation and Nonlinear-Model-Based Product Quality Control of a Crude Tower, *AIChE Journal*, Vol.41, 122-134, 1995.

[4] U.M.Diwekar, Unified Approach to solving Optimal design-Control Problems in Batch Distillation, *AIChE Journal*, Vol.38, 1551-1563, 1992.

[5] L.A.Gould, *Chemical Process Control: Theory and Applications*, Addison-Wesley Publishing Co., Massachusetts, 1969.

[6] S.E.Gallun, *Solution Procedures for Nonideal Equilibrium Stage Processes at Steady State Described by Algebraic or Differential-Algebraic Equations*, Ph.D. thesis, Texas A&M University, 1979.

[7] J.H.Lee, P.Kesavan and M.Morari, Control Structure Selection and Robust Control System Design for a High-Purity Distillation Column, *IEEE Trans. Contr. Syst. Technol.*, vol.5, 402-416, 1997.

[8] W.L.Luyben, *Process Modeling, Simulation and Control for Chemical Engineers*, McGraw-Hill Inc., 1973.

[9] C.D.Holland, *Fundamentals and Modeling of Separation Processes*, Prentice-Hall International Inc., 1975.

[10] G.M.Howard, Unsteady State Behavior of Multicomponent Distillation Columns: Pert I: Simulation, *AIChE Journal*, Vol.16, 1022-1029, 1970.

[11] G.K.Kel'mans, A.S.Poznyak and A.V.Chernitser, "Local" Optimization Algoritms in Asymptotic Control of Nonlinear Dynamic Plants. Automation and Remote Control, Vol.38, No.11 pp.1639-1652, 1977.

[12] A.S.Poznyak, E.N.Sanchez and W.Yu, Nonlinear adaptive trajectory tracking using dynamic neural network, *Proc. of 16th American Control Conference,* ACC'97, USA, 1997.

[13] A.S.Poznyak and E.N.Sanchez, Nonlinear system approximation by neural networks: error stability analysis, *Intl. Journ. of Intell. Autom. and Soft Comput.*, Vol. 1, pp 247-258, 1995.

[14] A.S.Poznyak, Wen Yu and E.N.Sanchez, Control and Synchronization of Unknown Chaotic Systems Based on Dynamic Neural Networks, submitted to *Chaos: American Institute of Physics,* 1997.

[15] O.Rademaker, J.E.Rijnsdorp and A.Maarleveld, *Dynamic and Control of Continuous Distillation Units*, Elsevier Scientific Publishing Co., 1975.

[16] G.A.Rovithakis and M.A.Christodoulou, Adaptive control of unknown plants using dynamical neural networks, *IEEE Trans. Syst., Man and Cybern.*, vol. 24, pp 400-412, 1994.

[17] A.M.Shaw and F.J.Doyle III, Multivariable Nonlinear Control Applications for a High Purity Distillation Column Using a Recurrent Dynamic Neuron Model, *J.Proc. Cont*, Vol.7, 255-268, 1997.

[18] S.Skogestad, M.Morari and J.C.Doyle, Robust Control of Ill-Conditioned Plants: High-Purity Distillation, *IEEE Trans. Automat. Contr.*, vol.33, 1092-1105, 1988.

[19] F.Viel, E.Busvell and J.P.Gauthier, A Stable Control Structure for Binary Distillation Columns, *Int. J. Control*, Vol.67, 475-505, 1997.

[20] H.K.Wimmer, Monotonicity of Maximal Solutions of Algebraic Riccati Equations, *System and Control Letters*, Vol.5, pp317-319, 1985.

[21] Wen Yu, Alexander S. Poznyak and Jaime Alvarez, Neuro Control for Multicomponent Column, *14th IFAC World Congress*, Beijing China , 1999.

11

General Conclusions and future work

In this book, the authors discussed the application of dynamic neural networks for identification, state estimation and trajectory tracking of nonlinear systems.

In chapter one, a brief review of neural networks is given: First, a short look of biological neural networks is taken; then the different structures of the artificial ones are discussed. This chapter also assesses the importance of autonomous systems and the reasons to consider neural networks as a useful tool to implement this kind of systems, as well as the applications of neural networks to control.

Chapter two focuses in nonlinear system identification; it was assumed that the dynamic neural network and the nonlinear system to be identified have the same state space dimension; and the system state space completely measurable. Stability conditions for the identification error were determined by means of a Lyapunov like analysis; the proposed learning laws, including one for dynamic multilayer neural networks ensure the convergence of the identification error to zero for the model matching case, and to a bounded region in presence of unmodeled dynamics.

Continuing with the research topic of developing learning laws with increasing capabilities, in Chapter three a new learning law based on the sliding mode technique is introduced. This new learning law guarantees a bound for the identification error, even for uncertain nonlinear systems in presence of bounded disturbances.

In Chapter five, an adaptive technique is suggested to provide the passivity property for a class of partially known SISO nonlinear systems. A simple differential neural network (DNN), containing only two neurons, is used to identify the unknown nonlinear system. By means of a Lyapunov-like analysis we derive a new learning law for this DNN guarantying both successful identification and passivation effects. Based on this adaptive DNN model we design an adaptive feedback controller serving for wide class of nonlinear systems with a priory incomplete model description.

All mentioned results are limited to nonlinear systems whose state space is completely measurable. In order to relax this condition, a nonlinear observed, using dynamic neural network, is proposed in Chapter four. A very general class of continuous observable perturbed nonlinear system was considered. This observer has an extended Luneburger structure; the corresponding gain matrix was calculated by solving a matrix optimization problem. The design of this sub-optimal neuro-observer achieved a prespecified state estimation error accuracy; this estimation error turned out to be a linear combination of external disturbances power level and internal uncertainties. The neuro-observer weights are learned on-line using a new adaptive gradient-like technique.

Once it was possible to model nonlinear systems by a neural identifier or a neuro-observer, the main objective was to derive a control law, which is explained in Chapter six. There an optimal control law, in order to track a reference nonlinear model, is introduced. First, a neuro identifier was considered; then it was assumed that not all the system state was measurable. For both cases, the proposed control scheme, which is composed by the neuro-identifier or the neuro-observer and the optimal control law, ensures a bounded tracking error.

These six chapters constitute the theoretical part of the book; even if the applicability of the results is illustrated by examples; it is very important to test them with challenging nonlinear systems. So, the second part of the book is developed to applications to a sort of nonlinear plants.

In chapter seven, the identification and control of unknown chaotic dynamical systems is discussed. The goal was to drive the chaotic system to a fixed point or to a stable periodic orbit. The Lorenz equation, the Duffing one and Chua circuit were used as examples.

A robot manipulator, with two degrees of freedom, uncertainties in its parameters and unknown load and friction, was considered in Chapter eight. The proposed neuro-control scheme was applied and was more effective than other schemes such as sliding mode or linear compensators.

The last two chapters center in process identification and control. In Chapter nine the identification of a multicomponent nonstationary ozonization process, with partially measurable state, is addressed. A neuro-observer is used to estimate the concentration of each component; then based on the neuro-observer states, a particular projection least square algorithm estimated unknown constants of the chemical reactions This scheme. is more effective, regarding computing time, and less complex that others based on differential geometry or global optimization.

Finally in Chapter ten, the neuro-control of a multicomponent nonideal distillation column is discussed. Holdup, liquid and vapor flow rates were assumed to be time-varying(nonideal conditions). Simulations using a five components distillation column with fifteen trays shows the effectiveness of the proposed neuro-control scheme.

Even if the area of neuro-control is maturing in recent years, there are still missing analysis of their properties, particularly concerning rigorous convergence proofs. This book contributes to establish this analysis for nonlinear system identification, state estimation, and trajectory tracking using differential neural networks. In order to guarantee error boundness new weight learning laws and dynamic neural networks structures were developed. The proposed neural schemes are very general in the sense that they are able to handle a large class of nonlinear systems even in presence of unmodeled dynamics and external perturbations.

In regards to future work and as a source for inspiration, we propose the suggestions as below:

- Stochastic continuous time nonlinear systems. They are almost no results concerning the identification and control of this kind of systems using neural networks. Some of the mathematical techniques to be taken into account are: Ito integrals, Girzanov Transformation and Zakai equations.

- Extension of the sliding mode learning law to the case of noise presence in both the dynamic system and the measurements.

- Discrete time nonlinear systems. This is also a very promising field; even if these systems are fundamental for real time applications, there exist very few results concerning on line dynamic neural network weights adaptation for identification and control.

- Application of concepts such as passivity, input-to-state, and input-to-output stability to the analysis of control schemes based on dynamic neural networks. These concepts have been seldom used in the existing analyses.

- Application of Hamilton-Jacobi-Issacs (HJI) equation to derive robust control law for tracking of nonlinear systems. Recent results implement robust controller without explicitly solving the partial differential equation appearing when HJI equation is used for robust control synthesis. Instead of the resulting feedback structure includes a gain matrix, which depends linearly on the gradient of an unknown solution of the corresponding HJI equation.

Finally, we strongly believe that this book can help the new generations of scientists to realize successfully the ideas discussed above, in their theoretical study and practical activities.

12

Appendix A: Some Useful Mathematical Facts

12.1 Basic Matrix Inequality

Lemma 12.1 *For any matrices $X \in \Re^{n\times k}$, $Y \in \Re^{n\times k}$ and any positive defined matrix $\Lambda = \Lambda^T > 0$, $\Lambda \in \Re^{n\times n}$ the following matrix inequality hold:*

$$X^T Y + Y^T X \leq X^T \Lambda X + Y^T \Lambda^{-1} Y, \qquad (12.1)$$

$$(X+Y)^T (X+Y) \leq X^T (I+\Lambda) X + Y^T \left(I+\Lambda^{-1}\right) Y. \qquad (12.2)$$

Proof. Define

$$H := X^T \Lambda X + Y^T \Lambda^{-1} Y - X^T Y - Y^T X.$$

Then for any vector v we can introduce the vectors

$$v_1 := \Lambda^{1/2} X v \text{ and } v_2 := \Lambda^{-1/2} Y v.$$

Based on this notations we derive:

$$v^T H v = v_1^T v_1 + v_2^T v_2 - v_1^T v_2 - v_2^T v_1 = \|v_1 - v_2\|^2 \geq 0$$

or, in matrix form:

$$H \geq 0,$$

that is equivalent to (12.1). The inequality (12.2) is direct consequence of (12.1). ∎

12.2 Barbalat's Lemma

Lemma 12.2 *[2] If $f : \Re^+ \to \Re^+$ is uniformly continuous for $t \geq 0$, and if the limit of the integral*

$$\lim_{t\to\infty} \int_0^t |f(\tau)|\, d\tau$$

exist and is finite, then

$$\lim_{t\to\infty} f(t) = 0.$$

Proof. Let $\lim_{t\to\infty} f(t) \neq 0$. Then there exist an infinite unbounded sequence $\{t_n\}$ and $\varepsilon > 0$, such that $|f(t_i)| > \varepsilon$. Since f is uniformly continuous,

$$|f(t) - f(t_i)| \leq k|t - t_i|, \quad \forall t, t_i \in \Re^+$$

for some constant $k > 0$. Also

$$|f(t)| \geq \varepsilon - |f(t) - f(t_i)| \geq \varepsilon - k|t - t_i|$$

Integrating the previous equation over an interval $[t_i, t_i + \delta]$, where $\delta > 0$

$$\int_{t_i}^{t_i+\delta} |f(\tau)| \, d\tau \geq \varepsilon\delta - k\delta^2/2$$

Choosing $\delta = \varepsilon/k$, we have

$$\int_{t_i}^{t_i+\delta} |f(\tau)| \, d\tau \geq \varepsilon\delta/2, \quad \forall t_i$$

This contradicts the assumption that $\lim_{t\to\infty} \int_0^t |f(\tau)| \, d\tau$ is finite. ■

Corollary 12.1 *If $g \in L_2 \cap L_\infty$, and $\dot{g}$ is bounded, then*

$$\lim_{t\to\infty} g(t) = 0$$

Proof. Choose $f(t) = g^2(t)$. Then $f(t)$ satisfies the condition of previous lemma, the result follows. ■

12.3 Frequency Condition for Existence of Positive Solution to Matrix Algebraic Riccati Equation

Let us assume that the system under study is time- invariant. The nominal system matrix in this case is given by $A_0(t) = A_0 = Const$. If this is the case , the solution of the differential matrix Riccati equations resulting in the control theory, can be found in the set of constant matrices $P(t) = P = Const$. The matrix Riccati equations is an algebraic equations. In this case, the following theorem due to Willems [3] can be very helpful to state the existence conditions of the positive defined matrix solution of this equation.

Theorem 12.1 *The matrix Riccati equation*

$$PA_0 + A_0^T P + PRP + Q = 0 \tag{12.3}$$

with constant parameters $A_0 \in \Re^{n\times n}$, $0 < R = R^T \in \Re^{n\times n}$ *and* $0 < Q = Q^T \in \Re^{n\times n}$ *with*

- $\operatorname{Re}\lambda_j(A_0) < 0 \quad \forall j = 1, \ldots, n,$
- *the pair* $(A_0, R^{1/2})$ *is stabilizable,*
- *the pair* $(Q^{1/2}, A_0)$ *is observable,*

has a unique positive definite solution $0 < P = P^T$ *if the following condition is satisfied:*

$$S(\omega) \triangleq I - \left[R^{1/2}\right]^T \left[-i\omega I - A_0^T\right]^{-1} Q \left[i\omega I - A_0\right]^{-1} \left[R^{1/2}\right] \geq 0 \tag{12.4}$$

for any $\omega \in (-\infty, \infty)$.

Proof. Straightforward from lemma 5 in [3]. ∎

Lemma 12.3 *Under the assumptions of the previous theorem, for* $R > 0$ *(in our case we deal with this situation) the function* $S(\omega)$ *satisfies the condition (12.4) if the following matrix inequality holds*

$$\frac{1}{4}\left(A_0^T R^{-1} - R^{-1} A_0\right) R \left(A_0^T R^{-1} - R^{-1} A_0\right)^T \leq A_0^T R^{-1} A_0 - Q \tag{12.5}$$

Proof. The condition $S(\omega) \geq 0$, (12.4), when $R > 0$, is equivalent to the following one

$$\left[-i\omega I - A_0^T\right] R^{-1} \left[i\omega I - A_0\right] \geq Q$$

or,

$$\omega^2 R^{-1} + i\omega \left(R^{-1} A_0 - A_0^T R^{-1}\right) \geq Q - A_0^T R^{-1} A_0$$

which can be rewritten as follows:

$$(u-iv)^T\left[\omega^2R^{-1}+i\omega\left(R^{-1}A_0-A_0^TR^{-1}\right)+A_0^TR^{-1}A_0\right](u+iv)\geq$$

$$\geq(u-iv)^T\left[Q-A_0^TR^{-1}A_0\right](u+iv)$$

The previous inequality can be rewritten as

$$\omega^2\left[\left(u,R^{-1}u\right)+\left(v,R^{-1}v\right)\right]+2\omega\left(u,Tv\right)\geq-\left(u,Gu\right)-\left(v,Gv\right)$$

where

$$\begin{aligned}T\ &:\ =A_0^TR^{-1}-R^{-1}A_0\\ G\ &:\ =A_0^TR^{-1}A_0-Q.\end{aligned}$$

which is true for any $\omega\in(-\infty,\infty)$. Minimizing the left hand side of this inequality with respect to ω, we obtain

$$\inf_{\omega\in(-\infty,\infty)}\left(\omega^2\left[\left(u,R^{-1}u\right)+\left(v,R^{-1}v\right)\right]+2\omega\left(u,Tv\right)\right)=$$

$$=-\frac{(u,Tv)^2}{(u,R^{-1}u)+(v,R^{-1}v)}\geq-\left(u,Gu\right)-\left(v,Gv\right)$$

or, in an other form,

$$\left[(u,Gu)+(v,Gv)\right]\left[\left(u,R^{-1}u\right)+\left(v,R^{-1}v\right)\right]\geq(u,Tv)^2$$

which should be valid for any real $u\in R^n$ and $v\in R^n$. Rewriting the last inequality for the new variable w given by

$$w:=\begin{pmatrix}u\\ v\end{pmatrix}\in R^{2n}$$

we can get the inequality

$$\left(w,\begin{bmatrix}G&0\\0&G\end{bmatrix}w\right)\left(w,\begin{bmatrix}R^{-1}&0\\0&R^{-1}\end{bmatrix}w\right)\geq\left(w,\begin{bmatrix}0&\frac{1}{2}T\\\frac{1}{2}T^T&G\end{bmatrix}w\right)^2\tag{12.6}$$

which should be valid for any $w \in R^{2n}$.

The inequality (12.5) is exactly equivalent to the following one:

$$\frac{1}{4} TRT^T \leq G$$

or, in the extended form,

$$\begin{bmatrix} 0 & \frac{1}{2}T \\ \frac{1}{2}T^T & G \end{bmatrix}^T \begin{bmatrix} R & 0 \\ 0 & R \end{bmatrix} \begin{bmatrix} 0 & \frac{1}{2}T \\ \frac{1}{2}T^T & G \end{bmatrix}$$

$$= \frac{1}{4} \begin{bmatrix} TRT^T & 0 \\ 0 & T^TRT \end{bmatrix} \leq \begin{bmatrix} G & 0 \\ 0 & G \end{bmatrix}$$

In view of this and using Cauchy -Bounyakovskii -Shwartz inequality, it follows

$$\left(w, \begin{bmatrix} 0 & \frac{1}{2}T \\ \frac{1}{2}T^T & G \end{bmatrix} w \right)^2$$

$$= \left(\begin{bmatrix} R^{-1/2} & 0 \\ 0 & R^{-1/2} \end{bmatrix} w, \begin{bmatrix} R^{1/2} & 0 \\ 0 & R^{1/2} \end{bmatrix} \begin{bmatrix} 0 & \frac{1}{2}T \\ \frac{1}{2}T^T & G \end{bmatrix} w \right)^2$$

$$\leq \left(w, \begin{bmatrix} R^{-1} & 0 \\ 0 & R^{-1} \end{bmatrix} w \right) \left(w, \begin{bmatrix} 0 & \frac{1}{2}T \\ \frac{1}{2}T^T & G \end{bmatrix}^T \begin{bmatrix} R & 0 \\ 0 & R \end{bmatrix} \begin{bmatrix} 0 & \frac{1}{2}T \\ \frac{1}{2}T^T & G \end{bmatrix} w \right)$$

$$\leq \left(w, \begin{bmatrix} R^{-1} & 0 \\ 0 & R^{-1} \end{bmatrix} w \right) \left(w, \begin{bmatrix} G & 0 \\ 0 & G \end{bmatrix} w \right)$$

which is true for any $w \in R^{2n}$. So, (12.6) is proven. ■

The advantages of this result are obvious. We do not need to check if a particular triplet A_0, R and Q satisfies condition (12.4) over all frequencies. We only need to look among all matrices A_0, R and Q satisfying (12.5) in order to assure that the matrix Riccati equation (12.3) has a definite positive solution.

12.4 Conditions for Existence of Positive Solution to Matrix Differential Riccati Equation

Lemma 12.4 *Let us consider a matrix differential Riccati equations (with parameters continuous in time) given by*

$$-\dot{P}_1(t) = A_t^T P_1(t) + P_1(t)A_t + P_1(t)R_t P_1(t) + Q_t \tag{12.7}$$

and a matrix differential Riccati equations (with constant parameters)

$$0 = A^T P_2 + P_2 A + P_2 R P_2 + Q \tag{12.8}$$

with the initial conditions

$$P_1(0) > P_2 \tag{12.9}$$

and with the corresponding Hamiltonians given by

$$H_{1,t} := \begin{bmatrix} Q_t & A_t^T \\ A_t & R_t \end{bmatrix} \quad H_2 := \begin{bmatrix} Q & A^T \\ A & R \end{bmatrix}.$$

Then the stabilizability ($\exists K : \operatorname{Re}\lambda_i (A - KR) < 0, i = 1, ..., n$) *of the pair* (A, R) *and*

$$0 \le H_{1,t} \le H_2 \tag{12.10}$$

imply

$$P_1(t) > P_2(t) \qquad \forall t > 0 \tag{12.11}$$

Proof. Let us define $\Delta_t := P_1(t) - P_2$. Using condition (12.10), the block-presentation

$$H = \begin{bmatrix} H_{11} & H_{12} \\ H_{21} & H_{22} \end{bmatrix}$$

and rewriting the Riccati equation (12.7) in the Hamiltonian form as

$$-\dot{P}_1(t) = \begin{bmatrix} I & P_1(t) \end{bmatrix} H_{1,t} \begin{bmatrix} I \\ P_1(t) \end{bmatrix}$$

$$-\dot{P}_2(t) = \begin{bmatrix} I & P_2(t) \end{bmatrix} H_2 \begin{bmatrix} I \\ P_2(t) \end{bmatrix}, \; P_2(t) = P_2 = const$$

we derive:

$$\begin{aligned} -\dot{\Delta}_t &= \begin{bmatrix} I & [\Delta_t + P_2(t)] \end{bmatrix} H_{1,t} \begin{bmatrix} I \\ [\Delta_t + P_2(t)] \end{bmatrix} - \begin{bmatrix} I & P_2(t) \end{bmatrix} H_2 \begin{bmatrix} I \\ P_2(t) \end{bmatrix} \leq \\ &\leq \left(\begin{bmatrix} I & \Delta_t \end{bmatrix} + \begin{bmatrix} 0 & P_2(t) \end{bmatrix} \right) H_2 \left(\begin{bmatrix} I \\ \Delta_t \end{bmatrix} + \begin{bmatrix} I \\ P_2(t) \end{bmatrix} \right) \\ &\quad - \begin{bmatrix} I & P_2(t) \end{bmatrix} H_2 \begin{bmatrix} I \\ P_2(t) \end{bmatrix} \\ &= \left(A^T + P_2(t)R\right)\Delta_t + \Delta_t\left(A + RP_2(t)\right) + \Delta_t R \Delta_t = L_t - Q_0 \end{aligned}$$

where

$$L_t := \left(A^T + P_2R\right)\Delta_t + \Delta_t\left(A + RP_2\right) + \Delta_t R \Delta_t + Q_0$$

Basing on the Theorem 3 in [4], we conclude that the pair $\left(A^T + P_2R\right)$ is stabilizable if (A, R) is stabilizable too. So, for $t = 0$, we have $\Delta_{t=0} > 0$ and hence (see Lemma1 in [4]), there exit $Q_0 > 0$ such that $L_{t=0} = 0$, that leads to

$$\dot{\Delta}_t \geq Q_0 > 0$$

Taking into account that the solution of the differential Riccati equation with parameters continuous in time is also continuous function, we conclude that for time $t = 0$ there exists $\varepsilon > 0$ such that

$$Q_\tau > 0 \; \forall \tau \in [t, t + \varepsilon]$$

As a result, we obtain:

$$\Delta_{t+\varepsilon} = \Delta_t + \int_t^{t+\varepsilon} \dot{\Delta}_\tau \, d\tau \geq \Delta_t + \int_t^{t+\varepsilon} Q_0 d\tau \geq \Delta_t + Q_0\varepsilon > 0$$

that leads to the following

$$P_1(\tau) > P_2(\tau) \; \forall \tau \in [0, 0 + \varepsilon]$$

Iterating this procedure for the next time interval $[\varepsilon, 2\varepsilon]$ we state the final result (12.11). ■

12.5 Lemmas on Finite Argument Variations

Lemma 12.5 *For any differentiable vector function $g(x) \in R^m$, $x \in R^n$ that, in addition, satisfies*

1) or Lipschitz condition in global, i.e. there exist a positive constant L_g such that

$$\|g(x_1) - g(x_2)\| \le L_g \|x_1 - x_2\| \tag{12.12}$$

for any $x_1, x_2 \in R^n$,

2) or Lipschitz condition for gradient in global, i.e. there exist a positive constant $L_{\partial g}$ such that for any $x_1, x_2 \in R^n$

$$\|\nabla g(x_1) - \nabla g(x_2)\| \le L_{\partial g} \|x_1 - x_2\| \tag{12.13}$$

the following property holds for any $x, \Delta x \in R^n$:

1) or

$$\left\|g(x + \triangle x) - \left(g(x) + \nabla^T g(x)\triangle x\right)\right\| \le 2L_g \|\triangle x\| . \tag{12.14}$$

2) or

$$\left\|g(x + \triangle x) - \left(g(x) + \nabla^T g(x)\triangle x\right)\right\| \le \frac{L_{\partial g}}{2} \|\triangle x\|^2 . \tag{12.15}$$

Proof. Based on the integral identity

$$\int_0^1 (\nabla g(x + \theta\triangle x), \triangle x)\, d\theta = g(x + \theta\triangle x) \mid_{\theta=0}^{\theta=1} = g(x + \triangle x) - g(x)$$

that is valued for any vectors $x, \Delta x \in R^n$ we derive:

$$\begin{aligned} g(x + \triangle x) - g(x) &= \textstyle\int_0^1 (\nabla g(x + \theta\triangle x) - \nabla g(x) + \nabla g(x), \triangle x)\, d\theta \\ &= \textstyle\int_0^1 (\nabla g(x + \theta\triangle x) - \nabla g(x), \triangle x)\, d\theta + (\nabla g(x), \triangle x), \end{aligned}$$

as a result,

$$\begin{aligned} \|g(x + \triangle x) - g(x) - (\nabla g(x), \triangle x)\| &\le \textstyle\int_0^1 \|(\nabla g(x + \theta\triangle x) - \nabla g(x), \triangle x)\|\, d\theta \\ &\le \textstyle\int_0^1 \|\nabla g(x + \theta\triangle x) - \nabla g(x)\| \, \|\triangle x\|\, d\theta. \end{aligned}$$

1) Using (12.12) we may state that for any $x \in R^n$

$$\|\nabla g(x)\| \leq L_g \tag{12.16}$$

and applying this estimate to (12.16) we conclude that

$$\begin{aligned} \|g(x+\triangle x) - g(x) - (\nabla g(x), \triangle x)\| &\leq \textstyle\int_0^1 (\|\nabla g(x+\theta\triangle x)\| + \|\nabla g(x)\|) \|\triangle x\| \, d\theta \\ &\leq \textstyle\int_0^1 2L_g \|\triangle x\| \, d\theta = 2L_g \|\triangle x\| \end{aligned}$$

2)

$$\|g(x+\triangle x) - g(x) - (\nabla g(x), \triangle x)\| \leq \int_0^1 L_{\partial g}\theta \|\triangle x\|^2 \, d\theta = \frac{L_{\partial g}}{2} \|\triangle x\|^2$$

Lemma is proved. ■

Corollary 12.2 *In the assumption of this lemma the following presentation takes place:*

$$g(x+\triangle x) = g(x) + \nabla^T g(x)\triangle x + v_g \tag{12.17}$$

where the vector v_g can be estimated as follow:

$$\|v_g\| \leq 2L_g \|\triangle x\| \tag{12.18}$$

Proof. Defining the vector

$$v_g := g(x+\triangle x) - g(x) - \nabla^T g(x)\triangle x$$

and using the estimate (12.14) we obtain the result. ■

Lemma 12.6 *If we define a positive function $V(x)$, $x \in R^n$ as*

$$V(x) := \frac{1}{2} [\|x - x^*\| - \mu]_+^2$$

where $[\cdot]_+^2 := ([\cdot]_+)^2$, $[\cdot]_+$ is defined as

$$[z]_+ = \begin{cases} z & z \geq 0 \\ 0 & z < 0 \end{cases}$$

then the function $V(x)$ is differentiable and its gradient is

$$\nabla V(x) = [\|x - x^*\| - \mu]_+ \frac{x - x^*}{\|x - x^*\|}$$

with the Lipschitz constant equal to 1.

Proof. see [3] (Chapter 4, paragraph 2, exercise 1). ■

12.6 References

[1] B.T.Polyak, *Introduction to Optimization*, Optimization Software, Publication Division, New York, 1987.

[2] T.Kailath, *Linear System,* Englewood Cliffs, N.J.: Prentice-Hall, 1980.

[3] J.C.Willems, "Least squares optimal control and algebraic Riccati equations", *IEEE Trans. on Automatic Control*, Vol. 16, No 6, pp 621-634, 1971.

[4] H.K.Wimmer, Monotonicity of Maximal Solutions of Algebraic Riccati Equations, *System and Control Letters*, Vol.5, pp317-319, 1985

13

Appendix B: Elements of Qualitative Theory of ODE

13.1 Ordinary Differential Equations: Fundamental Properties

Ordinary differential equations (**ODE**) given in the general form

$$\dot{x}_t = f(t, x_t, u_t), \quad x_{t_0} = x_0, \quad t \in [t_0, T] \tag{13.1}$$

provide simple deterministic descriptions of the laws of motion of a wide class of real physical systems. Here $x_t \in R^n$ is a state space vector and $u_t \in U \subseteq R^k$ is a control action defined at a given subset U at time t.

13.1.1 Autonomous and Controlled Systems

Definition 10 *The system (13.1) is said to be*

1. ***free (or autonomous)*** *if the right-side hand does not depend on control, i.e., for all $t \in [t_0, T]$ and $x_t \in R^n$*

$$\frac{\partial}{\partial u} f(t, x_t, u) = 0,$$

2. ***forced (or controlled)*** *if the right-side hand depends on control.*

Let us consider the class of control strategies which in addition satisfy

$$u_t = u(t, x_t), \tag{13.2}$$

i.e., we consider the class of **nonlinear nonstationary feedback controllers.**

Definition 11 *A control $u_t = u(t, x_t)$ $(t \in [t_0, T])$ is said to be **admissible** if*

- *the vector function $u(t, x)$ is measurable (or, more restrictive, piecewise continuous) with respect to t for any $x \in R^n$,*

- *the function $u(t,x)$ satisfies Lipschitz continuity condition on x uniformly on t, i.e., there exists a nonnegative constant L_u such that all $t \in [t_0, T]$ and any $x, x' \in R^n$*

$$\|u(t,x) - u(t,x')\| \leq L_u \|x - x'\|,$$

- *at each time $t \in [t_0, T]$ its value belong to a given value set U, i.e.,*

$$u_t \in U \subseteq R^k.$$

We will denote the set of all admissible control strategies by U_{adm}.

Substituting (13.2) into (13.1) we get a free system described by

$$\dot{x}_t = f(t, x_t, u(t, x_t)) = F(t, x_t), \quad x_{t_0} = x_0, \quad t \in [t_0, T] \tag{13.3}$$

13.1.2 Existence of Solution for ODE with Continuous RHS

The following basic theorem verifies existence and uniqueness of solutions of the differential equation (13.3), in particular these conditions require the function F to be Lipschitz continuous and exhibit linear growth in x.

Theorem 13.1 *[3] Suppose that*

1. *The function $F(t,x)$ is measurable with respect to t for all $x \in R^n$.*
2. *There exist a nonnegative constant L such that for all $t \in [t_0, T]$ and any $x, x' \in R^n$*

$$\|F(t,x) - F(t,x')\| \leq L \|x - x'\|.$$

3. *There exist a nonnegative constant K such that for all $t \in [t_0, T]$ and $x \in R^n$*

$$\|F(t,x)\|^2 \leq K\left(1 + \|x\|^2\right).$$

Then there is a unique solution x_t defined on $[t_0, T]$ which is continuous on t and on $x_{t_0} = x_0$.

13.1.3 Existence of Solution for ODE with Discontinuous RHS

Several theories (such as Sliding Mode Control [8]) lead to the necessity of studying the differential equations with discontinuous right-side hand. One of the examples such sort of equations is as follows:

$$\begin{aligned}\dot{x}_{1,t} &= 4 + 2\ sign x_{2,t},\\ \dot{x}_{2,t} &= 2 - 4\ sign x_{2,t}.\end{aligned}$$

Following to the Filippov's theory [4] we will present the definition of the solution for these equations and will discuss their properties such as the uniqueness and the continuous dependence on the initial conditions.

Definition 12 *A vector function x_t, defined on the interval (t_0, T), is called a solution of the ODE (13.3) containing, may be discontinuous right-hand side, if*

- *it is absolutely continuous,*
- *for almost all $t \in (t_0, T)$ and any $\delta > 0$ the vector $\dot{x}_t$ satisfies*

$$M_x^- \{F(t,x)\} \le \dot{x}_t \le M_x^+ \{F(t,x)\}$$

where the components of the vectors $M_x^- \{F(t,x)\}, M_x^+ \{F(t,x)\}$ are defined by

$$\begin{aligned}M_x^+ \{F_i(t,x)\} &:= \lim_{\delta \to 0} ess \max F_i(t,\tilde{x}),\\ M_x^- \{F_i(t,x)\} &:= \lim_{\delta \to 0} ess \min_{\overline{x} \in U(x,\delta)} F_i(t,\tilde{x})\end{aligned}$$

($U(x,\delta)$ is a δ-neighborhood of the point x).

A. We say that the ODE (13.3) fulfills **the condition A** in the open or closed region $\mathcal{Q}$ of the extended space t, x if the function $F(t,x)$ is defined almost everywhere in $\mathcal{Q}$, is measurable, and for any bounded closed domain $\mathcal{D} \subseteq \mathcal{Q}$ there exists a summable function A_t such that almost everywhere in $\mathcal{D}$ we have

$$\|F(t,x)\| \le A_t.$$

Theorem 13.2 *(Filippov 1988) Suppose that ODE (13.3) satisfies the condition* $\boldsymbol{A}$. *In order that the continuous vector function* x_t *be a solution of this equation on the interval* (t_0, T), *it is necessary and sufficient that for arbitrary* t' *and* $t'' > t'$ *in this interval and for any vector* v *the following inequality be satisfied:*

$$v^T (x_{t''} - x_{t'}) \leq \int_{t'}^{t''} M_x^+ \left\{ v^T F(t, x) \right\} dt.$$

All solutions are uniformly continuous for those values of t *for which their graphs are contained in* $\mathcal{D}$.

Theorem 13.3 *(Uniqueness condition) Under the assumptions of the previous theorem we have uniqueness and continuous dependence of the solution on initial conditions if for almost all* (t, x) *and* (t, z) *(where* $\|x - z\| \leq \varepsilon$*) we have*

$$(x - z)^T (F(t, x) - F(t, z)) \leq K \|x - z\|^2, \ K > 0.$$

In general, ODE given by (13.3) cannot be solved by quadratures: an explicit analytical expressions for solutions as functions of one or more independent variables may not exist. Even when obtained, close-form expressions for solutions may be sufficiently complicated to prevent ascertaining fundamental solution properties such as boundness of solutions, stability and etc. The qualitative theory of ODE encompasses techniques and methods that permit investigating the general behavior of solutions directly based on the properties of the right-side hand of this equation and available information on the initial conditions. *Stability theory* based on *Lyapunov-like analysis* gives an example of such sort of technique.

13.2 Boundness of Solutions

Let $F(t, x)$ be a continuous (on t and x) R^n-valued function defined for all $t \in [t_0, T]$ and $x \in R^n$.

Definition 13 *A solution* $x(t, x_0, t_0)$ *of the initial value (Cauchy) problem (13.3) is said to be*

- ***bounded*** *if there exists a constant* $\beta = \beta(t_0, x_0)$ *such that for all* $t \in [t_0, T]$

$$\|x(t, x_0, t_0)\| \leq \beta,$$

- ***uniformly bounded*** *if a constant* β *is independent on* t_0 *and if for each* $\alpha > 0$ *there exists a constant* β_α *such that for all* $t \in [t_0, T]$, *all* t_0 *and all* $x_0 : \|x_0\| < \alpha$

$$\|x(t, x_0, t_0)\| \leq \beta_\alpha.$$

The following definitions introduce readers into the foundation of Qualitative Lyapunov Theory which is under discuss in this Appendix.

Definition 14 *A function* $V_1(x) : R^n \to R^1$ *is said to be* ***definite positive*** *in the set* $X_h = \{x \in R^n : \|x\| \leq h\}$ *if*

1. *it is continuous in* X_h;
2.

$$V_1(0) = 0 \text{ and } V_1(x) > 0 \quad \forall x \neq 0, x \in X_h.$$

Definition 15 *A function* $V(t, x) : [t_0, T] \times R^n \to R^1$ *is said to be* ***definite positive*** *in the set* $X_h = \{x \in R^n : \|x\| \leq h\}$ *if*

1. $V(t, 0) = 0$ *for all* $t \in [t_0, T]$,
2. *it is a continuous function at point* $x = 0$ *for all* $t \in [t_0, T]$,
3. *there exists a positive defined function* $V_1(x)$ *such that for all* $t \in [t_0, T]$ *and all* $x \in X_h$

$$V_1(x) \leq V(t, x).$$

Example 13.1

$$V_1(x) = x^2 \leq V(t, x) = x^2\left(1 + (t+1)^{-2}\right).$$

Theorem 13.4 *[9] Suppose for all $t \geq t_0$ and any big enough $x : \|x\| > K > 0$ there exists a definite positive continuously differentiable (on both arguments) function $V(t,x)$ which satisfies the conditions*

1.

$$V_1(\|x\|) \leq V(t,x) \leq V_2(\|x\|)$$

where $V_1(r), V_2(r)$ are definite positive function such that

$$\lim_{r\to\infty} V_1(r) = \infty,$$

2. on any trajectory of (13.3)

$$\dot{V}(t,x_t) = \frac{\partial}{\partial t}V(t,x_t) + (\nabla_x V(t,x_t), F(t,x_t)) \leq 0.$$

Then the solutions of (13.3) are uniformly bounded.

Example 13.2

$$\dot{x}_t = -\frac{\arctan(x)}{1+t^{-2}}, x_{t_0} = x_0 \neq 0, t_0 > 0.$$

Let us select

$$V(t,x) := x^2\left(1+t^{-2}\right).$$

Then

$$V_1(\|x\|) = |x|^2 \frac{t_0^2}{1+t_0^2} \leq V(t,x) \leq V_2(\|x\|) = |x|^2$$

and

$$\dot{V}(t,x) = \tfrac{\partial}{\partial t}V(t,x) + (\nabla_x V(t,x), F(t,x_t)) = \\ -2t^{-3}x^2 - 2x\arctan(x) \leq 0,$$

so, x_t is uniformly bounded.

13.3 Boundness of Solutions "On Average"

Definition 16 *We say that the system (13.3) has the solutions which are bounded* ***"on average"***, *if*

$$\lim_{t\to\infty} \frac{1}{t} \int_{\tau=0}^{t} \|x_\tau\|^2 \, d\tau < \infty. \tag{13.4}$$

Theorem 13.5 *Suppose for all $t \geq t_0$ and any big enough $x : \|x\| > K > 0$ there exists a definite positive continuously differentiable (on both arguments) function $V(t,x)$ which satisfies the conditions*

1.

$$V_1(\|x\|) \leq V(t,x) \leq V_2(\|x\|)$$

where $V_1(r), V_2(r)$ are definite positive function such that

$$\lim_{r\to\infty} V_1(r) = \infty,$$

2. on any trajectory of (13.3)

$$\dot{V}(t,x_t) = \frac{\partial}{\partial t} V(t,x_t) + (\nabla_x V(t,x_t), F(t,x_t)) \leq -(x_t, Qx_t) + \xi_t$$

where the function ξ_t is bounded on average, i.e.,

$$\beta := \lim_{t\to\infty} \frac{1}{t} \int_{\tau=0}^{t} \xi_\tau \, d\tau < \infty$$

and $Q = Q^T$ a strictly positive matrix.

Then the solutions of (13.3) are bounded on average and

$$\lim_{t\to\infty} \frac{1}{t} \int_{\tau=0}^{t} \|x_\tau\|^2 \, d\tau \leq \beta \lambda_{\min}^{-1}(Q).$$

Proof. It follows directly from the property 2.. Indeed, after the integrating this inequality we obtain

$$\int_{\tau=0}^{t} \dot{V}(\tau, x_\tau)\, d\tau = V(t, x_t) - V(0, x_0) \leq -\int_{\tau=0}^{t} (x_\tau, Qx_\tau)\, d\tau + \int_{\tau=0}^{t} \xi_\tau\, d\tau,$$

from which it follows that

$$\begin{aligned} \lambda_{\min}(Q) \int_{\tau=0}^{t} \|x_\tau\|^2 d\tau &\leq \int_{\tau=0}^{t} (x_\tau, Qx_\tau)\, d\tau \leq \int_{\tau=0}^{t} \xi_\tau\, d\tau - V(t, x_t) + V(0, x_0) \\ &\leq \int_{\tau=0}^{t} \xi_\tau\, d\tau + V(0, x_0). \end{aligned}$$

dividing both side of the last inequality on T and calculating the upper limits we get the result. Theorem is proved. ■

13.4 Stability "in Small" , Globally, "in Asymptotic" and Exponential

13.4.1 Stability of a particular process

Denote by

$$x_t^0 = x(t, x_{t_0}, t_0)$$

the particular solution of the ODE (13.3) generated by an initial value $x_{t_0}^0$.

Definition 17 *A particular solution x_t^0 of the ODE (13.3) is called stable* ***"in small"*** *or* ***in Lyapunov sense*** *if for any $t_0 \geq 0$ and any $\varepsilon > 0$ there exists $\delta(t_0, \varepsilon) > 0$ such that for any $\tilde{x}$ subjected to*

$$\|\tilde{x} - x_{t_0}\| \leq \delta(t_0, \varepsilon)$$

the corresponding solution $x(t, \tilde{x}, t_0)$ satisfies

$$\|x(t, \tilde{x}, t_0) - x_t^0\| \leq \varepsilon$$

for any $t \geq t_0$.

Let us define the new variable

$$y_t := x_t - x_t^0$$

which , evidently, satisfies

$$\dot{y}_t := \dot{x}_t - \dot{x}_t^0 = F\left(t, y_t + x_t^0\right) - F\left(t, x_t^0\right) := g\left(t, y_t\right). \tag{13.5}$$

For this new ODE the process $y_t = 0$ turns out to be stable "in small" and, starting from this moment, we can talk about the stability of the zero point $y = 0$ instead of talk about the stability of the process x_t^0: both of these notion are equal.

13.4.2 Different Types of Stability

Definition 18 *The origin point $y = 0$ for the ODE (13.5) is called*

- *to be stable **"in small"** or **in Lyapunov sense** if for any $t_0 \geq 0$ and any $\varepsilon > 0$ there exists $\delta(t_0, \varepsilon) > 0$ such that for any y_{t_0} subjected to*

$$\|y_{t_0}\| \leq \delta(t_0, \varepsilon)$$

  the corresponding solution $y(t, y_{t_0}, t_0)$ satisfies

$$\|y(t, y_{t_0}, t_0)\| \leq \varepsilon$$

  for any $t \geq t_0$;

- *to be **uniformly stable "in small"** if $\delta(t_0, \varepsilon)$ can be chosen independently of t_0, i.e.,*

$$\delta(t_0, \varepsilon) = \delta_1(\varepsilon);$$

- *to be **asymptotically stable** if it is stable **"in small"** and, in addition,*

$$\lim_{t \to \infty} y(t, y_{t_0}, t_0) = 0$$

  if $\|y_{t_0})\| \leq \delta(t_0, \varepsilon)$

- *to be **exponentially stable** if any solution of (13.5) satisfies*

$$a_1 \|y_{t_0}\| e^{-\alpha_1(t-t_0)} \leq \|y(t, y_{t_0}, t_0)\| \leq a_2 \|y_{t_0}\| e^{-\alpha_2(t-t_0)}$$

  for any $t \geq t_0$. Here, $a_1, a_2, \alpha_1, \alpha_2 > 0$.

13.4.3 Stability Domain

Definition 19 *An open, piecewise connected set $\mathcal{A}$ containing a small neighborhood of the origin $y = 0$ is named* **the set of asymptotic stability** *for the system given by (13.5) if for any $t_0 \geq 0$ and any $y_{t_0} \in \mathcal{A}$ the corresponding solution $y(t, y_{t_0}, t_0)$ satisfies*

$$\lim_{t \to \infty} y(t, y_{t_0}, t_0) = 0$$

Definition 20 *The origin point $y = 0$ for the ODE (13.5) is called*

- *to be* **globally asymptotically stable** *if*

$$\mathcal{A} \equiv R^n.$$

13.5 Sufficient Conditions

Theorem 13.6 *(1-st Lyapunov's theorem 1892, see [10], [6]) The trivial solution $y_t = 0$ for the ODE (13.5) is* **uniformly stable** *if there exists a definite positive function $V(t, y)$ which satisfies the following conditions:*

1. *it is continuously differentiable on t and x;*
2. *it is continuous at $x = 0$ uniformly on $t \geq 0$;*
3. *it fulfills the inequality*

$$\frac{d}{dt} V(t, y) \leq 0$$

for any $t \geq 0$ and any small enough y_{t_0} satisfying $\|y_{t_0}\| \leq \delta_1(\varepsilon)$.

Corollary 13.1 *The trivial solution $y_t = 0$ for the ODE given by*

$$\dot{y}_t = A y_t + h(y_t), \quad y_{t_0} = y_0$$

is **uniformly stable** *if*

1. *the matrix* A *is stable, i.e.*

$$\operatorname{Re}\lambda_i(A) < 0 \quad \forall i = 1, ..., n,$$

2. *the nonlinear vector function* $h(y)$ *satisfies*

$$\frac{h(y)}{\|y\|} \to 0 \text{ if } \|y\| \to 0,$$

i.e.,

$$h(y) = o(\|y\|).$$

Example 13.3 *The origin point* $x_1 = x_2 = 0$ *turns out to be uniformly stable for the following nonlinear system*

$$\begin{aligned} \dot{x}_1 &= -x_1 + \frac{x_1^4}{\ln(1 + x_1^2 + x_2^2)} \\ \dot{x}_2 &= -x_2 + \frac{x_2^4}{\ln(1 + x_1^2 + x_2^2)}. \end{aligned}$$

Theorem 13.7 *(2-nd Lyapunov's theorem 1892, see [10]) The trivial solution* $y_t = 0$ *for the ODE (13.5) is* ***uniformly asymptotically stable*** *if there exist two definite positive functions* $V(t, y)$ *and* $W(t, y)$ *such that*

1. $V(t, y)$ *is continuous on* x *at any* $x \neq 0$ *uniformly on* $t \geq 0$;

2. *on the trajectories of (13.5)* $V(t, y)$ *fulfills the inequality*

$$\frac{d}{dt}V(t, y) \leq -W(t, y)$$

for any $t \geq 0$ *and any small enough* y_{t_0} *satisfying* $\|y_{t_0}\| \leq \delta_1(\varepsilon)$.

Theorem 13.8 *(Krasovskii 1963 [7]) If there exists a positive defined matrix* B *with constant elements such that the characteristic roots of the matrix* M *given by*

$$\begin{aligned} M &= \tfrac{1}{2}\left(J^T(y)\, B + BJ(y)\right) \\ J(y) &:= \tfrac{\partial}{\partial y} F(y) \end{aligned}$$

are bounded above by a fixed negative bound $-c$ *for any* $\|y\| \leq \delta_1(\varepsilon)$ *then the origin point* $y = 0$ *is guaranteed to be* **asymptotically stable** *for the autonomous ODE given by (13.6). If this bound is valid for all* $y \in R^n$ *then the equilibrium point is* **globally asymptotically stable.**

Example 13.4 *Consider ODE*

$$\dot{x}_1 = f_1(x_1) + f_2(x_2)$$
$$\dot{x}_2 = x_1 + ax_2.$$

We have

$$J(x) = \begin{pmatrix} \frac{\partial}{\partial x_1} f_1(x_1) & \frac{\partial}{\partial x_2} f_2(x_2) \\ 1 & a \end{pmatrix}.$$

Choosing $B = I$ *we obtain the following asymptotic stability conditions*

$$2\frac{\partial}{\partial x_1} f_1(x_1) + 2a \leq -\delta_1 < 0$$
$$4a\frac{\partial}{\partial x_1} f_1(x_1) - \left(1 + \frac{\partial}{\partial x_2} f_2(x_2)\right)^2 \geq \delta_2 > 0$$

which define the estimate for the stability set $\mathcal{A}$.

Theorem 13.9 *(Antosevitch 1958 [1]) If*

1. *the origin* $y = 0$ *is uniformly stable in small for the system (13.3),*
2. *there exists a continuously differentiable function* $V(t, y)$ *satisfying*

   -
   $$V(t, y) \geq a(\|y\|)$$
   for any $t \geq t_0$ *and any* y *,*

   -
   $$V(t, 0) = 0$$
   for any $t \geq t_0$,

- $$\frac{d}{dt}V(t,y) \leq -b(\|y\|)$$

  for any $t \geq t_0$ *and any* y ,

where the continuous functions $a(z)$ *and* $b(z)$ *are monotonically increasing and fulfill*

$$a(0) = b(0) = 0, \quad za(z) \geq 0, \quad zb(z) \geq 0$$

then the origin point $y = 0$ *of the system (13.3) is* ***globally asymptotically stable****.*

Theorem 13.10 *(Halanay 1966 [5]) If*

1. *there exists a continuously differentiable function* $V(t,y)$ *satisfying*

- $$V(t,y) \geq a(\|y\|)$$

  for any $t \geq t_0$ *and any* y ,

- $$V(t,0) = 0$$

  for any $t \geq t_0$,

- $$\frac{d}{dt}V(t,y) \leq -c(V(t,y))$$

  for any $t \geq t_0$ *and any* y ,

where the continuous functions $a(z)$ *and* $c(z)$ *are monotonically increasing and fulfill*

$$a(0) = c(0) = 0, \quad za(z) \geq 0, \quad zc(z) \geq 0$$

then the origin point $y = 0$ of the system (13.3) is ***globally asymptotically stable.***

Theorem 13.11 *(Chetaev 1955 [2]) If*

1. *there exists a continuous function $k(t) \geq 0$,*

2. *there exists a continuously differentiable function $V(t, y)$ satisfying*

   - $$V(t,y) \geq k(t)\, a(\|y\|)$$

     for any $t \geq t_0$ and any y, where the continuous function $a(z)$ is monotonically increasing and fulfill

     $$a(0) = 0, \quad za(z) \geq 0,$$

   - $$\frac{d}{dt}V(t,y) \leq 0$$

     for any $t \geq t_0$ and any y,

   - $$\lim_{t\to\infty} k(t) = \infty$$

then the origin point $y = 0$ of the system (13.3) is ***globally asymptotically stable.***

13.6 Basic Criteria of Stability

When we talk about any theorem of *Criterion Type* we mean that this theorem obligatory provides the necessary and sufficient conditions simultaneously. Below we present the criteria of Stability based on Lyapunov function approach [10]. In the previous subsection we have presented several classical results which state the sufficient conditions to guarantee the stability of trajectories. All of these results are based on the notion of *Lyapunov function.* If it fulfills some specific conditions then stability property is guaranteed. The importance of the theorems presented below is connected with the following question: "We know that if for a given system there exists a Lyapunov function then this system is stable. But if a given system is stable, does there exist a Lyapunov function?" The answer is positive and is presented below.

Theorem 13.12 *(Criterion of the stability "in small") The trivial solution $y_t = 0$ for the ODE (13.5) is* **stable in small** *if and only if there exist a definite positive function $V(t,y)$ and strictly positive function $\mu(t_0)$ such that for any $t_0 \geq 0$ from*

$$\|y_{t_0}\| < \mu(t_0)$$

it follows that the function $V(t, y(t, y_{t_0}, t_0))$ is a non increasing function of t.

Theorem 13.13 *(Criterion of the asymptotic stability) The trivial solution $y_t = 0$ for the ODE (13.5) is* **asymptotically stable** *if and only if there exists a definite positive function $V(t,y)$ and strictly positive function $\nu(t_0)$ such that for any $t_0 \geq 0$ from*

$$\|y_{t_0}\| < \nu(t_0)$$

it follows that the function $V(t, y(t, y_{t_0}, t_0))$ is a monotonically decreasing up to zero function of t on the trajectories of (13.5), i.e.,

$$V(t, y(t, y_{t_0}, t_0)) \downarrow 0.$$

Theorem 13.14 *(Criterion of the exponential stability) The trivial solution $y_t = 0$ for the ODE (13.5) is* **exponentially stable** *if and only if there exist two functions $V(t,y)$ and $W(t,y)$ such that for any $t \geq 0$*

1.

$$\mu_1 \|y\|^2 \leq V(t,y) \leq \mu_2 \|y\|^2, \quad \mu_1 > 0,$$

2.

$$\nu_1 \|y\|^2 \leq W(t,y) \leq \nu_2 \|y\|^2, \quad \nu_1 > 0,$$

3. *on the trajectories of (13.5)* $V(t,y)$ *fulfills the equality*

$$\frac{d}{dt} V(t,y) = -W(t,y)$$

for any $t \geq 0$.

Remark 13.1 *The following function selection is possible:*

$$W(t,y) := \|y\|^2,$$
$$V(t, y_{t_0}) := \int_{\tau=t}^{\infty} W(\tau, y(\tau, y_{t_0}, t_0))\, d\tau.$$

Theorem 13.15 *(Zubov 1957)* $\mathcal{A}$ *is a* ***stability domain*** *of an autonomous ODE*

$$\dot{y}_t = f(y_t, u(y_t)) = F(y_t), \quad y_{t_0} = y_0, \tag{13.6}$$

if and only if there exist two functions $V(y)$ *and* $W(y)$ *such that*

1. $V(y)$ *is negative defined and continuous in* $\mathcal{A}$ *and, in addition,*

$$-1 < V(y) < 0$$

for any $y \in \mathcal{A} \setminus \{0\}$

2. $W(y)$ *is definite positive and continuous in* $\mathcal{A}$ *and for any positive* α *there exists a positive* β *such that the inequality* $\|y\| \geq \alpha$ *provides* $W(y) \geq \beta$,

3. *on the trajectories of (13.6)* $V(y)$ *fulfills the inequality*

$$\frac{d}{dt} V(y) = [1 + V(y)]\, W(y)$$

for any $t \geq 0$ *and* y *satisfying* $y \in \mathcal{A} \setminus \{0\}$.

Corollary 13.2 *The* ***boundary*** $\partial\mathcal{A}$ *of stability domain* $\mathcal{A}$ *consists of all points* y *fulfilling*

$$V(y) = -1.$$

For the system of ODE given by

$$\begin{aligned}\dot{x}_1 &= -x_1 + 2x_1x_2 \\ \dot{x}_2 &= -x_2\end{aligned}$$

we can select

$$W(x) = \|x\|^2 = x_1^2 + x_2^2,$$
$$V(x) = \exp\left\{-\int_{\tau=t}^{\infty} W(y(\tau, x, t_0))\, d\tau\right\} - 1$$

and show that

$$\partial\mathcal{A} = \{x : x_1x_2 = -1\}.$$

13.7 References

[1] Antosiewcz H.A. A survey of Lyapunov's second Method. *Contr.Non.Oscill.* 4, 141-166, 1958.

[2] Chetaev N.G. *Stability of Motion.* Nauka. Moscow, 1955.

[3] Coddington E.A. and N. Levinson. *Theory of ordinary Differential Equations,* Malabar, Fla: Krieger Publishing Company, New York, 1984.

[4] Filippov A.F. *Differential Equations with Discontinuous Righthand Sides.* Kluwer Acad. Publishers. Dordrecht-Boston-London. 1988.

[5] Halanay A. *Differential Equations: Stability, Oscillations. Time Lag.* Academic Press, New York, 1966.

[6] Hanh W. *Stability of Motion.* Springer-verlag, New York Inc., 1967.

[7] Krasovskii N.N. *Certain Problems of the Theory of Stability of Motion.* Nauka. Moscow, 1959 (in Russian), transl. Stanford, Cal. 1963.

[8] V.I.Utkin, *Sliding Modes in Optimization and Control*, Springer-Verlag, 1992.

[9] Yoshizava T. Lyapunov's functions and boundness of solutions. *Funkc. Eqvacioj.* 2, pp.95-142, 1959.

[10] Zubov V.I. *Mathematical Methods for the Study of Automatic Control System,* New York: The Macmillan Company, 1963

14

Appendix C: Locally Optimal Control and Optimization

In this Appendix we present some elements of Locally Optimal Control Theory [1], [2] which are used throughout of this book It turns out that in nonliinear case one should apply the gradient descent technique to realize this theory and calculate numerically a control to be applied. That is why the part of Optimization Theory related to the Gradient Projection Method [3] is discussed in details to clarify the convergence property of the numerical procedure used in this book for the realization of the Locally Optimal control strategies.

14.1 Idea of Locally Optimal Control Arising in Discrete Time Controlled Systems

Let us consider a discrete time nonlinear system given in a general form as

$$x_{t+1} = x_t + f\left(t, x_t, u_t\right), \ x_{t=0} = x_0 \tag{14.1}$$

where $x_t \in R^n$ is a state vector at time t $(t = 0, 1, 2, ...)$, $u_t \in U \subseteq R^k$ is a control action defined on a convex compact $U \subseteq R^k$ and $f : R^{1+n+k} \to R^n$ is a known nonlinear function characterizing the nonlinear dynamics.

The general goal is to minimize asymptotically the *global performance index* J defined as

$$\begin{gathered} J = \lim_{t\to\infty} J_t \\ J_t = \frac{1}{t}\sum_{s=1}^{t} Q\left(s, x_{s+1}, u_s\right) \end{gathered} \tag{14.2}$$

where J_t is the *local performance index* up to time t. The last one can be rewritten in recursive form as

$$J_t = J_{t-1}\left(1 - \frac{1}{t}\right) + \frac{1}{t}Q\left(t, x_{t+1}, u_t\right), \ J_0 = 0, \ t = 1, 2, ... \tag{14.3}$$

Definition 21 *A control sequence* $\{u_t\}$ *is said to be* ***admissible*** *if*

- $$u_t = u_t\left(t, x_\tau\left(\tau = 0, 1, ..., t-1\right)\right),$$

  i.e., it realizes a nonlinear feedback control and it is dependent on only on available information $x_\tau\left(\tau = 0, 1, ..., t-1\right)$,

- $$u_t \in U.$$

Definition 22 *An admissible strategy* $\{u_t\}$ *is named* ***locally optimal*** *if it satisfies*

$$u_t = \arg\min_{u \in U} Q\left(t, x_{t+1}, u\right). \tag{14.4}$$

In other words, according to this strategy we minimize our *current losses* $Q\left(t, x_{t+1}, u_t\right)$ at each time trying to obtain the goal (14.2). Sure, it is not the optimal strategy which should obligatory depends on past but also on further information as well (see [4]).

Corollary 14.1 *As it follows from (14.4), the locally optimal strategy can be expressed as*

$$u_t^{loc} = \arg\min_{u \in U} Q\left(t, x_t + f\left(t, x_t, u\right), u\right). \tag{14.5}$$

As it is shown in [2], in the particular case when we have no any constraints $\left(U = R^k\right)$, the loss function $Q\left(t, x_{t+1}, u\right)$ is a stationary quadratic form

$$Q\left(t, x, u\right) = x^T Q x + u^T R u, \; Q = Q^T \geq 0, \; R = R^T > 0 \tag{14.6}$$

and the given plant is stationary and linear, i.e.,

$$f\left(t, x_{t+1}, u_t\right) = A x_t + B u_t, \tag{14.7}$$

the locally optimal strategy (14.4) turns out to be globally optimal in the sense of the asymptotic goal (14.2). In this special case when $U = R^k$, the plant is linear

(14.7) and the loss function is a quadratic form (14.6), the nonlinear programming problem (14.5) can be solved analytically:

$$u_t^{loc} = -\left(R + B^T Q B\right)^{-1} B^T Q \left(I + A\right) x_t.$$

In general, this nonlinear programming problem (14.5) can be solve numerically based on the Projection Gradient Procedure given by

$$u_t^{(s)} = \pi_U \left\{ u_t^{(s-1)} - \gamma_s \frac{\partial}{\partial u} Q\left(t, x_t + f\left(t, x_t, u_t^{(s-1)}\right), u_t^{(s-1)}\right) \right\} \tag{14.8}$$

where $\pi_U \{\cdot\}$ is a projection operator to the convex compact U and $\{\gamma_s\}$ is a non-negative step size sequence which, under the appropriate selection [3], provides the convergence

$$\lim_{s\to\infty} u_t^{(s)} = u_t^{loc}.$$

14.2 Analogue of Locally Optimal Control for Continuous Time Controlled Systems

The wide class of continuous time nonlinear systems can be described as

$$\dot{x}_t = f\left(t, x_t, u_t\right), \; x_{t=0} = x_0 \tag{14.9}$$

where all variable are the same as in (14.1). For any given time sequence $\{t_s\}_{s=1,2,\ldots}$ this model can be rewritten in a discrete approximation form as follows

$$x_{t_{s+1}} = x_{t_s} + \int\limits_{\tau=t_s}^{t_{s+1}} f\left(\tau, x_\tau, u_\tau\right) d\tau, \; x_{t_{s=0}} = x_0,$$

As for the asymptotic goal defined by

$$J = \overline{\lim_{t\to\infty}} \frac{1}{t+\varepsilon} \int\limits_{\tau=0}^{t} Q\left(\tau, x_\tau, u_\tau\right) d\tau, \tag{14.10}$$

it can also be rewritten in the following manner

$$\begin{aligned} J &= \overline{\lim_{t\to\infty}} J_t \\ J_t &= \tfrac{1}{t+\varepsilon} \sum_{s=0}^{t} Q_s \\ Q_s &= \int_{\tau=t_s}^{t_{s+1}} Q(\tau, x_\tau, u_\tau)\, d\tau \end{aligned} \tag{14.11}$$

and, in turn, Q_s can be approximated as

$$Q_s = \Delta t\, Q\left(t_s, x_{t_{s+1}}, u_{t_s}\right) + o(\Delta t\ ).$$

In the similar way as in (14.3), the function J_t can be presented in recursive form

$$\begin{aligned} J_t = J_{t-1}\left(1 - \tfrac{1}{t+\varepsilon}\right) + \tfrac{1}{t+\varepsilon}\, Q_t, \\ J_0 = 0,\ t = 1, 2, ... \end{aligned} \tag{14.12}$$

Again, the *locally optimal control* can be defined as

$$u_{t_s}^{loc} = \arg\min_{u\in U} Q\left(t_s, x_{t_s} + \Delta t\ f\left(t_s, x_{t_s}, u\right), u\right) \tag{14.13}$$

that coincides with (14.5). For small enough Δt from (14.13) we obtain

$$\begin{aligned} u_{t_s}^{loc} &= \arg\min_{u\in U} \left[Q\left(t_s, x_{t_s} + \Delta t\ f\left(t_s, x_{t_s}, u\right), u\right) - Q\left(t_s, x_{t_s}, u\right) + Q\left(t_s, x_{t_s}, u\right)\right] \\ &= \arg\min_{u\in U} \left[\ \Delta t \left(\nabla_x Q\left(t_s, x_{t_s}, u_{t_{s-1}}\right), f\left(t_s, x_{t_s}, u\right)\right) + Q\left(t_s, x_{t_s}, u\right)\right] \end{aligned} \tag{14.14}$$

Selecting $t_s = t$, we obtain

$$u_t^{loc} = \arg\min_{u\in U} \left[\Delta t\ \ \left(\nabla_x Q\left(t, x_t, u\right), f\left(t, x_t, u\right)\right) + Q\left(t_s, x_{t_s}, u\right)\right]$$

and, for the case when the loss function is independent of u_t,

$$u_t^{loc} = \arg\min_{u\in U} \left[\ \left(\nabla_x Q\left(t, x_t\right), f\left(t, x_t, u\right)\right)\right] \tag{14.15}$$

14.3 Damping Strategies

Let us consider the dynamic system given by (14.9) with the integral loss function given in Lagrange form as

$$J_T(u) = \int_{t=0}^{T} Q(t, x_t, u_t)\, dt \tag{14.16}$$

which we would like to minimize selecting the control strategy $\{u_t\}_{t\in[0,T)}$ such a way that for any $t \in [0, T)$

$$u_t \in U \subseteq R^k \tag{14.17}$$

where U is a given convex set (not obligatory a compact).

Define the new variable $x_{0,t}$ as follows:

$$x_{0,t} = \int_{s=0}^{t} Q(s, x_s, u_s)\, ds$$

which satisfies the following differential equation

$$\dot{x}_{0,t} = Q(t, x_t, u_t), \;\; x_{0,t=0} = 0.$$

In view of this definition the performance index (14.16) can be rewritten in Mayer's form

$$J_T(u) = x_{0,t=T} \tag{14.18}$$

Let us consider the extended state vector $\mathbf{x}_t \in R^{n+1}$defined as

$$\mathbf{x}_t^T = (x_{1,t}, ..., x_{n,t}, x_{0,t}),$$

fulfilling the dynamic equations

$$\dot{\mathbf{x}}_t = F(t, \mathbf{x}_t, u_t) = \begin{pmatrix} f(t, x_t, u_t) \\ Q(t, x_t, u_t) \end{pmatrix}, \tag{14.19}$$

and an auxiliary "energetic function" $V(t,\mathbf{x})$ which is differentiable with the respect of both arguments. Calculating its derivative along the trajectories of the dynamic system (14.19), we derive

$$\frac{d}{dt}V(t,\mathbf{x}) = \frac{\partial}{\partial t}V(t,\mathbf{x}) + (\nabla_x V(t,\mathbf{x}), F(t,\mathbf{x}_t,u_t)) := w(t,\mathbf{x}_t,u_t). \tag{14.20}$$

Definition 23 *Any control strategy* $\{u_t\}_{t\in[0,T)}$ *satisfying the "damping condition"*

$$u_t^{damp}(t,,\mathbf{x}_t) = \arg\min_{u\in U} w(t,\mathbf{x}_t,u_t) \tag{14.21}$$

is said to be a ***damping strategy.***

Substituting $u_t^{damp}(t,,\mathbf{x}_t)$ in to (14.19) we obtain

$$\dot{\mathbf{x}}_t = F\left(t,\mathbf{x}_t,u_t^{damp}(t,,\mathbf{x}_t)\right) := \tilde{F}(t,\mathbf{x}_t). \tag{14.22}$$

Definition 24 *We say that a damping control strategy* $u_t^{damp}(t,,\mathbf{x}_t)$ *is* ***admissable*** *at the interval* $[0,T)$ *if the ODE (14.22) has a solution* $\mathbf{X}_t = \mathbf{X}(t,\mathbf{x}_{t=0})$ *within this time interval.*

Define the program (depending only on t) control as

$$u^*(t) := u_t^{damp}(t,,\mathbf{X}_t).$$

By the construction, for any $t\in[0,T)$ this function fulfills $u^*(t)\in U$.

It is evident that the quality of any damping control depends on the selection of energetic function $V(t,\mathbf{x})$. Below we are presenting two more frequently selection ways.

14.3.1 Optimal control

Theorem 14.1 *If the energetic function* $V(t,\mathbf{x})$ *satisfies the following conditions*

1.

$$V(T,\mathbf{x}) = x_0,$$

2. it is differential and for any $t \in [0, T)$

$$\min_{u \in U} \frac{d}{dt} V(t, \mathbf{x}) = \min_{u \in U} \left[\frac{\partial}{\partial t} V(t, \mathbf{x}) + (\nabla_x V(t, \mathbf{x}), F(t, \mathbf{x}_t, u)) \right] = 0, \quad (14.23)$$

3. the corresponding $u_t^{damp}(t, , \mathbf{x}_t)$ is admissable,

*then this damping strategy $u_t^{damp}(t, , \mathbf{x}_t)$ is **optimal**.*

Proof. From the conditions 2 and 3 of this theorem we have

$$\frac{d}{dt} V(t, \mathbf{X}_t) = 0$$

and, hence,

$$V(T, \mathbf{X}_T) = V(0, \mathbf{X}_0).$$

From the condition 1 of this theorem we conclude that

$$J_T = x_{0,t=T} = V(T, \mathbf{X}_T) = V(0, \mathbf{X}_0) = J_T^{opt}.$$

For any other admissable control u_t (which guarantees the existence of the solution of a corresponding close dynamic system and satisfying (14.17)) we get

$$\frac{d}{dt} V(t, \mathbf{X}_t) = w(t, \mathbf{x}_t, u_t) \geq \min_{u \in U} w(t, \mathbf{x}_t, u_t) \equiv 0$$

and, as a result,

$$V(T, \mathbf{X}_T) \geq V(0, \mathbf{X}_0)$$

and, finally,

$$J_T = x_{0,t=T} = V(T, \mathbf{X}_T) \geq V(0, \mathbf{X}_0) = J_T^{opt}$$

■

If we do not want to solve the Bellman partial differential equation (14.23) to find its solution $V(t, \mathbf{x})$, and only keep the condition

$$u := \arg\min_{u \in U} [(\nabla_x V(t, \mathbf{x}), F(t, \mathbf{x}_t, u))]$$

for the Lyapunov function $V(t, \mathbf{x}) = Q(t, x_t)$, we obtain the locally optimal strategy (14.15).

14.4 Gradient Descent Technique

Let us consider *the Gradient Method* [3] applied for the minimization of a strictly convex differential function $f(x)$ defined at a given convex set $X \subseteq R^n$. Assume, for the simplicity, that its minimal point is an internal point of X,i.e.,

$$x^* := \arg\min_{x \in X} f(x) \in \text{int } X$$

This method is described by the following recursive scheme

$$\begin{aligned} x_{k+1} &= \pi_X \{x_k - \gamma_k s_k\} \\ s_k &= \nabla f(x_k) + \xi_k, \ k = 0, 1... \end{aligned} \tag{14.24}$$

where the unmeasured disturbances (noises) ξ_k are assumed to be bounded

$$\|\xi_k\| \leq \varepsilon \tag{14.25}$$

and $\pi_X \{\cdot\}$ is a projection operator to the set X satisfying for any $x \in R^n$ and any $x' \in X$

$$\|\pi_X \{x\} - x'\| \leq \|x - x'\| \, .$$

The following assumptions, concerning to optimized function $f(x)$, are assumed to be valid:

A1. The optimized differential function $f(x)$ is *strictly convex* in R^n, i.e., there exists a positive constant $l > 0$ such that for all $x, x' \in R^n$

$$(\nabla f(x), x - x') \geq l \|x - x'\|^2$$

A2. The gradient $\nabla \ f(x)$ of the optimized differential function $f(x)$ satisfies Lipschitz condition, i.e. there exists a constant $L \in (0, \infty)$ such that for all $x, x' \in R^n$

$$\|\nabla f(x) - \nabla f(x')\| \leq L \|x - x'\|^2$$

The following theorem states the convergence of the projection gradient algorithm (14.24) to a neighborhood of the minimum point x^* .

Theorem 14.2 *Under the assumptions A1, A2 and (14.25) there exists a constant* $\bar{\gamma} > 0$ *such that for any* $0 < \gamma_k = \gamma < \bar{\gamma}$ *we have*

$$\|x_k - x^*\| \le \rho(\varepsilon) + q^k \|x_0 - x^*\|, \ \rho(\varepsilon) = O(\varepsilon)$$

Proof. Let us introduce the following Lyapunov function

$$V(x) = \frac{1}{2}\left(\|x_k - x^*\| - \frac{\varepsilon}{l}\right)_+^2$$

where $(\cdot)_+$ is the projection operator acting according to the rule

$$(x)_+ = \begin{cases} x & if \quad x \ge 0 \\ 0 & if \quad x < 0 \end{cases}$$

It is easy to check that the function $V(x)$ is differentiable,

$$(\nabla V(x_k), s_k) = \left(\|x_k - x^*\| - \frac{\varepsilon}{l}\right)_+ \left(\frac{x_k - x^*}{\|x_k - x^*\|}\right)$$

and $\nabla V(x)$ satisfies the Lipschitz condition with constant $L = 1$. In view of this, we derive:

$$\begin{gathered}(\nabla V(x_k), s_k) = \left(\|x_k - x^*\| - \tfrac{\varepsilon}{l}\right)_+ \tfrac{(\nabla f(x) + \xi_k, x_k - x^*)}{\|x_k - x^*\|} \ge \\ \left(\|x_k - x^*\| - \tfrac{\varepsilon}{l}\right)_+ (l\|x_k - x^*\| - \varepsilon) = 2lV(x_k)\end{gathered}$$

and

$$\begin{gathered}\|s_k\|^2 = \|\nabla f(x_k) + \xi_k\|^2 \le (L\|x_k - x^*\| + \varepsilon)^2 \le \\ a(\varepsilon) + bV(x_k) \le a + \tfrac{b}{2l}(\nabla V(x), s_k)\end{gathered}$$

Hence, applying Lemma 5 (part 2) from Appendix 1, we get

$$\begin{gathered}V(x_{k+1}) = \tfrac{1}{2}\left(\|\pi_X\{x_k - \gamma_k s_k\} - x^*\| - \tfrac{\varepsilon}{l}\right)_+^2 \le \\ \tfrac{1}{2}\left(\|x_k - x^* - \gamma_k s_k\| - \tfrac{\varepsilon}{l}\right)_+^2 = V(x_k - \gamma_k s_k) \le \\ V(x_k) - \gamma_k(\nabla V(x_k), s_k) + \tfrac{L}{2}\gamma_k^2\|s_k\|^2 \le \\ V(x_k) - \gamma_k(\nabla V(x_k), s_k) + \tfrac{L}{2}\gamma_k^2\left[a(\varepsilon) + \tfrac{b}{2l}(\nabla V(x), s_k)\right] \le \\ V(x_k) - \gamma_k\left(1 - L\tfrac{b}{2l}\gamma_k\right)(\nabla V(x_k), s_k) + \tfrac{L}{2}\gamma_k^2 a \le \\ V(x_k)\left[1 - 2l\gamma_k\left(1 - L\tfrac{b}{2l}\gamma_k\right)\right] + \tfrac{L}{2}\gamma_k^2 a(\varepsilon) \le \\ V(x_k)q + \tfrac{L}{2}\gamma a(\varepsilon)\end{gathered}$$

from which the result follows directly. ■

14.5 References

[1] G.K.Kel'mans and A.S.Poznyak, Algorithm for Control of Dynamic Systems on the Basis of Local Optimization, *Engineering Cybernetics*, No.5, 134-141, 1977.

[2] G.K.Kel'mans, A.S.Poznyak and V.Chernltser, "Local" Optimization Algorithm in Asymptotic Control of Nonlinear Dynamic Plants, *Automation and Remote Control*, Vol.38, No.11, 1639-1653, 1977.

[3] B.T.Polyak, *Introduction to Optimization*, Optimization Software, Publication Division, New York, 1987.

[4] Pontryagin L.S., Boltyansky V.G., Gamkrelidze R.V. and E.F. Mishenko. *Mathematical Theory of Optimal Processes.* Nauka, Moscow. 1969 (in Russian).

Index